dk online
science

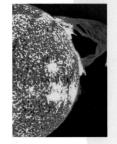     

encyclopedia

Google

CONTENTS

dk online
science

encyclopedia

LONDON, NEW YORK, MELBOURNE,
MUNICH and DELHI

Project Editors Fran Baines, Paula Borton, Gilly Cameron Cooper, Robert Dinwiddie, Jacqueline Fortey, Sarah Goulding, Margaret Hynes, Patricia Moss, Sue Nicholson, Nigel Ritchie, Richard Williams, Selina Wood, Jane Yorke

Project Designers Sarah Cowley, Yumiko Tahata, Ross George, Jim Green, Nick Harris, Adrienne Hutchinson, Alex Menday, Andrew Nash, Rebecca Painter, Johnny Pau, Owen Peyton Jones, Joanna Pocock, Smiljka Surla, Jacqui Swan, Clair Watson

Weblink editors Clare Lister, Mariza O'Keeffe, Steve Barker, John Bennett, Roger Brownlie, Clare Hibbert, Phil Hunt

Illustrators Lee Gibbons, Nick Gopalla, Robin Hunter, Andrew Kerr, Patrick Mulrey, Darren Poore

Managing Editor Camilla Hallinan

Managing Art Editor Sophia M Tampakopoulos Turner

Digital Content Manager Fergus Day

Picture Research Bridget Tily, Martin Copeland, Sean Hunter, Michelle Faram, Liz Moore, Alison Prior, Amanda Russell, Fran Vargo

DTP Co-ordinator Toby Beedell
DTP Designers Jay Jackson, Sarah Nunan, Sarah Pfitzner
Cartography Simon Mumford

Picture Librarian Gemma Woodward
Jacket Design Yumiko Tahata
Production Kate Oliver

Category Publisher Sue Grabham

Art Director Simon Webb

Revised and updated editon
Editor Steven Carton   **Designer** Stefan Podhorodecki

   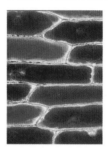  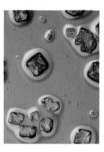

Contributors and Consultants Roger Bridgman, Kim Bryan, Dr Sue Davidson, Helen Dowling, Graham Farmelo, Dr Philip Gates, Dr Jen Green, Derek Harvey, Robin Kerrod, Dr Jacqueline Mitton, Alan Q Morton, John Nicholson, Christopher Oxlade, Dr Penny Preston, Professor Robert Spicer, Carole Stott, John Stringer, Chris Woodford

First published in hardback in Great Britain in 2004 as e.encyclopedia science
This revised and updated paperback edition first published in 2009 by
by Dorling Kindersley Limited, 80 Strand, London WC2R 0RL

Colour reproduction by Colourscan, Singapore
Printed and bound by LEO, China

Discover more at
www.dk.com

ABBREVIATIONS

METRIC		IMPERIAL		DATES	
mm	millimetre	in	inches	c.	*circa* (about)
cm	centimetre	ft	feet	BC	before Christ
m	metre	yd	yards	AD	*anno Domini* (in the year of Our Lord), after the birth of Christ
km	kilometre				
sq km	square kilometres	sq miles	square miles		
km²	square kilometres	miles²	square miles	b.	born
kph	kilometres per hour	mph	miles per hour	d.	died
°C	degrees Celsius	°F	degrees Fahrenheit	r.	reigned
g	grams	oz	ounces		
kg	kilograms	lb	pounds		
				billion = thousand million	

FROM THE BOOK TO THE NET AND BACK AGAIN

The e.science encyclopedia has its own website, created by DK and Google™. When you look up a subject in the book, the article gives you key facts and displays a keyword that links you to extra information online. Just follow these easy steps.

1   **Enter this website address**   Address : @ http://www.science.dkonline.com/

2   **Find the keyword in the book**

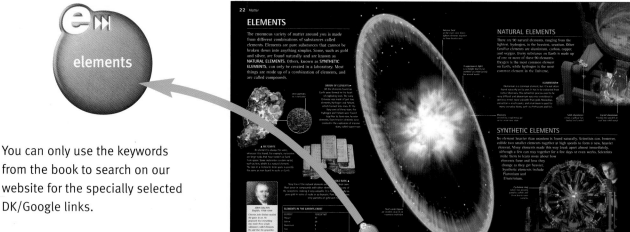

You can only use the keywords from the book to search on our website for the specially selected DK/Google links.

3   **Enter the keyword, eg elements**

elements

Be safe while you are online:

- Always get permission from an adult before connecting to the internet.

- Never give out personal information about yourself.

- Never arrange to meet someone you have talked to online.

- If a site asks you to log in with your name or email address, ask permission from an adult first.

- Do not reply to emails from strangers – tell an adult.

Parents: Dorling Kindersley actively and regularly reviews and updates the links. However, content may change. Dorling Kindersley is not responsible for any site but its own. We recommend that children are supervised while online, that they do not use Chat Rooms, and that filtering software is used to block unsuitable material.

4 Click on your chosen link...

▶▶ **Watch a video about superheavy elements**

Links include:

▶▶ animations   ▶▶ interactive quizzes

▶▶ videos   ▶▶ databases

▶▶ sound buttons   ▶▶ timelines

▶▶ virtual tours   ▶▶ realtime reports

Let Google direct you to more great sites about your subject

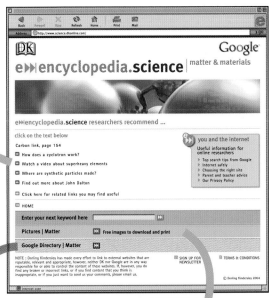

Download fantastic pictures!

The pictures are free of charge, but can be used for personal non-commercial use only.

Atom structure

5 Go back to the book for your next subject...

Headword identifies main entry

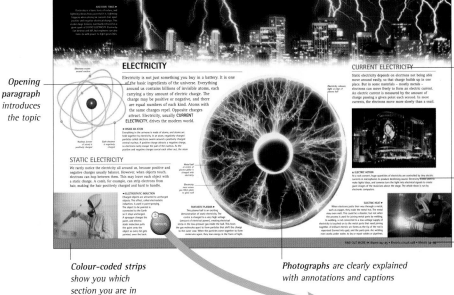

Opening paragraph introduces the topic

Colour-coded strips show you which section you are in

Photographs are clearly explained with annotations and captions

Sub-entry with its own definition, captions, and annotations

Captions give in-depth information about the topic

Find-out-more cross-references to other, related headwords

You will find:

■ e-links

■ data boxes

■ biographies

■ timelines

■ cross-references

■ full index

...and enter a new keyword online   `electricity`

MATTER & MATERIALS

MATTER

Everything you can hold, taste, or smell is made of matter. Matter makes up everything you can see, including clothes, water, food, plants, and animals. It even makes up some things you cannot see, such as air or the smell of perfume. You can describe a type of matter by its **MATERIAL PROPERTIES** such as its colour or how hard it is. Matter is made up of **PARTICLES** so tiny that only the most powerful microscope can see them.

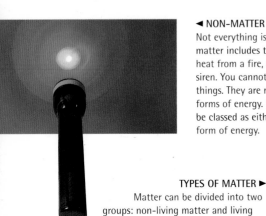

◄ NON-MATTER

Not everything is made of matter. Non-matter includes the light from a torch, the heat from a fire, and the sound of a police siren. You cannot hold, taste, or smell these things. They are not types of matter, but forms of energy. Everything that exists can be classed as either a type of matter or a form of energy.

STATES OF MATTER ►

All matter on Earth is in one of three different states (forms): solid, liquid, or gas. Solids, such as the firefighter's visor, keep their shape. Liquids, such as water, don't keep their shape, but always take up the same amount of space. Gases, such as the gases in the smoke, flow to fill whatever space is available.

TYPES OF MATTER ►

Matter can be divided into two groups: non-living matter and living matter. Non-living matter does not move on its own, grow, or reproduce. The rocks that make up the Earth are examples of non-living matter. All living things, including animals and plants, are living matter.

NON-LIVING MATTER   LIVING MATTER

MATERIAL PROPERTIES

Different types of matter have different material properties that make them useful for different jobs. A plastic hosepipe is flexible, so it can be pointed in any direction. A perspex visor is transparent, so the wearer can see straight through it. A fire-fighter's suit is shiny so it can reflect heat and light. Flexibility, transparency, and shininess are three examples of material properties.

◄ COLOUR

Colour is a very obvious material property. The bright colours of this Queen Alexandra's Birdwing butterfly warn off predators and help it to attract a mate. Matter can be brightly coloured, dull, or transparent. Glass is an example of a transparent material.

SHININESS ►

Like most metals, the glittering, stainless steel metal of the Walt Disney Concert Hall in Los Angeles, USA, is a very shiny material. Shininess is the ability of a material to reflect light. Shiny materials, such as stainless steel, reflect light very well.

DENSITY ►

Density is the amount of matter packed into a space. Lead, for example, is very dense. A small cube of lead has a lot of matter packed into a small space, so it feels very heavy. Flour is not as dense. To balance a set of scales, only two small lead weights are needed in comparison to a much larger pile of flour.

Lead weights are more dense than flour

PARTICLES IN
GASEOUS STATE

*Gases in the smoke
are released from
burning objects*

*Visor is made from
a transparent solid
so the firefighter
can see through it*

*Water in liquid
state flows out
of the hose*

PARTICLES IN
SOLID STATE

PARTICLES IN
LIQUID STATE

PARTICLES

All matter is made of incredibly tiny particles called atoms. Atoms are far too small to see with our eyes, but scientists have worked out how small they are. There are many kinds of atom. Sand grains are made of two kinds of atom: oxygen and silicon. People are made of about 28 different kinds of atom. Material properties depend on the kinds of atom the material is made from.

DEMOCRITUS
Greek, 460–c. 370 BC

Democritus was one of the first philosophers (thinkers) to say that everything was made up of particles too small to be seen. He believed these particles could not be destroyed or split. Democritus said that all changes in the world could be explained as changes in the way particles are packed together.

▲ SAND PARTICLES
Grains of sand look like pieces of gravel when viewed through a microscope. They have different shapes and sizes. Each grain contains millions of atoms, too small to see with a microscope. A sand grain the size of the full stop at the end of this sentence would contain about 10 million million million atoms.

◀ ATOMS
We cannot really see atoms with microscopes. The best we can do is image them, by bouncing light off the particles. A computer translates the light beams into an image. Scanning tunnelling microscopes (STM) and atomic force microscopes (AFM) do this. This STM image shows atoms in a section of a grain of sand.

FIND OUT MORE ▶▶ Atoms **24–25** • Gases **15** • Gravity **72** • Liquids **14** • Solids **12–13**

SOLIDS

Solids are one of the three states of matter and, unlike liquids or gases, they have a definite shape that is not easy to change. Different solids have particular properties such as stretch, **STRENGTH,** or hardness that make them useful for different jobs. Most solids are made up of tiny crystals. This is because their particles are arranged in a regular pattern, called a **CRYSTALLINE STRUCTURE.**

Criss cross of steel beams strengthen road

◄ PARTICLE STRUCTURE
Solids behave as they do because of the way their particles are arranged. The particles of a solid are linked by strong forces, which pull the particles tightly together. So, although the particles can vibrate, they cannot move about easily. This arrangement explains why solids usually keep their shape and feel firm.

◄ CHANGING SHAPE
Some solids, such as the metal in this car bonnet and the plastic bumper, can be hammered or squashed into many different shapes without breaking. They are known as malleable materials. Other solids, such as biscuits or glass, will not bend when hammered or squashed, but will break and split. These materials are brittle.

Metal body crumples under impact

Plastic bumper bends and changes shape

▲ METAL BRIDGE EXPANSION GAPS
Metal bridges always have spaces, known as expansion gaps, built into them. This is because metals expand when heated. Heat makes the metal particles vibrate more energetically and so take up more space. The gaps allow the metal to expand in hot weather without the road buckling.

STRETCH

Some solids, such as the metal copper, can be pulled and stretched easily into extremely thin wires. They are known as ductile materials. They have this property because their particles are not held in a rigid structure, but are arranged in rows that can slide past one another. Copper can be stretched into a thread half the width of a human hair, and is used in many kinds of wiring, including electrical and telephone wiring.

Copper wire

UNSTRETCHED SOLID

STRETCHED SOLID

Metal clip springs back to remembered shape

Metal clip moves with the broken bone, so ballerina can still dance

◄ SHAPE MEMORY METAL
Shape memory metals can remember their shape. When brought to a certain temperature, these metals can be set to a shape that they never forget. They have many uses, including repairing broken bones. Even if the bones move, the metal always returns to its original shape, bringing the bones back to their correct position.

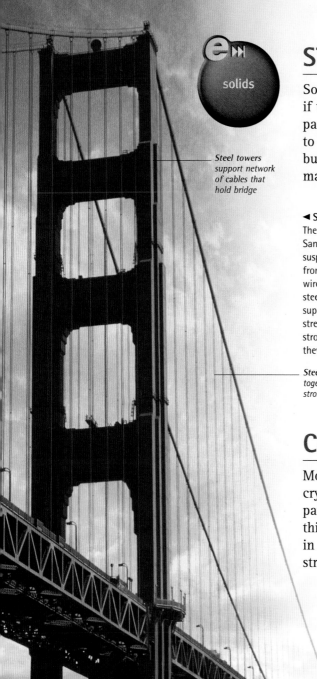

Steel towers support network of cables that hold bridge

STRENGTH

Some solids, such as steel or concrete, are difficult to break, even if they are made to carry a heavy weight. This is because their particles are bound together very strongly. Such materials are said to have high strength and are used to construct bridges and buildings. Strength is a different property from hardness. Hard materials cannot be bent or scratched easily.

◄ SUSPENSION BRIDGE
The Golden Gate Bridge in San Francisco, USA, is a suspension bridge. It is made from thousands of steel wires bound into cables, steel towers, and concrete supports. The bridge's strength comes from the strong materials and the way they have been combined.

Steel wires bound together to make strong cables

MOHS HARDNESS SCALE

MINERAL	MOH
Diamond	10
Corundum	9
Topaz	8
Quartz	7
Feldspar	6
Apatite	5
Fluorite	4
Calcite	3
Gypsum	2
Talc	1

Hardness is a measure of how easily a material can be scratched. Mohs hardness scale arranges 10 minerals from 1 to 10. The higher the number, the harder the mineral. Each mineral in the scale will scratch all those below it. Other materials can be compared to these minerals. Copper, for example, has a hardness of 2.5.

DIAMOND

CRYSTALLINE STRUCTURE

Most solids, such as metals, salt, and sugar, are made up of tiny crystals. Their particles are arranged in regular three-dimensional patterns such as cubes or hexagonal shapes. Not all solids are like this, however. The particles of glass, for example, are not arranged in a regular pattern, and so glass does not have a crystalline structure. Its structure is described as amorphous.

SALT'S STRUCTURE ►
Table salt is made up of thousands of tiny crystals, too small to see without the help of a scanning electron microscope. Salt crystals are cubes. Here, they have been tinted turquoise so their structure can be seen clearly. Crystals form many geometric shapes including cubes, pyramids, hexagons, and prisms.

Six-sided crystal with sharp points at both ends

QUARTZ CRYSTAL ►
Quartz crystals are big enough to see with your eyes. It is the most common mineral and can be found in many rocks. Most of the sand on Earth is made up of grains of quartz. In its pure form, it is transparent, but impurities can transform (change) it into many different colours.

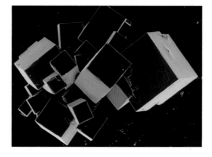

Concrete support forms stable platform for pillar to sit on

FIND OUT MORE ►► • Elasticity **69** • Metals **34–35** • Microscopes **116**

LIQUIDS

As water flows along a river, it constantly changes its shape to fit the space available. This is because water is a liquid, and liquids flow and do not have a fixed shape. Instead, they take on the shape of whatever container they are in. If you pour a liquid from a glass onto a plate, the volume of liquid (the space it takes up) stays the same, but its shape changes.

Viscous crude oil sticks to tube

Rate of flow is slow because the liquid has high viscosity

◄ POWER OF FLOW
A fast-flowing liquid, such as the water rushing over this waterfall, has a lot of energy. The power of flowing water can be used to turn wheels to drive machinery and even create electricity. Fast-moving liquids, such as tidal waves, can also cause a lot of damage.

◄ VISCOSITY
A measure of how fast or slowly a liquid can flow is its viscosity. Crude oil, for example, is a liquid that does not flow very easily. It is said to have high viscosity. Heating crude oil lowers its viscosity and enables it to flow more freely through pipes. Other liquids, such as water, flow easily without being heated. Water has low viscosity.

◄ LIQUID PARTICLES
The forces between liquid particles are weaker than the forces between solid particles. This means that liquid particles are further apart and can move about more easily. Since the particles can move, the liquid can flow and take the shape of its container.

Liquid mercury forms small balls on a surface

Crude oil drips very slowly

▲ COHESION
Mercury is a liquid metal that is poisonous. When mercury is dropped onto a surface, it rolls off in little balls. This is because the forces between the mercury particles are very strong, so the particles clump together. This force between particles of the same type is called cohesion. Water particles do not have such strong cohesion, so they wet surfaces.

liquids

◄ VOLUME
Although they look very different, these two containers contain the same volume of liquid. The volume of a liquid is the amount of space it takes up. Although liquids change their shape when moved from one container to another, their volume always stays the same. For this reason, liquids are usually measured by their volume, in litres or gallons.

Light legs of pond-skater do not break water surface

Volume of liquid in a tall container is identical to volume in a short container

▲ SURFACE TENSION
Some insects, such as pond-skaters, are able to walk on water without sinking. This is because the forces of attraction between the water particles pull the particles at the surface together. This creates a tension, called surface tension, that makes the water surface behave as if an invisible, stretchy skin covers it.

Liquid has changed shape but not volume

FIND OUT MORE ►► Changing State 16–17 • Energy 76–77 • Forces 64–65

GASES

Gases are all around us, but although many, such as perfume, can be smelt, most gases are invisible. Like liquids, gases can flow but, unlike solids or liquids, gases will not stay where they are put. They have no set shape or volume, and they expand in every direction to fill completely whatever container they are put into. If the container has no lid, the gas escapes.

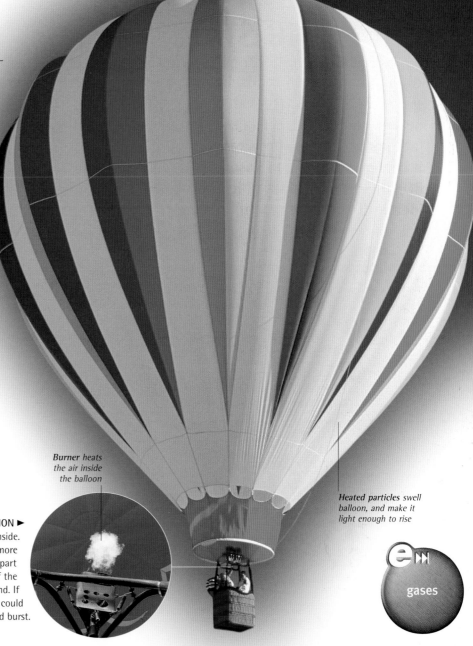

Burner heats the air inside the balloon

Heated particles swell balloon, and make it light enough to rise

◄ GAS PARTICLES
Gas particles move around at over 1,600 km (1,000 miles) per hour. The particles are widely spaced and can move freely in any direction. Gases can spread out to fill whatever container they are put into. When gas particles collide, the forces between them are not strong enough to keep them together – instead they bounce apart.

EXPANSION ►
In hot air balloons, a burner heats the air inside. This causes the particles of air to gain more energy and so they move faster and farther apart from one another, pushing at the sides of the balloon. Heat always causes gases to expand. If you left a balloon near a fire, the air inside could expand so much that the balloon would burst.

Gas particles forced closer together

◄ COMPRESSION
Gases can be easily squashed, or compressed. When you push a bicycle pump, for example, you are squeezing the air inside into a smaller space. The air particles are forced closer together, and bang against each other and against the sides of the pump.

Wall of pump warms up as gas particles bang against it

PRESSURE ►
Why does a champagne cork explode out of a shaken bottle? The champagne inside the bottle contains lots of tiny bubbles of gas. Shaking the bottle releases the gas, and the high-speed gas particles bang against the cork. This creates an enormous pressure on the cork, and eventually forces the cork out of the bottle.

▲ VAPOUR
Vapour is a gas that has evaporated from a liquid before the liquid has reached its boiling point. Water, for example, boils to form a gas at 100°C (212°F). But, even at much lower temperatures, some water particles escape from the liquid to form a gas, called vapour, that mixes with the air. When vapour cools slightly, the gas forms droplets seen as mist.

gases

FIND OUT MORE ►► Changing States 16–17 • Pressure 74–75 • Water 40–41

CHANGING STATES

All matter exists as solids, liquids, or gases. These are called the states of matter. Matter can change from one state to another if heated or cooled. If ice (a solid) is heated it changes to water (a liquid). This change is called **MELTING**. If water is heated, it changes to steam (a gas). This change is called **BOILING**. The particles of ice, water, and steam are identical, but arranged differently.

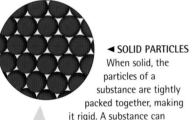

◄ **SOLID PARTICLES**
When solid, the particles of a substance are tightly packed together, making it rigid. A substance can change from a solid state to a liquid state, and from a liquid state to a solid state.

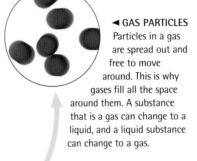

◄ **GAS PARTICLES**
Particles in a gas are spread out and free to move around. This is why gases fill all the space around them. A substance that is a gas can change to a liquid, and a liquid substance can change to a gas.

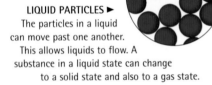

LIQUID PARTICLES ►
The particles in a liquid can move past one another. This allows liquids to flow. A substance in a liquid state can change to a solid state and also to a gas state.

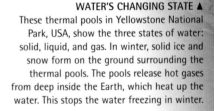

WATER'S CHANGING STATE ▲
These thermal pools in Yellowstone National Park, USA, show the three states of water: solid, liquid, and gas. In winter, solid ice and snow form on the ground surrounding the thermal pools. The pools release hot gases from deep inside the Earth, which heat up the water. This stops the water freezing in winter.

FREEZING

MELTING

MELTING

When a solid is heated, the particles are given more energy and start to vibrate faster. At a certain temperature, the particles vibrate so much that their ordered structure breaks down. At this point the solid melts into liquid. The temperature at which this change from solid to liquid happens is called the melting point. Each solid has a set melting point at normal air pressure. At lower air pressure, such as up a mountain, the melting point lowers.

◄ **FREEZING**
Lava is liquid rock, which erupts through a volcano at temperatures as high as 1,500°C (2,732°F) through a volcano. However, the red-hot lava cools as it meets the Earth's surface, and turns back into solid rock again. This change from liquid to solid is called freezing or solidifying. It is the opposite process to melting.

changing states

▲ SOLAR PLASMA
The glowing corona of the Sun visible during a total eclipse is made of a fourth state of matter called plasma. Plasma is formed when a lot more energy is given to a gas, such as by heating the gas or passing electricity through it. This extra energy splits the particles of the gas into even smaller pieces so hot that they glow.

HOAR FROST ►
Hoar frost creates fine needles of solid ice on leaves. If the temperature is below -9.5°C (15°F), water vapour (gas) in the air changes straight to solid ice on leaves, without going through the liquid phase. This is known as sublimation. Most gases, when cooled, turn to liquids first, and then to solids if cooled further.

BOILING

When a liquid is heated, the particles are given more energy. They start to move faster and further apart. At a certain temperature, the particles break free of one another and the liquid turns to gas. This is the boiling point. The boiling point of a substance is always the same; it does not vary.

CONDENSATION

STEAM

BOILING

INVISIBLE STEAM ►
Water boils when it reaches its boiling point of 100°C (212°F). This is the temperature at which water turns to steam. Steam is an invisible gas. When it reaches the lid it cools back to a liquid.

▼ EVAPORATION
Even without boiling water in a kettle, some of the liquid water changes to gas. This is evaporation. It occurs when a liquid turns into a gas far below its boiling point. There are always some particles in a liquid that have enough energy to break free from the rest to become a gas.

CONDENSATION ►
Dewdrops are often found on a spider's web early in the morning after a cold night. Water that is present as a gas in the air cools down and changes into tiny drops of liquid water on leaves and windows. This change from gas to liquid is called condensation.

FIND OUT MORE ►► Gases 15 • Liquids 14 • Solids 12–13 • Water 40–41

MIXTURES

Almost everything is made of different substances mixed together. Things are only easy to recognize as mixtures if the **PARTICLE SIZE** of each substance is big enough to see. The flakes, nuts, and raisins in a bowl of cereal are a mixture that is easy to see. A fruit drink, though, doesn't look like a mixture because the particles of fruit and water are so small. It is a type of mixture called a **SOLUTION**, made of different, very tiny particles dissolved (evenly spread out) in water.

PARTICLE SIZE

There are many different types of mixtures, which are divided into groups based on how small their particles are. A mixture such as sand has a large particle size. Mud stirred in water is a type of mixture called a suspension; the particles are too small to see when mixed, but they eventually settle out. A mixture such as fog (water and air) is called a colloid; its particles are too small ever to settle out.

◄ COARSE MIXTURES
The particles of some mixtures are large enough to see without a microscope. When you look closely at a handful of sand, for example, you can make out the different coloured grains mixed together. Some sands have smaller grains than others. The smaller the grain size, the softer and more powdery the sand feels.

Sand *contains quartz*

COMMON PARTICLES ►
Rock, sand, and seawater are all mixtures of the same substances – such as the minerals feldspar, mica, and quartz – but in different particle sizes. Rock contains these substances in chunks or veins; sand has them as small grains; and seawater contains them as tiny dissolved particles that are invisible to the eye. Rain and rivers dissolve the minerals as they wash over the rock on their way to the sea.

Seawater contains minerals dissolved from rock

mixtures

Dye particles scattered through glass marble

COLLOID ►
A colloid is a mixture containing tiny particles of one substance scattered throughout another substance, such as dye particles mixed with glass in a marble. The particles are smaller than those in a suspension, but larger than those in a solution. The particles are so small and light, they do not ever settle out.

Dust particles suspended in the air

▲ SUSPENSION
When a volcano releases a huge cloud of dust, the dust is actually a mixture of solid ash (powder from burnt substances) and gases, such as carbon dioxide. This cloud of dust is suspended in the air for a while, but eventually the fine ash particles clump together, fall to Earth, and cover the ground. The volcanic dust cloud is an example of a suspension.

EMULSION ►
Milk is made up of tiny globules of fat scattered throughout water. It is an example of an emulsion, a special type of colloid in which oils or fats are mixed with water to create a creamy liquid or paste. Other examples of emulsions are mayonnaise, emulsion paints, lipsticks, and face creams.

Fat particles dispersed in water

Rock is the original source for minerals

Sand formed from rock particles weathered into tiny grains

MINERAL MIXTURES

All rocks are mixtures of naturally occurring substances called minerals. Granite is a common rock made of three differently coloured minerals called feldspar, mica, and quartz. The pink grains in granite are feldspar, the black grains are mica, and the light grey, glass-like grains are quartz. Granite is usually about 75% feldspar, 5% mica, and 20% quartz. These proportions can vary and the rock often contains small amounts of other minerals as well.

FELDSPAR   MICA   QUARTZ

SOLUTIONS

A solution is a mixture in which the different particles are tiny and are mixed completely evenly. Solutions are often made by dissolving a solid, such as sugar, into a liquid, such as water. The sugar is called the solute and the water is called the solvent. Water is the most common solute. Solutions can also be a liquid dissolved in another liquid, for example antiseptic liquid. This is water and alcohol. Or they can be a gas dissolved in another gas, such as oxygen dissolved in nitrogen in the air.

SOLID SOLUTION ▶
Wood's Metal is found in automatic fire sprinklers. It is an alloy (mixture of metals) containing bismuth, lead, tin, and cadmium. This mix of metals has a low melting point of 71°C (158°F). It is used as a sensor in automatic fire sprinklers; if the temperature gets too high, the metal alloy melts and releases the water.

◄ GAS SOLUTION
Another type of solution occurs when a gas is dissolved in a liquid. When a vitamin tablet is dissolved in water, carbon dioxide is produced to help the tablet dissolve quickly. Sparkling water is also a solution of carbon dioxide in water. When the gas is in solution, you cannot see it. It is only when the gas comes out of solution and bubbles to the surface of the liquid that you can see the gas that was once dissolved.

FIND OUT MORE ▶▶ • Erosion 222–223 • Metals 34–35 • Rocks 218–219 • Separating Mixtures 18–19

SEPARATING MIXTURES

The substances in a mixture are separated by the differences in their physical properties, such as their particle size. The more different the properties are, the easier it is to separate the substances. Tea leaves do not dissolve in water, so you can use a strainer to **FILTER** them. The particles in other mixtures can be far smaller; in **CHROMATOGRAPHY**, microscopic substances are separated by how easily each sticks to another substance.

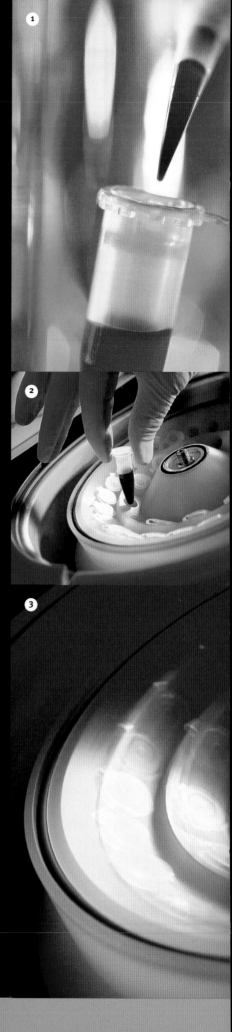

DECANTING GOLD ▶
To search for tiny particles of gold in rivers, a mixture of sand, mud, and gravel is scooped up in a pan and swirled around. Gold particles are heavier than the other particles, so they settle to the bottom of the pan. The lighter particles stay suspended in the water, and are decanted (poured off). This technique of panning for gold is called decanting. Cream is also separated from milk by decanting – the cream is less dense than the milk.

Thermometer shows temperature of evaporating gas

◀ DISTILLATION
In distillation, a mixture of liquids is heated in a flask. The liquid with the lower boiling point evaporates (changes to a vapour) first, and is condensed (changes back to a liquid) and collected. The liquid with the higher boiling point and any solid particles are left behind in the flask. Fractional distillation separates liquids one by one as they boil. The oil industry separates crude oil using this technique.

separating mixtures

Vapour enters the condenser's inner tube

Vapour is cooled by cold water surrounding the inner tube

Water leaves the condenser's outer tube

Mixture contains a solution of different substances

Cold water enters the condenser's outer tube

Bunsen burner heats the mixture

Condensed liquid drips into the flask

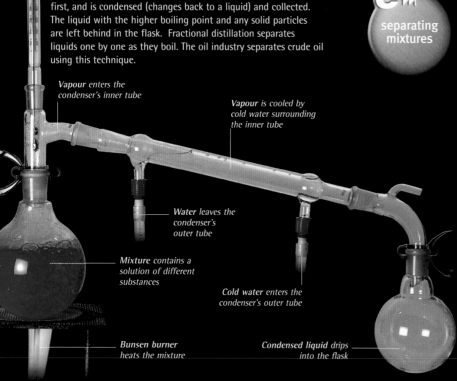

FILTRATION

When the substances in a mixture have different particle sizes, they are separated by filtration. The mixture is poured through a sieve or filter. The smaller particles slip through the holes, but the larger particles do not. Filtration is the first stage in water recycling. Chemists use filters called zeolites, which have holes so tiny that they can remove microscopic particles from water.

Dirty water is a mixture of solid particles and liquid

Largest particles are trapped in the gravel layer

Filtered water contains only liquid

▲ FILTERING DIRTY WATER
You can turn dirty water into clear water using a filter. Place a container with a hole in the bottom inside another container and line it with filter paper. Fill the container with layers of charcoal, sand, and gravel. Pour dirty water into the container. The layers will filter out smaller and smaller particles of dirt. The result is clearer (but not necessarily drinkable) water.

◄ CENTRIFUGING A MIXTURE
A centrifuge is like an extra-fast spin dryer. It spins a liquid so quickly that the particles separate out. The heavier particles sink to the bottom and the lighter particles collect at the top. Doctors separate blood samples for analysis (study) using a centrifuge.

1 A microtube is filled with the blood to be separated. It is the heavier, red blood cells that give blood its red colour.

2 The microtube is placed in a secure holder in the centrifuge. A centrifuge holds up to 50 microtubes.

3 The cover of the centrifuge is shut firmly, and the centrifuge spins at around 4,000 revolutions per minute.

4 The red blood cells settle to the bottom of the tube, and the yellow, liquid plasma rises to the surface.

Red blood cells can be frozen for later use

CHROMATOGRAPHY

Scientists separate many liquid mixtures using chromatography. The mixture is dissolved in a liquid or a gas to make a solution. The solution is put on a solid material and the substances that dissolved most easily travel farthest up the solid material. The separated substances form bands of colour called chromatograms. Food scientists study chromatograms to discover which colourings a food contains.

Clip holds filter paper in place

Pole laid along top of jars to support clips

Blue dye travels to top of paper

Water moves up the paper, carrying the colouring with it

Bottom edge of filter paper placed in water

PAPER CHROMATOGRAPHY ▲
Food scientists separate food colouring for analysis using paper chromatography. A drop of colouring is put onto filter paper. The edge of the filter paper is dipped in water. As the water flows up through the paper, it carries the colours with it. Some colours travel faster than others, so the substances split into different coloured bands.

THIN LAYER CHROMATOGRAPHY ►
Genetic scientists use thin layer chromatography (TLC) to study the substances that make up our genes. In TLC, the solid material is a plate of glass or plastic coated with a chemical, usually aluminium oxide or silicon oxide. When the liquid mixture travels up the plate, some of the substances move farther up the plate than others. The substances appear as spots on the plate. Scientists study genes to learn about inherited characteristics.

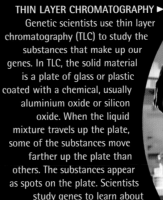

FIND OUT MORE ▶▶ Chemical Industry 50–51 • Genetics 364–365 • Mixtures 18–19 • Sediments 225

The enormous variety of matter around you is made from different combinations of substances called elements. Elements are pure substances that cannot be broken down into anything simpler. Some, such as gold and silver, are found on their own. Most elements, however, are combined in twos, threes, and more to make compounds. The **NATURAL ELEMENTS** are found on Earth. **SYNTHETIC ELEMENTS** are created in laboratories and are often short-lived.

ORIGIN OF ELEMENTS ►
All the elements on Earth were formed in the heart of exploding stars. The early Universe was made of just two elements, hydrogen and helium, which formed into stars. At the fiery core of these stars, the hydrogen and helium were forced together to form new, heavier elements. Even heavier elements were created in the explosions of massive stars, called supernovas.

Iron filings

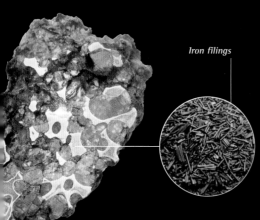

▲ METEORITE
An element is always the same, wherever it is found. For example, meteorites are large rocks that have landed on Earth from space. Some meteorites contain metal, such as iron, which is a natural element. The iron in a meteorite from space is exactly the same as iron found in rocks on Earth.

PURE GOLD BARS ▲
Very few of the natural elements are found on their own. Most occur in compounds with other elements. The metal gold is one of the exceptions. It is found as pure gold in veins (small cracks) of rocks or as deposits (nuggets) in the earth. Pure gold contains only particles of gold and nothing else.

JOHN DALTON
English, 1766–1844

Chemist John Dalton studied the gases in air. He proposed that everything was made from simple substances called elements. He said that the properties (characteristics) of every particle of one element are identical, and are different to the properties of any other element. This is how elements are defined today.

ELEMENTS IN THE EARTH'S CRUST

ELEMENT	PERCENTAGE
Oxygen	47
Silicon	28
Aluminium	8
Iron	5
Calcium	3.5
Sodium	3
Potassium	2.5
Magnesium	2
All other elements	1

Star's outer layers are pushed away in an enormous explosion

elements

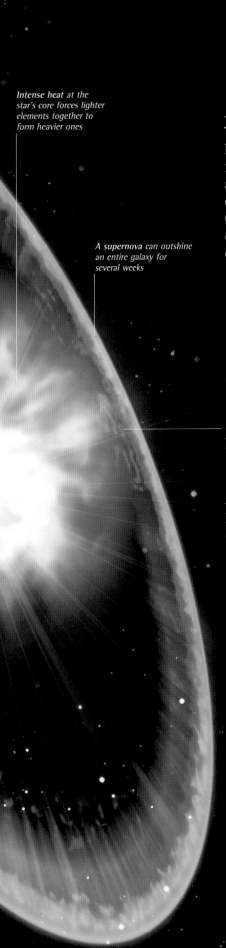

Intense heat at the star's core forces lighter elements together to form heavier ones

A supernova can outshine an entire galaxy for several weeks

Elements created by a supernova go on to create new stars

NATURAL ELEMENTS

There are 90 natural elements, ranging from the lightest, hydrogen, to the heaviest, uranium. Other familiar elements are aluminium, carbon, copper, and oxygen. Every substance on Earth is made up of one or more of these 90 elements. Oxygen is the most common element on Earth. Hydrogen is the most common element in the Universe.

ALUMINIUM ▶
Aluminium is a common element, but it is never found naturally on its own. It has to be extracted from rocks called minerals. This extraction process used to be very difficult and aluminium was once considered a precious metal, more valuable than gold. Nowadays, extraction is much easier, and aluminium is used for many everyday items, such as drinks cans and foil.

Solid aluminium where a spillage has landed and cooled

Liquid aluminium flowing into moulds to cool into solid metal

SYNTHETIC ELEMENTS

No element heavier than uranium is found naturally. Scientists can, however, collide two smaller elements together at high speeds to form a new, heavier element. Many elements made this way break apart almost immediately, although a few can stay together for a few days or even weeks. Scientists make them to learn more about how elements form and how they change as they get heavier. Synthetic elements include plutonium and einsteinium.

Cyclotron ring where fast-moving particles collide and form heavier, new particles

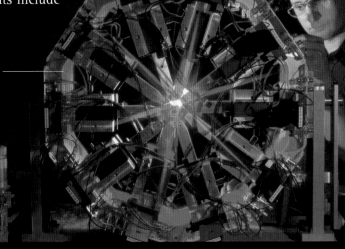

▲ CYCLOTRON
Scientists create synthetic elements in a cyclotron. The cyclotron contains a circular track, into which particles are released. The particles are speeded up to extremely high speeds, before being allowed to collide with a target of another element to form new elements. In the largest cyclotrons, the ring is many kilometres wide and the particles speed at 225,000 kph (13,809 mph).

FIND OUT MORE ▶▶ Atoms **24** • Hydrogen **38** • Metals **34** • Oxygen **39** • Periodic Table **26** • Supernovas **169**

An atom is the smallest part of an element that can exist on its own. Copper, for example, is made from copper atoms, which are different to the oxygen atoms that make up oxygen. Atoms are so tiny that even the full stop at the end of this sentence has a width of around 20 million atoms. Inside each atom are even smaller particles, called subatomic particles. These include a nucleus, which contains protons and neutrons, and electrons that whizz around the nucleus.

◄ FOOTBALL STADIUM
Imagine an atom magnified to the size of a football stadium. The nucleus of the atom would be the size of a pea in the centre of the stadium, and the electrons would be whizzing around the outer stands. Everything in between would be empty space.

NIELS BOHR
Danish, 1885–1962

In 1913, Bohr published his model of atomic structure in which electrons travelled in orbits around the central nucleus. He also introduced the idea of electron shells, saying that the properties of an atom depended on how its electrons were arranged in the shells. In 1922, Bohr was awarded the Nobel Prize for Physics.

NUCLEUS ►
The nucleus is a tightly bound cluster of protons and neutrons. This carbon atom nucleus has 6 protons and 6 neutrons. Protons have a positive electric charge and neutrons have no charge. Positively charged protons would normally repel each other, but the nucleus is held together by a powerful force called the strong nuclear force.

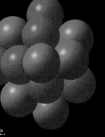

NUCLEUS

PROTON

NEUTRON

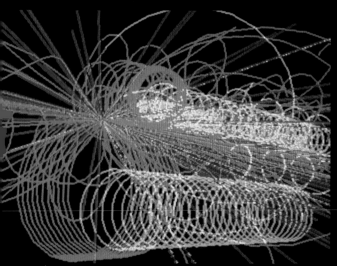

◄ PARTICLE TRACKS
Nuclear scientists smash subatomic particles together at very high speed in a machine called a collider to discover what the particles are made of. This breaks them up into even smaller particles. Their collisions leave tracks, which are processed into an image by a computer. Each particle has its own distinctive track.

Each electron's orbit is different

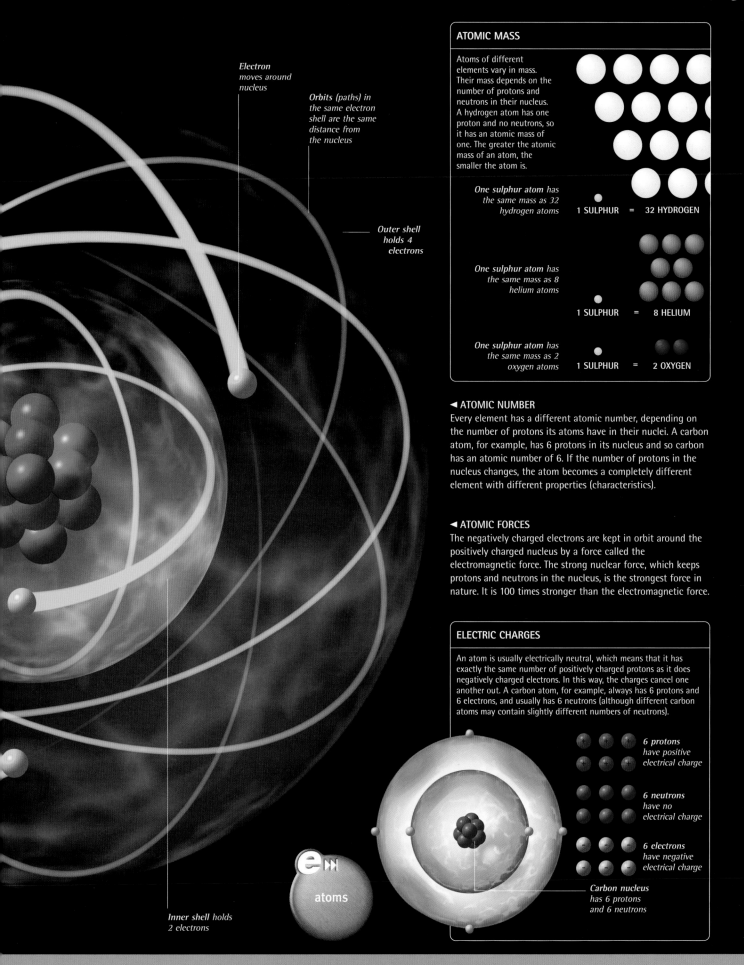

*Electron
moves around
nucleus*

*Orbits (paths) in
the same electron
shell are the same
distance from
the nucleus*

*Outer shell
holds 4
electrons*

ATOMIC MASS

Atoms of different
elements vary in mass.
Their mass depends on the
number of protons and
neutrons in their nucleus.
A hydrogen atom has one
proton and no neutrons, so
it has an atomic mass of
one. The greater the atomic
mass of an atom, the
smaller the atom is.

*One sulphur atom has
the same mass as 32
hydrogen atoms*

1 SULPHUR = 32 HYDROGEN

*One sulphur atom has
the same mass as 8
helium atoms*

1 SULPHUR = 8 HELIUM

*One sulphur atom has
the same mass as 2
oxygen atoms*

1 SULPHUR = 2 OXYGEN

◄ ATOMIC NUMBER

Every element has a different atomic number, depending on
the number of protons its atoms have in their nuclei. A carbon
atom, for example, has 6 protons in its nucleus and so carbon
has an atomic number of 6. If the number of protons in the
nucleus changes, the atom becomes a completely different
element with different properties (characteristics).

◄ ATOMIC FORCES

The negatively charged electrons are kept in orbit around the
positively charged nucleus by a force called the
electromagnetic force. The strong nuclear force, which keeps
protons and neutrons in the nucleus, is the strongest force in
nature. It is 100 times stronger than the electromagnetic force.

ELECTRIC CHARGES

An atom is usually electrically neutral, which means that it has
exactly the same number of positively charged protons as it does
negatively charged electrons. In this way, the charges cancel one
another out. A carbon atom, for example, always has 6 protons and
6 electrons, and usually has 6 neutrons (although different carbon
atoms may contain slightly different numbers of neutrons).

*6 protons
have positive
electrical charge*

*6 neutrons
have no
electrical charge*

*6 electrons
have negative
electrical charge*

*Carbon nucleus
has 6 protons
and 6 neutrons*

atoms

*Inner shell holds
2 electrons*

FIND OUT MORE ►► Elements **22–23** • Forces **64–65** • Matter **10–11** • Periodic Table **26–27**

PERIODIC TABLE

DIMITRI MENDELEYEV
Russian, 1834-1907

This chemist was convinced there was an order to the elements. He collected information on each one and, in 1869, he published a table of elements on which the modern periodic table is based. He left gaps for elements he predicted would be found, such as gallium, germanium, and scandium.

At first glance, the periodic table looks very complex. In fact it is a large grid of every element that exists. The elements are arranged in order of their atomic number. The atomic number is the number of protons each atom has in its nucleus. By arranging the elements in this way, those with similar properties (characteristics) are grouped together. As with any grid, the periodic table has rows running left to right, and columns running up and down. The rows are called PERIODS and the columns are called GROUPS.

KEY

The elements in the Periodic Table can be colour coded to show the nine different groupings. Hydrogen does not belong to any one group.

- Alkali metals
- Alkali-earth metals
- Transition metals
- Rare earths
- Radioactive rare earths
- Other metals
- Semimetals
- Non-metals
- Noble gases
- Hydrogen

◄ READING THE PERIODIC TABLE

Hydrogen (H) is the first element in the periodic table because it has just one proton in its nucleus. Helium (He) is second, because it has two protons, and so on. The periodic table can be coloured-coded. Often, each group is given a particular colour so that it is easy to pick out all the elements that belong to a particular group.

1 **H** Hydrogen 1																	2 **He** Helium 4
3 **Li** Lithium 7	4 **Be** Beryllium 9											5 **B** Boron 11	6 **C** Carbon 12	7 **N** Nitrogen 14	8 **O** Oxygen 16	9 **F** Fluorine 19	10 **Ne** Neon 20
11 **Na** Sodium 23	12 **Mg** Magnesium 24											13 **Al** Aluminium 27	14 **Si** Silicon 28	15 **P** Phosphorus 31	16 **S** Sulphur 32	17 **Cl** Chlorine 35	18 **Ar** Argon 40
19 **K** Potassium 39	20 **Ca** Calcium 40	21 **Sc** Scandium 45	22 **Ti** Titanium 48	23 **V** Vanadium 51	24 **Cr** Chromium 52	25 **Mn** Manganese 55	26 **Fe** Iron 56	27 **Co** Cobalt 59	28 **Ni** Nickel 58	29 **Cu** Copper 63	30 **Zn** Zinc 64	31 **Ga** Gallium 69	32 **Ge** Germanium 74	33 **As** Arsenic 75	34 **Se** Selenium 80	35 **Br** Bromine 79	36 **Kr** Krypton 84
37 **Rb** Rubidium 85	38 **Sr** Strontium 88	39 **Y** Yttrium 89	40 **Zr** Zirconium 90	41 **Nb** Niobium 93	42 **Mo** Molybdenum 98	43 **Tc** Technetium 97	44 **Ru** Ruthenium 102	45 **Rh** Rhodium 103	46 **Pd** Palladium 106	47 **Ag** Silver 107	48 **Cd** Cadmium 114	49 **In** Indium 115	50 **Sn** Tin 120	51 **Sb** Antimony 121	52 **Te** Tellurium 130	53 **I** Iodine 127	54 **Xe** Xenon 132
55 **Cs** Caesium 133	56 **Ba** Barium 138	57-71	72 **Hf** Hafnium 180	73 **Ta** Tantalum 181	74 **W** Tungsten 184	75 **Re** Rhenium 187	76 **Os** Osmium 192	77 **Ir** Iridium 193	78 **Pt** Platinum 195	79 **Au** Gold 197	80 **Hg** Mercury 202	81 **Tl** Thallium 205	82 **Pb** Lead 208	83 **Bi** Bismuth 209	84 **Po** Polonium 209	85 **At** Astatine 210	86 **Rn** Radon 222
87 **Fr** Francium 223	88 **Ra** Radium 226	89-103	104 **Rf** Rutherfordium 261	105 **Db** Dubnium 262	106 **Sg** Seaborgium 266	107 **Bh** Bohrium 264	108 **Hs** Hassium 277	109 **Mt** Meitnerium 268	110 **Ds** Darmstadtium 271	111 **Rg** Roentgenium 272	112 **Uub** Ununbium 285	113 **Uut** Ununtrium 284	114 **Uuq** Ununquadium 289	115 **Uup** Ununpentium 288	116 **Uuh** Ununhexium 292	117 **Uus** Ununseptium (undiscovered)	118 **Uuo** Ununoctium 294

57 **La** Lanthanum 139	58 **Ce** Cerium 140	59 **Pr** Praseodymium 141	60 **Nd** Neodymium 142	61 **Pm** Promethium 145	62 **Sm** Samarium 152	63 **Eu** Europium 153	64 **Gd** Gadolinium 158	65 **Tb** Terbium 159	66 **Dy** Dysprosium 164	67 **Ho** Holmium 165	68 **Er** Erbium 168	69 **Tm** Thulium 169	70 **Yb** Ytterbium 174	71 **Lu** Lutetium 175
89 **Ac** Actinium 227	90 **Th** Thorium 232	91 **Pa** Protactinium 231	92 **U** Uranium 238	93 **Np** Neptunium 237	94 **Pu** Plutonium 244	95 **Am** Americium 243	96 **Cm** Curium 247	97 **Bk** Berkelium 247	98 **Cf** Californium 251	99 **Es** Einsteinium 254	100 **Fm** Fermium 257	101 **Md** Mendelevium 258	102 **No** Nobelium 255	103 **Lr** Lawrencium 256

periodic table

Atomic number is the number of protons in the atom's nucleus ⎯ **32**

Symbol is used as a shorthand in chemical equations ⎯ **Ge**
Germanium
74

Mass number is the number of protons and neutrons in the nucleus

◄ SYMBOL

As well as a name, each element has a symbol, a shorthand way of writing the element in chemical equations. Often this is the first letter or two of the element's name, but it can come from a Latin name. Each also has an atomic number and a mass number.

GALLIUM ►

One element that Mendeleyev left a gap for in his periodic table was gallium (element 31). Mendeleyev called it eka-aluminium because he predicted it would have similar properties to aluminium. In 1875, French scientist Lecoq de Boisbaudran discovered gallium. It has the exact properties that Mendeleyev predicted. Gallium is a soft, silvery metal with a melting point of 29.8°C (85.6°F).

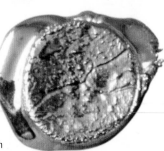

GROUPS

There are 18 groups (columns) in the periodic table. Group 1 (also known as the alkali metals) is the column on the far left of the table. Elements in the same group have similar, but not identical characteristics. This is because they all have the same number of electrons in their outermost shell. You can tell a lot about an element just by knowing which group it is in.

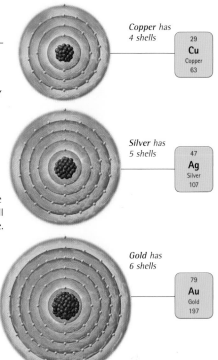

Copper has 4 shells

29 Cu Copper 63

Silver has 5 shells

47 Ag Silver 107

Gold has 6 shells

79 Au Gold 197

INCREASING SIZE ►
As you move down one element in a group, there is a large jump in the number of protons and neutrons in the nucleus, and a new shell of electrons is added. The extra particles make the atom heavier and the extra shell of electrons makes the atom take up more space.

◄ METAL IN SPACE
An astronaut's visor is gold-plated to reflect sunlight. This shiny, hard-wearing metal does not corrode (rust), making it ideal for use in space, where materials cannot be replaced easily. Gold, copper, and silver belong to group 11. Group 11 metals are also called coinage metals, because they are used to make coins.

PERIODS

The properties of the elements across a period (row) change gradually. The first and last elements are very different. The first is a reactive solid – it catches fire when it mixes with oxygen – and the last is an unreactive gas. However, they have the same number of electron shells. All the elements in the third period, for example, have three shells for their electrons.

DECREASING SIZE ▼
As you go across a period, the atoms get slightly heavier, but they also get smaller. This is because the number of electron shells stays the same across the period, but the number of protons in the nucleus increases. The stronger, attractive force from the positively charged protons sucks the negatively charged electrons tighter into the centre.

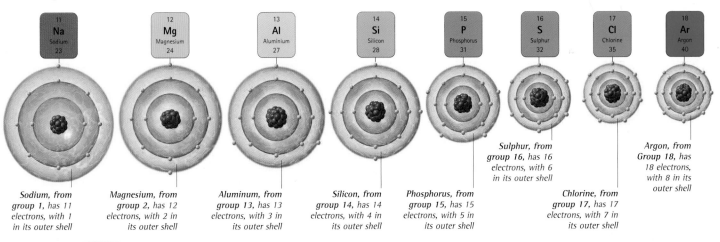

Sodium, from group 1, has 11 electrons, with 1 in its outer shell

Magnesium, from group 2, has 12 electrons, with 2 in its outer shell

Aluminum, from group 13, has 13 electrons, with 3 in its outer shell

Silicon, from group 14, has 14 electrons, with 4 in its outer shell

Phosphorus, from group 15, has 15 electrons, with 5 in its outer shell

Sulphur, from group 16, has 16 electrons, with 6 in its outer shell

Chlorine, from group 17, has 17 electrons, with 7 in its outer shell

Argon, from Group 18, has 18 electrons, with 8 in its outer shell

▲ FIZZING SODIUM
Sodium is a highly reactive metal. This means that it reacts with water and burns violently in air. Because of this, sodium always combines with other elements, and is not found on its own in nature.

◄ PHOSPHORUS
Phosphorus is a non-metal element. There are two kinds. Red phosphorus on the sides of matchboxes help matches ignite. Yellow phosphorus is a more reactive, waxy solid, which catches fire in air. It glows in the dark—an effect called phosphorescence.

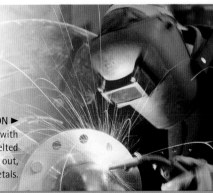

UNREACTIVE ARGON ►
Argon is very unreactive and does not combine with other elements. In arc welding, metals are melted surrounded by argon gas. The argon keeps oxygen out, so that oxygen cannot react with the melted metals.

FIND OUT MORE ►► Atoms 24–25 • Chemical Reactions 30–31 • Elements 22–23 • Metals 34–35

MOLECULES

Most atoms join up with other atoms through chemical **BONDS** to form larger particles called molecules. They can join up with atoms of the same element or with atoms of different elements. Substances whose molecules contain different types of atom are called compounds. Chemical reactions can **CHANGE MOLECULES** and when this happens, new molecules and therefore new compounds are formed.

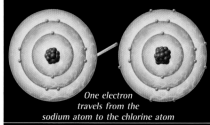

molecules

VARIETY OF MOLECULES ►
Molecules can be simple or complex. They can even be made up of just one atom. The element argon is a one-atom molecule. Other molecules can consist of two atoms of the same element. The oxygen molecule is made up of two oxygen atoms bonded together. However, in certain circumstances, three oxygen atoms bond together, forming a molecule called ozone.

A

Argon

O_2

Oxygen

Oxygen

H_2O

Hydrogen

SIMPLE MOLECULE ▲
Water molecules (H_2O) are very simple. They are made of two hydrogen (H) atoms bonded to one oxygen (O) atom. All water molecules are the same, but they are different from the molecules of any other substance. A water molecule is the smallest possible piece of water. You can break it up into smaller pieces, but they wouldn't be water anymore. The symbols that scientists use to represent molecules are called chemical formulae.

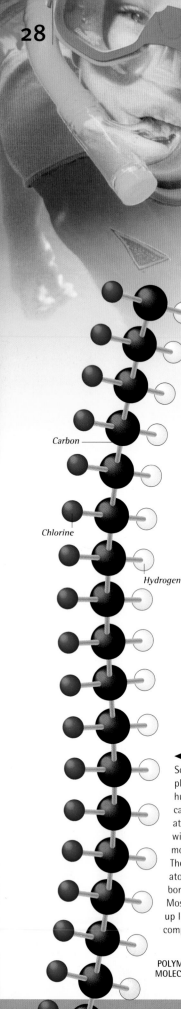

Carbon

Chlorine

Hydrogen

◄ COMPLEX MOLECULE
Some molecules, such as the plastic in a snorkel, contain hundreds or even thousands of carbon, hydrogen, and chlorine atoms joined together in long, winding chains. Such complex molecules are called polymers. They are possible because carbon atoms are able to form very stable bonds with other carbon atoms. Most of the molecules that make up living things are made of complex polymers.

POLYMER MOLECULE

BONDING

When atoms join together to form molecules, they are held together by chemical bonds. These bonds form as a result of the sharing or exchange of electrons between the atoms. It is only the electrons in the outermost shell that ever get involved in bonding. Different atoms use these electrons to form one of three different types of bond: ionic bonds, covalent bonds, or metallic bonds.

DIFFERENT KINDS OF BONDS BETWEEN ATOMS

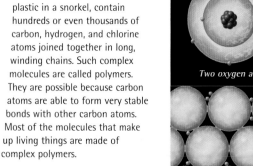

One electron travels from the sodium atom to the chlorine atom

Two oxygen atoms share four electrons

IONIC BONDS
In ionic bonds, electrons are transferred from one atom to another. When sodium and chlorine combine to form sodium chloride (salt), sodium loses an electron and becomes positively charged; chlorine takes that electron and becomes negatively charged. Ionic bonds are difficult to break. Ionic compounds are usually solids with high melting points.

COVALENT BONDS
In a covalent bond, electrons are shared between two atoms. When two oxygen atoms bond together to form an oxygen molecule, they share four electrons – two from each oxygen atom. Other examples of covalent bonding are water (H_2O), and carbon dioxide (CO_2). Covalent compounds are usually liquids or gases with low melting points.

METALLIC BONDS
Metal atoms are bonded to each other through metallic bonding. In this type of bonding, all the atoms lose electrons, which float around in a common pool. The electrons in this pool can move around freely, which is why metals can transfer heat or electricity so well. If one part of the metal is heated, the electrons carry the heat quickly to other parts.

CHANGING MOLECULES

All around you, molecules are changing and rearranging their atoms in chemical reactions to form new molecules and new compounds. When you breathe in oxygen, it goes through a chemical change inside your body and forms a new compound, carbon dioxide, which you breathe out. Catalysts are special types of molecules that speed up chemical reactions, but do not actually change themselves. They are used, for example, in catalytic converters in cars.

Nitrogen oxide

Carbon monoxide

Hydrocarbon

Exhaust gases from the engine contain harmful pollutants

Catalyst made of platinum and rhodium metals

▲ SFX REACTION
A special effects explosion is a chemical reaction that releases energy. Pyrotechnic experts want each explosion to be unique, so they use different types and amounts of explosives. In every chemical reaction, some bonds between atoms are broken and new ones are made. Energy is needed to break a bond, but energy is released when a bond is made. Depending on the number and type of bonds broken and made, a reaction may take in or give out energy.

Bubbles of gas in the dough make it expand

Yeast mixture froths as carbon dioxide is produced

Flour and water kneaded into yeast mixture to form dough

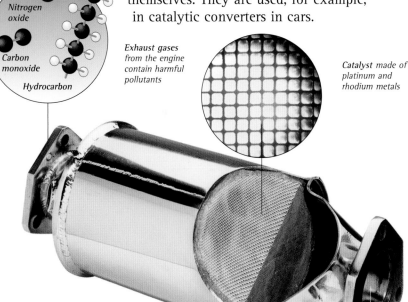

▲ CATALYTIC CONVERTER
When a car engine burns petrol, it releases harmful gases. Cars fitted with a catalytic converter change the harmful gases into safer gases. When they enter the catalytic converter, the gases form temporary bonds with the surface of the catalyst. This brings them into close contact with each other and allows new, safer gases to form.

Reactions in the converter produce relatively harmless gases

Water

Carbon dioxide

Nitrogen

CHEMICAL EQUATION FOR YEAST REACTION:

$$C_6H_{12}O_6 \xrightarrow{yeast} 2C_2H_5OH + 2CO_2$$

ENZYMES IN THE KITCHEN ▲
Enzymes are catalysts found in nature. For example, it is the enzymes in yeast that cause bread dough to rise. When yeast is mixed with warm water and sugar it starts to grow and bubbles of carbon dioxide gas are produced. When the yeast mixture is added to flour and water to make a dough, the dough rises. Heating bakes the bread and kills the yeast. Scientists use chemical equations to show how molecules change in a chemical reaction.

FIND OUT MORE ▶▶ Atoms 24–25 • Chemical Reactions 30–31 • Elements 22–23

CHEMICAL REACTIONS

In a chemical reaction, the molecules of one substance break apart and join together with those of another substance to create a different compound (combination of molecules). Many chemical reactions are **NON-REVERSIBLE CHANGES.** You cannot turn a baked cake back into its raw ingredients. Some chemical reactions can be reversed, and re-formed into the original substances. These are **REVERSIBLE CHANGES.**

Frozen lolly is chemically identical to melted liquid

CHEMICAL CHANGE ▶
When the iron and magnesium in a firework burn, they react with oxygen and produce ash and smoke. They also release spectacular heat, light, and noise. Chemical changes produce new materials. They also usually give out or take in energy such as heat or light because chemical bonds have been broken and made.

◀ PHYSICAL CHANGE
A melting ice lolly is an example of a physical change, not a chemical change. The liquid ice lolly is not a new material, just a different form of the old one. Physical changes do not create new substances and no chemical bonds are broken or made. Melting, freezing, tearing, bending, and crushing are all physical changes that alter a substance's appearance but not its chemical properties.

e▶▶
chemical reactions

Ship's hull has rusted under water

◀ CONSERVING MATTER
When iron rusts, it reacts with oxygen in water or in air to create a new compound called iron oxide (rust). As in every chemical reaction, no mass is lost or gained. The same atoms from the original material are in the new materials, but in different places. If you weighed the iron oxide in this rusting ship, it would weigh the same as the original iron and oxygen.

| *4 iron atoms* | *6 oxygen atoms* | *4 iron atoms* | *6 oxygen atoms* |

$$4Fe + 3O_2 \longrightarrow 2Fe_2 2O_3$$

Rust forms a loose, flaky layer

CHEMICAL EQUATION FOR RUSTING

NON-REVERSIBLE CHANGE

Many chemical reactions are non-reversible changes. This means they are permanent changes that cannot be undone. You cannot turn the new materials made back into the original materials again. Rusting is a non-reversible change. However, if rust is mixed with magnesium powder another chemical reaction occurs and iron can be extracted from the rust.

BURNING ▶
Burning is a non-reversible chemical change. When you burn wood, the carbon in the wood reacts with oxygen in the air to create ash and smoke, and energy in the form of light and heat. This is a permanent change that cannot be undone – you cannot turn ashes back into wood.

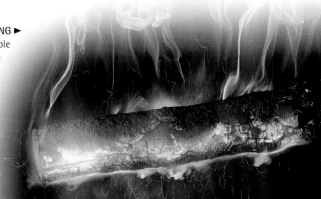

◀ DECOMPOSITION
Decomposing (rotting) of food is a non-reversible reaction. Tiny living things called micro-organisms feed on the food and turn it into other substances, including nitrogen compounds and carbon dioxide. It is impossible to re-create fresh food from rotten food. This process is called decomposition because complex compounds are splitting up into simpler compounds.

FRESH RED PEPPER   ROTTEN RED PEPPER

REVERSIBLE CHANGE

A few chemical reactions can be reversed – the original materials can be re-created from the new materials. These reactions are called reversible changes. They have a forward reaction and a backward reaction. Both reactions are actually happening at the same time but, depending on the conditions, one will be stronger than the other.

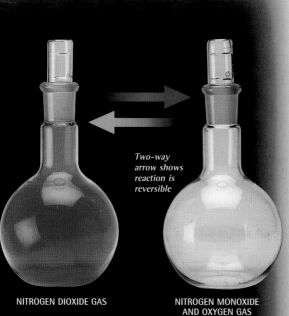

Two-way arrow shows reaction is reversible

NITROGEN DIOXIDE GAS   NITROGEN MONOXIDE AND OXYGEN GAS

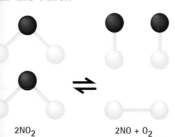

$2NO_2$ $2NO + O_2$

◀ NITROGEN DIOXIDE
When the gas nitrogen dioxide is heated, a forward chemical reaction changes the brown nitrogen dioxide gas into two colourless gases – nitrogen monoxide and oxygen. However, if these colourless gases are cooled, they will re-form into brown nitrogen dioxide gas. This is called a backward chemical reaction.

FIND OUT MORE ▶▶ Atoms **24–25** • Mixtures **18–19** • Molecules **28–29**

▲ CITRIC ACID
Lemons and other citrus fruits taste sour because they contain citric acid. Citric acid is used to add a tangy taste to food and soft drinks. Citrus fruits also contain another acid, called ascorbic acid or Vitamin C, which we need for healthy skin and gums.

ACIDS

The sour taste of food such as lemons is due to acids. Acids in food are weak, but they can sting if they touch a cut on your skin. Strong acids, such as sulphuric acid in car batteries, are much more dangerous as they can burn through materials. Acid compounds all contain hydrogen. They dissolve in water to produce particles called hydrogen ions. The more hydrogen ions an acid contains, the stronger an acid it is.

ACIDRAIN DAMAGE ▶
The pock-marked appearance of some statues is caused by acids in rainwater attacking the stone. Rainwater is always slightly acidic because carbon dioxide in the air dissolves in water to form carbonic acid. In addition, industrial areas give off pollutants, such as sulphur dioxide. These pollutants react with water in clouds to form strong acids that react with stone, especially limestone.

Unexposed stone not affected by acid rain

Exposed stone flaking off with acid rain damage

Hole caused by acids in rain reacting with limestone

Acidic liquid turns pH paper pink

▲ ACID BATH
Baths of strong acid are used to clean machine parts, such as this jet engine rotor bearing. Acids corrode (eat away) metals. Each metal part is immersed in acid for a set time to remove the top layer of metal, along with any rust or dirt. Each part is then washed thoroughly to ensure that no acid remains on the metal and continues to corrode it.

acids

pH SCALE ▶
Scientists used the pH scale to measure the strength of acids and of bases called alkalis. pH stands for power of hydrogen, and it measures how many hydrogen ions a liquid contains. The lower the pH, the more acidic a liquid is. Any liquid with a pH of above 7 is alkaline.

Strong acids
Acids such as hydrochloric acid and sulphuric acid are very strong and have a pH of about 1.

pH1   pH2   pH

Gland in the slug's skin produces sulphuric acid

BASES

Many cleaning products, such as soap and oven cleaner, are bases. Bases neutralize (cancel out) acids. Alkalis are bases that dissolve in water. Strong bases, such bleach, are corrosive and burn skin. Bases contain particles called hydroxide ions. The more hydroxide ions a base contains, the stronger it is.

LIMESTONE ▲
Limestone is an important base that is dug from the earth in quarries. It comes from calcium carbonate, which formed millions of years ago from the compressed remains of sea-shells and other marine life. Once quarried, limestone is crushed and used to make cement, fertilizers, paints, and ceramics.

▲ SEA SLUG
This frilled nudibranch sea slug oozes a strong acid called sulphuric acid to protect itself. The sulphuric acid makes the nudibranch poisonous and taste awful, so it does not have many predators. Ants and stinging nettles contain an acid called methanoic acid, which they use to protect themselves.

◄ NEUTRAL WASP STING
People used to think that a wasp sting contained a base. In fact, it contains a complex protein, which is neutral. This means that the sting is neither a base nor an acid. The wasp punctures the skin with its hollow stinger. It then pumps the protein through the stinger and into the wound. The protein contains poisons, which cause pain and swelling.

HARMFUL CHEMICALS

bases

CORROSIVE CHEMICALS

▲ HAZARD SIGNS
Strong acids and bases are extremely poisonous, corrosive, and cause bad burns, so their containers are labelled with hazard symbols. Some give information about how to handle the chemicals safely. The symbols are also displayed on the tankers that transport acids and bases, so emergency services know how to handle the substances in the case of an accident or spillage.

NEUTRALIZATION ►
As a result of acid rain, many lakes in Scandinavia have a high acid content. Acid water can poison wildlife. The acid is neutralized by spraying powdered limestone in the lake. When an acid reacts with a base, it forms water and a compound called a salt.

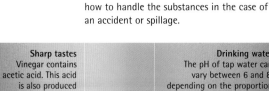

Sharp tastes		**Drinking water**	**Pure water**	**Soap**		**Cleaning fluids**
Vinegar contains acetic acid. This acid is also produced when wine is exposed to the air.		The pH of tap water can vary between 6 and 8, depending on the proportion of gases and minerals dissolved in it.	Pure water is neutral – it is not acidic or alkaline	Soaps are made by combining a weak acid with a strong base, and so they are mildly alkaline with a pH of around 8.		Household cleaning fluids such as bleach and the caustic soda in oven cleaner have a pH of around 10.
pH4	**pH5**	**pH6**	**pH7**	**pH8**	**pH9**	**pH10**

FIND OUT MORE ►► Defence **320–321** • Erosion **222–223** • Insects **297** • Pollution **250**

METALS

Almost three-quarters of all elements are metals, such as gold and silver. There are also some elements we may not think of as metals, such as the calcium in our bones, and the sodium in table salt (sodium chloride). Metals are defined by their **METALLIC PROPERTIES**, such as high melting points. Mixtures of metals are called **ALLOYS**. Solder is an alloy that is used to join metals in plumbing and electrical wiring. It is mainly tin with lead or silver.

Molten gold pours through steel tubes in exact measures

Melting point of gold is 1,063°C (1,945°F)

◄ GOLD IN QUARTZ
Some metals, such as gold, are found naturally as pure metals in rocks. Gold is unreactive, so it does not combine with other elements. Most metals are more reactive and are found combined with other elements in rocks. Iron, for example, is usually combined with oxygen. The rocks in which metals are found are called ores.

Quartz rock is the ore

Pure gold has not reacted with any of the other elements

EXTRACTING GOLD ►
To extract gold from its ore, huge grinders crush the ore to a fine powder. The powder is mixed with a solution of cyanide. Only the gold from the ore dissolves in the solution. Powdered zinc is added to bring the gold out of the solution. The gold is melted down and poured into moulds.

Crucible tips when full and pours gold into moulds

METALLIC PROPERTIES

Metals are usually shiny solids with high melting points and are very good conductors of heat and electricity. They are malleable, so they can be beaten into sheets, and ductile, which means they can be drawn into wires. Most are strong and cannot be broken easily. Of course, there are exceptions: mercury, for example, is a metal that has a low boiling point and is liquid at room temperature.

Gold cools in moulds to form ingots

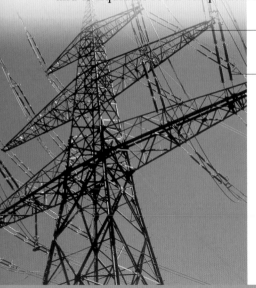

Steel pylon strong enough to support the cables

Cables contain copper wires that conduct electricity

◄ ELECTRICAL CONDUCTORS
The transmission lines (electric cables) that bring electricity to our homes, schools, and offices all rely on copper. Copper is a red-orange metal that is one of the best electrical conductors. Metals conduct electricity well because when metal atoms bond (join together), the electrons in their outer shells move freely. If electricity passes through one part of the metal, the electrons carry the electricity quickly to other parts.

ALLOYS

Alloys are mixtures of metals with properties that make them more useful than pure metals. A mixture of chromium and iron resists rusting much better than iron on its own. Most alloys are made of two or more metals, but some contain a non-metal. Steel is an alloy of iron and carbon. Alloys are made by melting the different materials together. Changing the proportions of the materials can change the properties of the alloy.

Copper and tin particles

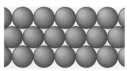

ATOMS IN A PURE METAL   ATOMS IN AN ALLOY

▲ ARRANGEMENT OF ATOMS

In a pure metal, the identical atoms are arranged in layers that can slide over one another. This is why pure metals are often soft and malleable. In an alloy, the different-sized atoms disrupt the regular arrangement, making it more difficult for the layers to slide over one another. The alloy is therefore harder and less malleable than the pure metal.

metals

METAL GROUPS

Metals are classified according to where they are found in the periodic table. Each group has a set of properties that make the metals useful for different purposes.

ALKALI METALS

These include potassium and sodium, and form Group I of the periodic table. They are extremely reactive metals: they react strongly with water to form strong alkalis.

ALKALINE-EARTH METALS

These elements make up Group II of the periodic table. They combine with many elements in the Earth's crust. Their oxides react with water to form alkalis.

TRANSITION METALS

This group includes copper, silver, and gold. They are hard and shiny, have high melting points, and are good conductors of heat and electricity.

OTHER METALS

Also called poor metals, these metals are fairly soft and melt easily. They include tin and aluminium and are often used in alloys. Bronze is an alloy of tin and copper.

BRONZE PANEL ▲

Bronze is an alloy of 90 per cent copper and 10 per cent tin. Molten bronze is poured into moulds to create objects with fine detail, such as this bronze panel from north Africa. Bronze was first made 6,000 years ago from minerals containing copper and tin. This alloy is much harder than pure copper. Bronze was so widely used for so many years that this period is called the Bronze Age.

Filament is made of tungsten

Potassium bubbles in the filament prolong its life

Wires that carry the current are a copper and nickel alloy, which is a good electrical conductor

Support wires are made of molybdenum

◄ LIGHT BULB ALLOYS

A light bulb is made from many different types of metals and alloys. Tungsten is a metal with a melting point of 3,422°C (6,192°F) and is used as the filament. When electricity flows through the filament, it heats up and gives out light. Light bulbs with a high wattage can overheat, so a heat deflector made of aluminium is placed in the neck of the bulb to diffuse the gases.

Base is made of an alloy of copper and zinc

FIND OUT MORE ►► Atoms **24–25** • Electricity Supply **131** • Mixtures **18–19** • Molecules **28–29** • Periodic Table **18–19**

NON-METAL ELEMENTS

The metal elements in the periodic table have easily defined properties. The remaining elements, however, have very different properties. They consist of a group of unreactive gases called the **NOBLE GASES**, a group of reactive elements known as the **HALOGENS**, and a set of elements referred to as non-metals. In addition, a few elements have properties that place them in between metals and non-metals. They are called the **SEMI-METALS**.

▲ SULPHUR CRYSTALS
Deposits of the non-metal sulphur are found as deep as 300 m (1,000 ft) below ground. Combined with other elements, sulphur is also found in rocks and minerals, such as gypsum.

▲ MAKING SULPHURIC ACID
Sulphur crystals are ground to a powder at sulphur processing plants. The powder is sprayed into a furnace where it reacts with oxygen, forming sulphur dioxide. More oxygen is reacted with the sulphur dioxide to make sulphur trioxide, which is dissolved in water to make oleum.

▲ TRANSPORTING SULPHURIC ACID
Oleum is concentrated sulphuric acid. It is transported to manufacturing plants in tankers. Here, water is added to the oleum in precise measures to make the correct concentration of sulphuric acid. Sulphuric acid is used in the manufacture of detergents, paints, medicines, plastics, and synthetic fabrics.

SEMI-METALS

SOLAR PANELS ►
The Hubble Space Telescope is constantly orbiting (circling) the Earth. Electricity is needed to power the telescope. Huge solar panels made of silicon create electricity from sunlight. The panels rotate, so they always face the Sun. This means the maximum amount of sunlight is used to create electricity.

The elements known as semi-metals have some of the properties of metals and some of the properties of non-metals. Arsenic, for example, has the shininess of a metal but does not conduct heat or electricity very well. Other semi-metals, such as silicon and germanium, are semi-conductors. This means that they can conduct electricity, but only under special conditions. This property makes them very useful in solar panels and computers.

nonmetals

◄ ELECTRICITY IN SPACE
Hubble's solar panels are made from thousands of tiny silicon cells. When sunlight hits a cell, it is absorbed by the silicon. This alters the movement of electrons within the silicon atoms so that a tiny current of electricity is created. The thousands of solar cells within a panel create enough electricity to power the Hubble Space Telescope and the computers inside the telescope. The computers send images of the Universe back to Earth.

SILICON

HALOGEN

At first sight, the halogens don't seem very alike. For example, fluorine is a yellow gas and iodine is a shiny, black solid. However, they are all highly reactive and are quick to combine with other elements to form salts, such as table salt (sodium chloride). They also have important uses. Chlorine is used to disinfect water, and compounds of fluorine – fluorides – are added to toothpaste to prevent tooth decay.

◄ BROMINE GAS
Bromine is the only liquid non-metal element. It is a reddish-brown colour and evaporates quickly to form a choking, poisonous gas. Bromine is found in seawater and mineral springs in the form of salts, called bromides. Bromine compounds are used in photography, as mild sedatives, and in the manufacture of flameproof coatings and dyes.

Black areas are formed when a lot of X-rays hit the silver bromide

SILVER BROMIDE IN X-RAYS ►
In X-ray photography, a plastic film is coated with a paste of a bromine compound called silver bromide. When X-ray light strikes the film, the silver bromide breaks apart and pure silver atoms are left on the film. The more intense the light, the more silver atoms are formed and the darker that part of the image becomes.

Solid bone blocks X-rays

White areas are formed when no X-rays hit the silver bromide

NOBLE GASES

Group 18 of the periodic table contains the noble gases. These six unreactive gases do not combine with other elements, so they are usually found on their own. Nearly 1 per cent of air is argon. Traces of neon, helium, krypton, radon, and xenon are also found in air. Argon is used in light bulbs, xenon is used in lighthouse arc lamps, and helium is used to fill airships and hot-air balloons.

Red light is created by passing electricity through neon gas

◄ NEON LIGHTING
A neon light is a tube containing a noble gas, but not always neon. When electricity is passed through the tube, the atoms of the noble gas emit (give out) light of different colours. Helium emits a yellow light, neon a red light, argon a blue light, and krypton a purple light. Other colours are created by giving the glass tube different coloured coatings.

Blue light is created by passing electricity through argon gas

Green light is created by passing electricity through a mixture of helium and argon gases

FIND OUT MORE ►► Electricity **126–127** • Gases **15** • Periodic Table **26–27** • Space Observatories **196–197**

| 1 |
| H |
| Hydrogen |
| 1 |

HYDROGEN

You cannot see, taste, or smell hydrogen, yet this element makes up over 90 per cent of matter. The Sun and stars are made of hydrogen gas. On Earth, hydrogen forms compounds (mixture of elements), and is found in almost every living thing. Hydrogen gas is used to make chemicals such as ammonia, which is needed to make fertilizers. Hydrogen is also used to increase the amount of petrol produced from crude oil.

Shuttle powered by three shuttle main engines

Exhaust gases from solid rocket boosters give shuttle initial thrust

SPACE SHUTTLE ▶
The space shuttle uses liquid hydrogen fuel because hydrogen gives out a lot of power for very little weight. Hydrogen, like all fuel, needs oxygen to burn, so the shuttle has a tank of liquid hydrogen and a tank of liquid oxygen. A fine mist of the two liquids is sprayed into the engines and ignited (set alight). The hydrogen explodes, sending steam out of the nozzles and helping to thrust the shuttle into space.

Steam explodes out of nozzle at over 10,000 kph (6,000 mph)

▲ HYDROGEN IN STARS
Stars are fuelled by hydrogen. At incredibly high temperatures inside stars, hydrogen atoms smash into one another and fuse (join) together to create helium atoms. These reactions give out a huge amount of energy as light and heat. Hydrogen atoms were probably the first atoms to form in the Universe and fuse together to create other, heavier atoms.

▲ HYDROGENATION
Margarine is made by passing bubbles of hydrogen gas through hot vegetable oil. Extra hydrogen atoms bond (join) with the oil molecules, and the oil changes from a liquid to a more solid form. This process is called hydrogenation. If oil is fully hydrogenated, it becomes completely solid; by stopping part way, it becomes a semi-solid.

hydrogen

▲ HYDROGEN-FUELLED CAR
Scientists are developing hydrogen-powered cars. The cars contain tanks of hydrogen that combine with oxygen from the air to drive them. Hydrogen-fuelled cars produce water instead of polluting exhaust gases. They are not mass-produced, because scientists have not developed a compact and lightweight method for storing hydrogen yet.

ANTOINE LAVOISIER
French, 1743–1794

This chemist is often known as the father of modern chemistry. He studied the "inflammable air" that was discovered by English scientist Henry Cavendish (1731–1810). Lavoisier discovered that this gas combines with oxygen to make water. He named the gas hydrogen, which is Greek for water-former.

FIND OUT MORE ▶▶ Engines 92 • Molecules 28–29 • Nuclear Energy 85 • Space Travel 190–191 • Stars 166–167

8
O
Oxygen
16

OXYGEN

On Earth, oxygen is more common than any other element. It is an invisible, odourless gas that makes up 21 per cent of air. Oxygen is found in water, minerals, and almost all living things. It is essential to life. Ordinary oxygen molecules contain two oxygen atoms. Ozone, a three-atom form, is found high up in the atmosphere. Oxygen moves through the environment via the **OXYGEN CYCLE**.

◄ OXYGEN FOR LIFE
Divers wear a SCUBA (self-contained underwater breathing apparatus), so they can breathe under water. The SCUBA contains a cylinder of compressed air, which divers carry on their backs. The air is compressed (or squeezed) into the cylinder to increase the amount of air the divers can carry. Divers breathe through a regulator, which decompresses the air as it comes out of the cylinder.

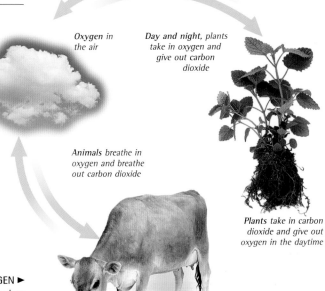

BURNING FUSE ▲
This burning fuse is reacting with oxygen. The reaction gives out energy in the form of heat and light. Oxygen is needed to make things burn. The more oxygen there is, the faster an object burns. This fuse is burning with oxygen in the air. Fireworks burn even more fiercely, because oxygen-rich compounds are added to their fuses, which mixes with oxygen in the air.

JOSEPH PRIESTLEY
British, 1733-1804

In 1774, this chemist announced his discovery of oxygen. He didn't realize that Swedish chemist Carl Scheele (1742–1786) had found it first, a year or two previously. They both showed that air is not one element. Priestley also discovered how to combine carbon dioxide with water to make fizzy water.

OXYGEN CYCLE

Almost all living things, including humans, need oxygen to survive. Both plants and animals take in oxygen from their surroundings to release energy. Underwater plants and animals cannot use the oxygen in air – instead they use oxygen dissolved in water. The oxygen cycle continuously circulates oxygen through the environment, so it is always available to all living things.

Oxygen in the air

Day and night, plants take in oxygen and give out carbon dioxide

Animals breathe in oxygen and breathe out carbon dioxide

Plants take in carbon dioxide and give out oxygen in the daytime

oxygen

CHANGING OXYGEN ►
Plants are able to use the energy of sunlight to convert carbon dioxide (CO_2) and water (H_2O) into carbohydrates and oxygen (O_2) in a process called photosynthesis. This oxygen is taken in by plants and animals to provide energy, releasing carbon dioxide and water. This process is called respiration.

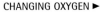

FIND OUT MORE ►► Atmosphere 234–235 • Atoms 24–25 • Photosynthesis 258 • Respiratory System 354–355

WATER

The simple combination of two hydrogen atoms and one oxygen atom creates a water molecule (H_2O). Water is the most common compound on Earth, making up over half the weight of living things. It is vital to life, bringing nutrients to and taking away waste from every living cell. Water molecules are attracted to one another through **HYDROGEN BONDS**, and this gives water some unusual but useful properties.

ABUNDANCE OF WATER ►
Water covers around 70 per cent of the Earth's surface. This is why Earth looks blue from space, and why it is often called the Blue Planet. Water is liquid in the oceans and forms solid ice caps at the ice caps. Water vapour is a gas in air. Humid places, such as rainforests, have a lot of water vapour. The human body is about 60 per cent water and a ripe tomato contains over 95 per cent water.

Oceans, rivers, and lakes cover three-quarters of the Earth's surface

Arctic ice cap is approximately the same size as the United States

Gaseous water is found as vapour everywhere

DRINKING WATER ►
At room temperature, pure water is a colourless liquid with a neutral pH – it is not an acid or a base. But most water is not pure. Hard water contains calcium and magnesium minerals, which have dissolved in the water as it flows over rocks. Soap does not lather well in hard water – the minerals react with the soap to form a scum. Hard water is softened by boiling or by passing it through a water softener.

Dissolved potassium permanganate crystals turn the water purple

◄ UNIVERSAL SOLVENT
The chemical potassium permanganate dissolves in water to form a pink liquid. More substances dissolve in water than in any other liquid. Its molecules are small and have a slight electrical charge, so they can move around and interact with other particles. If water did not have this property, life could not exist. Water is nature's carrier. Dissolved gases, such as oxygen and carbon dioxide, are carried by water to and from all living cells.

HYDROGEN BOND

Water molecules have an attraction to other water molecules. This attraction is called the hydrogen bond. It is a fairly weak bond compared to the bonds within a water molecule, but it is still strong enough to give water some unusual properties. For example, water is a liquid at room temperature; other molecules of a similar size are gases. It is also less dense as a solid than as a liquid.

Droplet of water forms when steam touches a cool surface

Steam is an invisible gas

◄ MOLECULAR STRUCTURE

In a water molecule, electrons are pulled closer to the oxygen atom than the hydrogen atoms. So the oxygen atom has a small negative charge, and the hydrogen atoms have a small positive charge. The slightly positively charged hydrogen atoms of one water molecule are attracted to the slightly negatively charged oxygen atoms of another water molecule. This attraction is the hydrogen bond.

WATER MOLECULES IN LIQUID STATE

Water molecules slide over each other in liquid water

Heat makes water molecules move quickly

Water evaporates from the leaves, forcing more water molecules upwards

Water molecules travel through the stem to the leaves

◄ CAPILLARY ACTION

Plants use capillary action to bring water up from their roots to their leaves. Water molecules move up through tubelike xylem cells in the plants. The slightly charged water molecules are attracted to the xylem walls and this attraction drags the molecules upwards.

BOILING WATER ▲

Water boils at 100˚C (212˚F). This is almost 200˚C (424˚F) higher than the boiling points of other similar-sized molecules, such as hydrogen sulphide. Water's high boiling point can be explained by its hydrogen bonds. Extra heat is needed to break the hydrogen bonds, so a water molecule can break free of other water molecules and leave the liquid's surface as steam, which is a gas.

Water molecules travel through the root to the stem

Pericycle (outer layer of xylem) supports the cells

Xylem cell carries water up the root

Root hair takes in water from the soil

LOW DENSITY SOLID ▲

Floating ice insulates and protects this seal from the freezing air above. When most liquids freeze they become denser, but when water freezes it becomes less dense, which is why ice floats on top of water. The hydrogen bonds hold the water molecules apart in a rigid, ringlike structure.

WATER MOLECULES IN SOLID STATE

FIND OUT MORE ►► Atoms **24–25** • Changing States **16–17** • Molecules **28–29** • Transpiration **259**

| 7 |
| N |
| Nitrogen |
| 14 |

NITROGEN

Nitrogen is needed to make proteins, which are vital to life. Plants and animals recycle nitrogen through the air and soil in a process called the **NITROGEN CYCLE**. As a gas, nitrogen makes up 78 per cent of air. At everyday temperatures it is very unreactive. It is used in place of air in crisp packets, for example, so the contents do not go stale. Nitrogen is also used to make industrial chemicals such as fertilizers and explosives.

Gloves protect hand from extremely cold liquid nitrogen

Mist is tiny droplets of water, cooled to a liquid by cold nitrogen gas

Nitrogen is present as nitrogen gas in the air

Nitric acid falls to the ground dissolved in rainwater

Nitrates are essential for plants

Nitrites are essential for micro-organisms in the soil

▲ LIGHTNING

The heat produced by lightning forces nitrogen molecules in the air to split. Nitrogen atoms bond with oxygen to form nitrogen oxides, which dissolve in water to create nitric acid. Weak nitric acid falls to the soil, where it splits apart to form the compounds nitrates and nitrites. These compounds are essential to life for plants and micro-organisms.

LIQUID NITROGEN ▲

When nitrogen gas is cooled to –196˚C (–320˚F), it turns to a liquid. Liquid nitrogen is so cold that it can freeze a substance in seconds. In hospitals, it is used to preserve blood and body parts for transplant. The material to be preserved is placed in a special, sealed container that is filled with liquid nitrogen. Because nitrogen is so unreactive, it does not alter the preserved materials in any way.

nitrogen

FERTILIZING SOIL ▼

Farmers often use fertilizers to help their crops grow well. Many fertilizers contain nitrogen in the form of nitrates, because this is the form that plants can use. Natural fertilizers are made from compost and manure. Synthetic fertilizers are made by combining nitrogen from the air with hydrogen from natural gas.

Fine mist of liquid fertilizer sprayed onto crop

KIJ 5544

NITROGEN CYCLE

All living things need nitrogen, but most cannot use nitrogen gas directly from the air. The nitrogen has to be fixed (combined) with other elements to form nitrites and nitrates. This is done by lightning and by nitrogen-fixing bacteria. The nitrates are taken up by plants, which are eaten by animals. This starts the continual cycle of nitrogen called the nitrogen cycle.

▲ EXPLOSIVES

Nitrogen compounds are used to make explosives. These compounds contain chemicals that break apart easily to release huge volumes of gases extremely quickly. They can be used in a controlled way to demolish a building without harming other buildings nearby. The explosive TNT (trinitrotoluene) releases hydrogen, carbon monoxide, and nitrogen, and carbon powder, which produces black smoke.

Fertilizers contain between 15–80% nitrogen

Plants take in nitrogen compounds through their roots

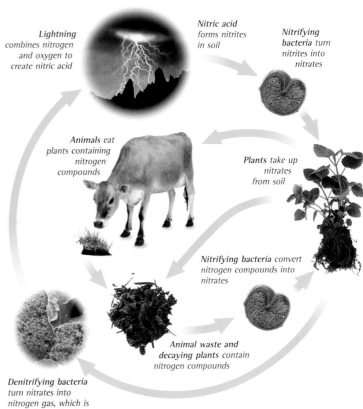

Lightning combines nitrogen and oxygen to create nitric acid

Nitric acid forms nitrites in soil

Nitrifying bacteria turn nitrites into nitrates

Animals eat plants containing nitrogen compounds

Plants take up nitrates from soil

Nitrifying bacteria convert nitrogen compounds into nitrates

Animal waste and decaying plants contain nitrogen compounds

Denitrifying bacteria turn nitrates into nitrogen gas, which is released into the air

MOVEMENT OF ATOMS ▲

Nitrogen from the air is fixed to make nitrates in the soil by nitrifying bacteria. The nitrates are taken up by plants to build plant protein. When an animal eats a plant, it turns the plant protein into animal protein. Denitrifying bacteria convert the nitrogen contained in animal waste and in decaying plant and animal material back into nitrogen gas again.

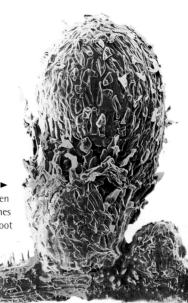

NITRIFYING BACTERIA IN ROOT NODULES ►

Nitrifying bacteria are a key part of the nitrogen cycle. Some live in the root nodules of legumes (peas and beans), like this nodule from the root of a pea plant. Others live free in the soil. Bacteria in the soil make nitrates from nitrites and other nitrogen molecules. Bacteria in legume root nodules take up nitrates from the soil.

FIND OUT MORE ►► Molecules **28–29** • Periodic Table **26–27**

6
C
Carbon
12

CARBON

Carbon is the sixth most common element in the universe and is the main element in every living thing on Earth. Carbon atoms are passed between living things through the **CARBON CYCLE**. Carbon is present as carbon dioxide in the air, and makes up a large part of coal, crude oil, and natural gas. Pure carbon is very rare in nature, although it can be found in one of several different forms, or **ALLOTROPES**.

▼ CARBON AS FUEL

Anything that burns well usually contains carbon. Coal, charcoal, wood, and paper are packed full of carbon. Carbon atoms joined together store a lot of energy. When carbon burns, each carbon atom breaks away from its surrounding atoms and reacts with oxygen in the air to form carbon dioxide. The stored energy is released as heat.

Coal starts to burn at 400°C (848°F)

carbon

ALLOTROPE

The atoms of some elements can link up in different ways to create different forms called allotropes. Carbon is found in three allotropes: diamond, graphite, and fullerene. Each allotrope has very different physical properties. Graphite, diamond, and fullerene contain only carbon atoms, but the atoms are arranged differently in each allotrope.

FULLERENES ▶

In a fullerene, the carbon atoms link together to form a ball-shaped cage. Fullerenes may contain 100, 80, or 60 carbon atoms. This fullerene contains 80 atoms. The first fullerene discovered was buckminsterfullerene in the 1980s, which has 60 carbon atoms. It is named after Buckminster Fuller, an American architect who designed buildings similar in shape to the fullerene molecule.

◀ GRAPHITE

Some lubricating engine oils and all pencil leads contain graphite. Graphite has layers of carbon atoms that can slide across one another. There are strong bonds between the carbon atoms of each layer, but weak bonds between the different layers. Because the layers can move over one another, graphite is quite a soft material.

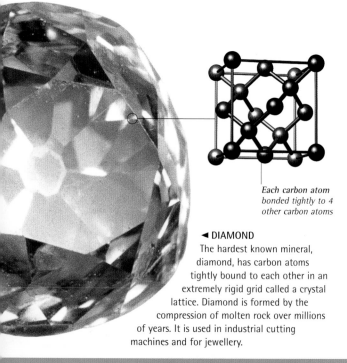

Each carbon atom bonded tightly to 4 other carbon atoms

◀ DIAMOND

The hardest known mineral, diamond, has carbon atoms tightly bound to each other in an extremely rigid grid called a crystal lattice. Diamond is formed by the compression of molten rock over millions of years. It is used in industrial cutting machines and for jewellery.

Layer of carbon atoms slides over layer below

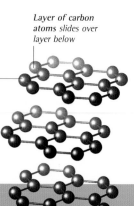

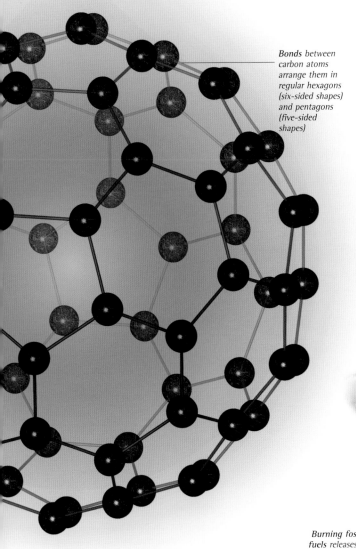

Bonds between carbon atoms arrange them in regular hexagons (six-sided shapes) and pentagons (five-sided shapes)

CARBON CYCLE

Carbon atoms continually circulate through the air, animals, plants, and the soil. This recycling of carbon atoms in nature is called the carbon cycle. The bodies of all living things contain carbon. The carbon comes originally from carbon dioxide gas in the air. Green plants and some bacteria take in the carbon dioxide and use it to make food. When animals eat plants, they take in some of the carbon. Carbon dioxide goes back into the air when living things breathe out, and when they produce waste, die, and decay.

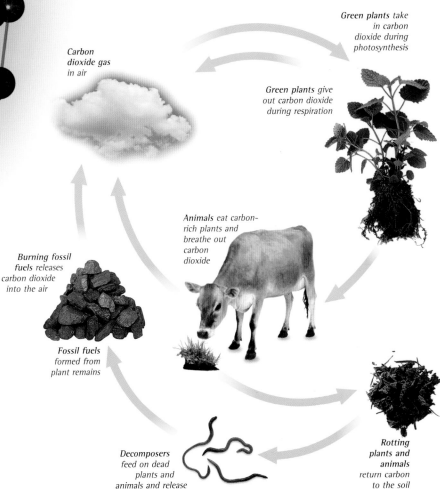

Green plants take in carbon dioxide during photosynthesis

Carbon dioxide gas in air

Green plants give out carbon dioxide during respiration

Animals eat carbon-rich plants and breathe out carbon dioxide

Burning fossil fuels releases carbon dioxide into the air

Fossil fuels formed from plant remains

Decomposers feed on dead plants and animals and release carbon dioxide

Rotting plants and animals return carbon to the soil

MOVEMENT OF ATOMS ▲
Green plants use carbon dioxide from the air to make food. When an animal eats a plant, it uses the carbon to build body tissue. When the animal breathes out, it returns carbon into the air as carbon dioxide. When the animal dies and decays, the carbon in its body returns to the soil. Decomposers such as worms, bacteria, and fungi, feed on the decaying remains of animals. As they feed, the decomposers breathe out carbon dioxide into the air. Green plants then take in carbon dioxide from the air, and the cycle is repeated.

GEODESIC DOME ▲
The stable structure of fullerenes works well on large-scale buildings. In the 1940s, architect Buckminster Fuller designed a type of building called a geodesic dome. It is made of a network of triangles that together form a sphere. This shape is very stable and encloses a lot of space with little building material, making it strong but light.

FIND OUT MORE ▸▸ Biochemistry 46–47 • Molecules 28–29 • Organic Chemistry 48–49

BIOCHEMISTRY

The study of the chemical processes of all living things is called biochemistry. These processes include respiration (breathing) and the digestion of food. Carbon atoms can combine in so many ways that living things are mainly made up of molecules containing carbon. The molecule **DNA** carries the chemical instructions that allow living things to create and make copies of their molecules and reproduce.

biochemistry

FOOD FOR ENERGY ▶
Like all living things, orang-utans need food for the energy to make all the other body processes happen, such as growth, movement, and repair. These complicated chemical reactions are called metabolism in animals. Plants make their food, through a process called photosynthesis.

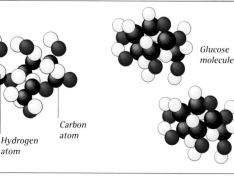

Oxygen atom

Hydrogen atom

Carbon atom

Glucose molecule

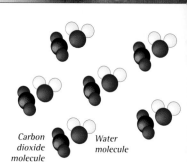

Carbon dioxide molecule

Water molecule

▲ BREAKING DOWN CARBOHYDRATES
Many foods, including apples, contain molecules called carbohydrates. When broken down through the process of digestion, carbohydrates release a lot of energy. Food contains two other kinds of molecule: fats and proteins. Fats are another good source of energy, and proteins are important for growth.

▲ DIGESTION
The carbohydrate sugar contains 12 carbon atoms, 22 hydrogen atoms, and 11 oxygen atoms. Once eaten, the molecules are broken down to form simpler glucose molecules through the chemical process of digestion. Digestive enzymes, such as amylase, speed up this process.

▲ RESPIRATION
Glucose molecules pass into the bloodstream and are carried to cells around the body. Every cell uses glucose molecules in a chemical process called respiration. In respiration, the bonds within the glucose molecule break, releasing the energy in the molecule in a form our bodies can use.

▲ ENERGY RELEASED
Glucose molecules react with oxygen in the air we breathe to release energy and create carbon dioxide (CO_2) and water (H_2O) molecules. Processes that release energy, such as respiration, are called catabolic reactions. Other processes that take in energy, such as building proteins, are called anabolic reactions.

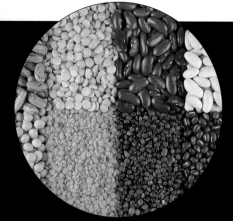

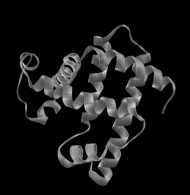

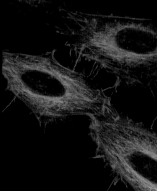

▲ FOOD FOR BUILDING MOLECULES
Living things do not only break down molecules, but they also build up complex molecules, such as muscle proteins. Protein molecules are needed for growth. They are made from amino acids, which come from protein-rich food, such as beans, pulses, and meat.

▲ AMINO ACID BUILDING BLOCKS
Amino acid molecules, such as this histidine molecule, are made of carbon, hydrogen, oxygen, and nitrogen atoms. They join up to form protein molecules, the building blocks of our bodies. Our 20 different amino acids make thousands of proteins.

▲ PROTEIN MOLECULE
This ribbon model of a muscle protein molecule is made from a chain of amino acids linked together. Some proteins are a few amino acids long, but others are made up of thousands. Amino acid chains fold and twist in complex ways, giving each protein a unique 3-D shape.

▲ PROTEIN CELLS
The cells of our skin, blood, hair, and muscles are all made up of proteins. There are thousands of different proteins in our bodies, including the enzymes that drive every reaction and the antibodies that fight disease.

DNA

Probably the most amazing molecule in our bodies is one called deoxyribonucleic acid, or DNA for short. This molecule contains the genes (instructions) for making every different type of protein in our bodies. Almost every cell in our bodies contains DNA, divided up into 46 parts called chromosomes. Each cell uses just the part of the instructions it needs. For example, only a muscle cell makes muscle proteins.

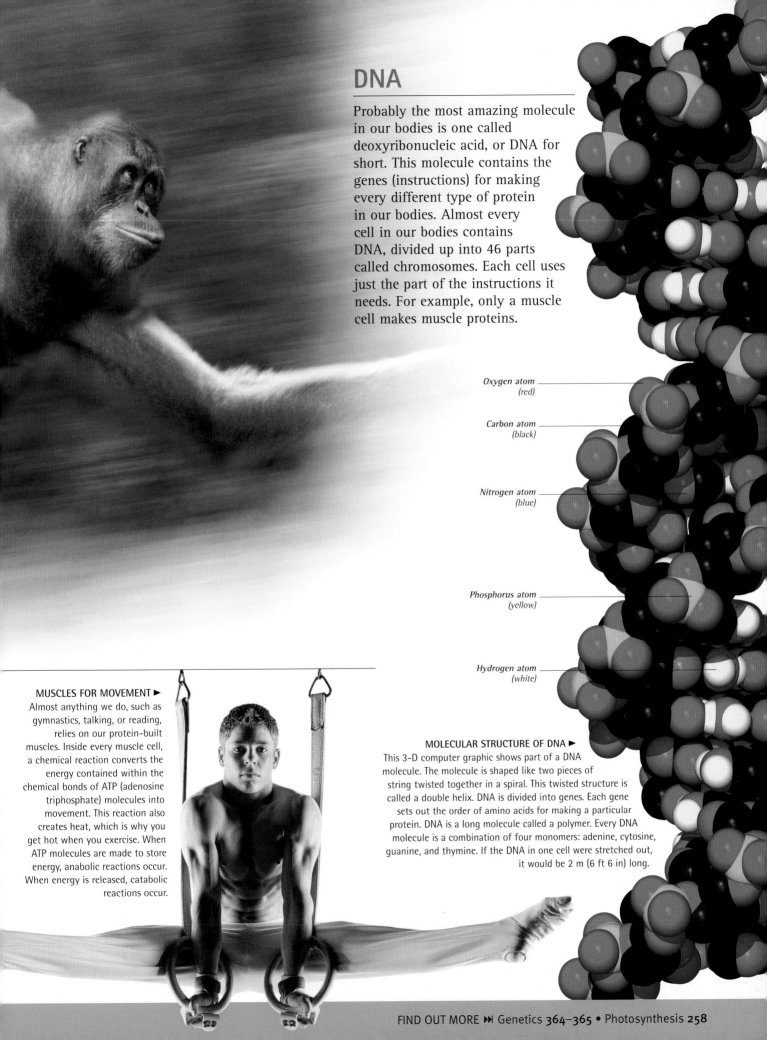

Oxygen atom
(red)

Carbon atom
(black)

Nitrogen atom
(blue)

Phosphorus atom
(yellow)

Hydrogen atom
(white)

MUSCLES FOR MOVEMENT ▶
Almost anything we do, such as gymnastics, talking, or reading, relies on our protein-built muscles. Inside every muscle cell, a chemical reaction converts the energy contained within the chemical bonds of ATP (adenosine triphosphate) molecules into movement. This reaction also creates heat, which is why you get hot when you exercise. When ATP molecules are made to store energy, anabolic reactions occur. When energy is released, catabolic reactions occur.

MOLECULAR STRUCTURE OF DNA ▶
This 3-D computer graphic shows part of a DNA molecule. The molecule is shaped like two pieces of string twisted together in a spiral. This twisted structure is called a double helix. DNA is divided into genes. Each gene sets out the order of amino acids for making a particular protein. DNA is a long molecule called a polymer. Every DNA molecule is a combination of four monomers: adenine, cytosine, guanine, and thymine. If the DNA in one cell were stretched out, it would be 2 m (6 ft 6 in) long.

FIND OUT MORE ▶▶ Genetics 364–365 • Photosynthesis 258

ORGANIC CHEMISTRY

The study of all compounds that contain carbon is called organic chemistry. Carbon atoms are unique. They can combine with each other to make molecules that contain hundreds, even thousands, of carbon atoms. There are more **CARBON COMPOUNDS** than compounds of all the other elements put together. **CARBON TECHNOLOGY** uses carbon compounds to make many modern materials, from the interiors of aircraft to medicines.

organic chemistry

◀ **CARBON IN ALL LIVING THINGS**
From butterfly wings to the petals of a flower, all living things are made of carbon compounds. All the processes that happen in living things – such as digestion, movement, and growth – are chemical reactions involving carbon compounds. It is the ability of carbon to make so many different compounds that results in the rich diversity of life on Earth.

CARBON COMPOUNDS

Many carbon compounds contain the same few elements, but in different quantities and arranged in different ways. The most important elements to join with carbon are hydrogen, oxygen, and nitrogen. Carbon atoms can form chains of just carbon and hydrogen, which are called hydrocarbons. They can also form rings of carbon, called aromatics.

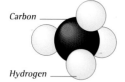

Carbon

Hydrogen

METHANE MOLECULE

BUTANE MOLECULE

BENZENE MOLECULE

Ethyl butanoate is one of many esters in the skin of an apple

Sorbitol is one of the alcohols in apple flesh

Carboxylic acid in the apple flesh reacts with alcohol to create esters

▲ **SIMPLE HYDROCARBONS**
Methane is a hydrocarbon. It contains one carbon atom bonded to four hydrogen atoms. The prefix "meth–" always refers to compounds whose molecules contain only one carbon atom. Methane is a natural gas. It is used in domestic central heating.

▲ **CHAINS OF CARBON**
Butane gas is a slightly more complex hydrocarbon than methane. Butane contains four carbon atoms and ten hydrogen atoms. The prefix 'but–' always refers to compounds whose molecules contain four carbon atoms in a chain.

▲ **RINGS OF CARBON**
A benzene molecule is made of a ring of six carbon atoms, each of which is bonded to a hydrogen atom. This gas is used to make dyes and pigments. Compounds whose molecules are made of carbon rings are called aromatics as they have distinctive smells.

ALCOHOLS AND ESTERS ▲
A carbon compound called an ester gives an apple its distinctive smell. Esters are liquids with a sweet, fruity smell, and evaporate quickly. They are made when alcohol reacts with an acid. Alcohols and esters contain carbon, hydrogen, and oxygen atoms.

CARBON TECHNOLOGY

The carbon industry is one of the largest and most important industries because so many products contain organic (carbon) compounds. Carbon technology is vital to the production of medicines, paints, synthetic fabrics, food flavourings, plastics, cosmetics, and glues. The raw materials that are the basis for these products come from coal, crude oil, and natural gas.

CARBON FIBRES ▲
Polyacrylonitrile (PAN) is heated to 3,000°C (5,432°F) to create thin filaments of carbon fibre. This material is fireproof and five times lighter than steel, yet twice as strong. Carbon fibre has many uses, such as in lightweight sports equipment, car body panels, construction pipes, and on the wings and nose of space shuttles.

CARBON FIBRE BIKE FRAME ▶
Racing bikes are often made from carbon fibre because it is strong and light, and can be moulded into complex shapes. The carbon fibres are woven into a cloth which is then cut and layered in a mould. The moulded part is filled with a chemical called a resin and then baked in an oven to form the hard, tough carbon fibre material.

Seat post is a composite of carbon fibre and Kevlar®

Layers of carbon fibres strengthen wheel

Carbon fibre frame is eight times lighter than a steel frame

Rubber tyre contains carbon

▲ MEDICINES
New medicines are made to treat specific illnesses by combining organic (carbon) compounds in new ways. Some are similar in structure to compounds found naturally in our bodies or in plants. New medicines undergo a series of tests to ensure they do not have any poisonous effects.

▲ PAINT
Paint pigments and the dyes that colour our clothes are mostly organic compounds. Pigments coat the surface of a material. Dyes bond with the molecule of the fabric they are colouring. The molecules of pigments and dyes often contain many rings of carbon atoms.

▲ PLASTICS
All plastics are organic compounds, made from recycled plastic or from the products of coal, oil, and natural gas. From flexible bags to hard chairs, plastics are light and cheap to make. Their molecules are made of long chains of carbon atoms called polymers.

▲ COSMETICS
Organic chemicals such as oils and pigments are mixed with talcs, clays, and metal compounds to make cosmetics, such as nail polish, eye shadow, lipstick, and blusher. As with medicines, every new cosmetic has to go through rigorous tests to make sure it does not harm our skin.

FIND OUT MORE ▶▶ Biochemistry 46–47 • Carbon 44–45 • Chemical Industry 50–51 • Plastics 52–53 • Space Travel 190–191

CHEMICAL INDUSTRY

Materials from the chemical industry are all around us. They include chemicals to make paint for cars, plastic for computers, and **PHARMACEUTICALS** (medicines). Chemical engineers start with cheap, raw (natural) materials, such as **PETROCHEMICALS**, seawater, and minerals. They separate materials by using physical processes, such as evaporation, and by using chemical reactions. These processes take place in factories, called chemical plants.

SODIUM CHLORIDE
NaCl

◄ EXTRACTING SALT
The elements that make up salt – sodium and chlorine – are used to make paint, soap, fertilizer, detergents, and paper. Salt is collected by evaporating water from a salt solution. In hot countries, seawater is fed into wide, shallow pools called salt pans. The water evaporates in the sun, leaving salt. The salt is then transported to factories all over the world.

SPLITTING SALT AND WATER ►
At huge chemical plants, in a process called electrolysis, an electric current is passed through brine (concentrated salt water). This breaks up brine into its elements (parts). Chlorine from the salt and hydrogen from the water are released. Sodium and hydroxide ions are left behind, as sodium hydroxide. This alkali is used to make soap, paper, and some pigments. Chlorine is used to make plastics, and hydrogen to make fertilizers.

Water pipes are
coloured green

Chlorine gas is collected
in yellow pipes

Hydrogen gas is
collected in red pipes

SODIUM
HYDROXIDE
NaOH

CHLORINE GAS
Cl$_2$

MAKING PIGMENTS ►
Many pigments are made using sodium hydroxide. Pigments are coloured compounds (chemical mixtures) that do not dissolve in water. They are fine powders that mix easily to colour paint and printing ink. Modern pigments are made by chemists in huge plants. Pigments are often made by mixing solutions of chemicals. The mixture is filtered, dried, and crushed to a fine powder by a series of heavy rollers.

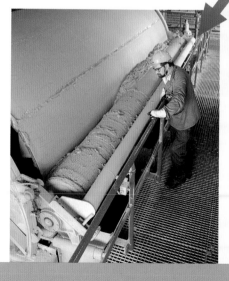

PAINT MANUFACTURE ►
To make paints, pigments are mixed with a sticky liquid called a binder in huge vats. A chemical called a wetting agent is stirred in by a machine with rotating blades. The wetting agent makes the paint flow easily. Water-based paints use water as their wetting agent. Gloss paints use a chemical called white spirit. When an object is painted, the wetting agent evaporates and the binder hardens.

PETROCHEMICALS

Crude oil is a sticky, dark liquid found under the ground or sea. Each drop of oil contains hundreds of hydrocarbon (hydrogen and carbon) compounds, called petrochemicals. Chemists separate the different hydrocarbons into fractions by heating them at an oil refinery. Thousands of products are made from hydrocarbons.

PHARMACEUTICALS

The pharmaceutical industry creates thousands of medicines to help prevent and fight disease. Chemists create synthetic substances in a laboratory to target specific illnesses. These substances are made artificially by heat and chemical reactions. After thorough testing, the substances are made into medicines.

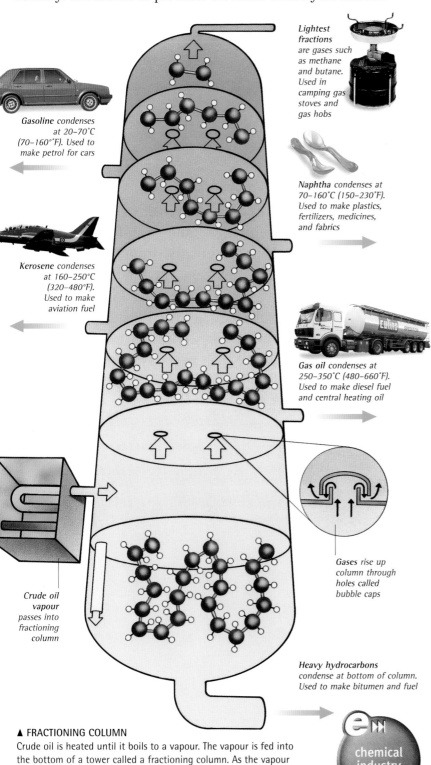

Gasoline condenses at 20–70°C (70–160°F). Used to make petrol for cars

Kerosene condenses at 160–250°C (320–480°F). Used to make aviation fuel

Crude oil vapour passes into fractioning column

Lightest fractions are gases such as methane and butane. Used in camping gas stoves and gas hobs

Naphtha condenses at 70–160°C (150–230°F). Used to make plastics, fertilizers, medicines, and fabrics

Gas oil condenses at 250–350°C (480–660°F). Used to make diesel fuel and central heating oil

Gases rise up column through holes called bubble caps

Heavy hydrocarbons condense at bottom of column. Used to make bitumen and fuel

▲ FRACTIONING COLUMN
Crude oil is heated until it boils to a vapour. The vapour is fed into the bottom of a tower called a fractioning column. As the vapour rises up the column, it cools. When a fraction (one hydrocarbon compound) in the vapour cools to its boiling point, it condenses to a liquid and is piped to another part of the refinery for processing.

chemical industry

RESEARCH AND DEVELOPMENT ▶
A research chemist models molecules on a computer. The computer has a large database of how atoms bond and react with each other. This helps the chemist to model a molecule that has the right kind of shape and structure to interact with chemicals inside our bodies. The chemist has in-depth knowledge of a particular illness and the body chemicals involved.

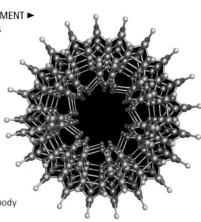

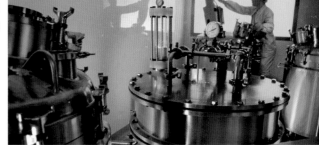

▲ MAKING MEDICINES
Chemists combine and test thousands of molecules to create compounds for a new medicine. Chemists make and test the compounds in small amounts for three years of laboratory trials. The compounds that show promise and pass safety tests are then tested on people for five years in clinical trials.

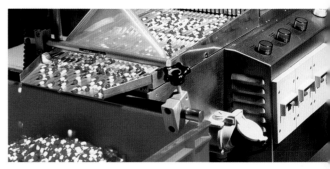

▲ DRUG PRODUCTION
After clinical trials, the compounds that treat an illness most successfully are developed into new medicines. Tablets or capsules are a useful way of distributing the medicine because they are easy to store and take. Some capsules have a gelatine coating that dissolves in the stomach, releasing the medicine.

FIND OUT MORE ▶▶ Changing States **16–17** • Earth's Resources **248–249** • Molecules **28–29** • Separating Mixtures **20–21**

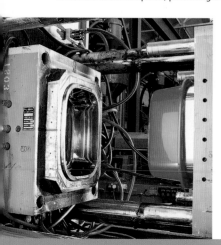

PLASTICS

Plastics are used to make a wide range of materials, including furniture, computers, and toys. Plastics are not found in nature, but are created from the products of coal, oil, and natural gas. They are made up of carbon, hydrogen, and other atoms linked in long-chained molecules called **POLYMERS**. Plastics are so useful because they are strong, light and can withstand heat and chemicals better than many materials. They can also be moulded into practically any shape or size.

TRANSPARENT PLASTIC ▲
The Eden Project in Cornwall, UK, is a collection of huge greenhouses. Each hexagonal section is made from an air-filled bag of a plastic called ethyltetrafluorethylene (EFTE). EFTE is 100 times lighter than glass and is also non-stick, so dirt washes off whenever it rains.

▲ INSULATING PLASTIC
Modern plastics can be created with precise properties that suit a particular use. Mylar® is a plastic used to insulate space shuttles. It is a shiny, strong, and light polyester film made in very thin sheets. It reflects the intense heat generated when a space shuttle re-enters Earth's atmosphere, protecting the craft and its crew.

◄ INJECTION MOULDING
Plastic bowls are often shaped by injection moulding. Plastic pellets are heated until they have melted. The liquid plastic is then pushed, or injected, into a mould, held in place by a clamp. After the plastic in the mould has cooled and hardened, the finished product is pushed out. This process is used to mass-produce items such as bowls, butter tubs, and yoghurt pots.

COLOURED PLASTIC ►
Plastics have revolutionized design, offering colours and shapes never available before. Plastic is liquid when first made and is then moulded or pushed into shape. Mixing dyes into the liquid creates translucent coloured plastics. Mixing in pigments creates opaque coloured plastics.

Red pigment colours chair bright red

Chair moulded from one piece of plastic

plastics

COMMON PLASTICS

PVC
Polyvinylchloride, or PVC for short, is the plastic used to make credit cards and waterproof clothes. It is tough, flexible, cheap to produce, and easy to print on.

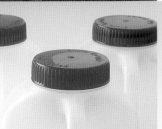

POLYTHENE
There are two types of polythene. Low-density polythene is used to make lightweight plastic bags. High-density polythene is stronger and is used to make plastic milk bottles.

PS
Polystyrene can be either rigid or foamed. Rigid polystyrene is used to make toys and containers. Foamed polystyrene is used for fast food packaging.

PET
Polyethylene terephthalate, more commonly called PET, is a strong plastic used to make fizzy drinks bottles. It can be recycled into carpets and ribbons for video cassettes.

PP
Polypropylene, or PP, is a plastic with a relatively high melting point of 160°C (320°F). It is used to make camera film and dishwasher-proof plastic objects.

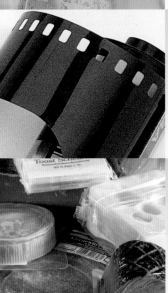

PA
Polyamide, or PA for short, is the plastic used to package oily food such as cheese and meats. It is also known as nylon, and is used in clothing, ropes, carpets, and bristles for brushes.

POLYMERS

All plastics are synthetic polymers. Polymers are substances whose molecules are made of simpler molecules called monomers joined together in long, winding chains. The monomers contain carbon and hydrogen, and sometimes other elements such as oxygen and nitrogen. Synthetic polymers can be divided into two groups, thermoplastics and thermosets.

Thermoplastic polymer chains are not linked to each other

◄ THERMOPLASTICS
These plastic balls are made from a thermoplastic polymer called PVC. In thermoplastic polymers, the molecules are arranged in long chains and there are no links between the chains. When heated, the chains can easily slide over one another. Thermoplastics can be melted and re-solidified many times over. This makes them easy to recycle.

Thermoset polymer chains are linked together to form a strong network

Polyurethane wheel is strong and light

◄ THERMOSETS
Thermosets, such as the polyurethane used for these skateboard wheels, cannot be melted and re-moulded. Rather than melting, like thermoplastics, thermosets will blister and burn when heated. They do this because their polymer chains are linked to other chains in a rigid network and cannot flow freely past one another.

FIND OUT MORE ►► Composites 57 • Elasticity 69 • Molecules 28–29 • Synthetic Fabrics 56

GLASS

First made over 5,000 years ago, glass is a thick liquid that never completely sets (hardens). That is why old window panes are thicker at the bottom than at the top. Glass is still in widespread use because it is transparent (see-through), strong, and can be melted and recycled endlessly. Molten glass can be shaped in many ways, including flat panels for windows and threads for optic fibres. **GLASS TECHNOLOGY** is so advanced that glass can be made fire-resistant and shatterproof.

glass

◀ GLASS ARCHITECTURE
Over 7,000 diamond-shaped glass panels make up the Swiss Re Tower in London, UK. Each panel is flat, but when so many are put together in a steel lattice, they create a curved building. Light floods in through the floor-to-ceiling windows, which give workers an uninterrupted 360° view of London.

▲ COLOURING GLASS
Coloured glass is created by dissolving metal compounds into melted sand. Different metal compounds create different colours. For example, selenium sulphide makes glass red. Iron and chromium compounds produce a deep green glass.

MAKING GLASS

MOLTEN GLASS ▶
Sand, broken glass, soda, and limestone are heated in a furnace. At around 1,500°C (2,732°F), the mixture melts to form molten (liquid) glass, which is cut into individual globules of glass called gobs.

SHAPING ▶
The gobs are dropped into bottle moulds. Compressed air blows the glass against the mould walls. The bottles are removed from the moulds and reheated slightly to remove imperfections.

COOLING ▶
The bottles are cooled slowly on a moving conveyor belt under carefully controlled conditions. This ensures that no dust is trapped in them, and that the glass does not shatter.

GLASS TECHNOLOGY

Material scientists have developed and improved the properties of glass to suit a range of uses. Heat-proof oven doors are made by adding chemicals to molten glass so that the glass lets light but not heat through. Car windscreens are made shatterproof by cooling molten glass rapidly with jets of air. Test tubes and other glass apparatus used in science labs need to withstand the heat of a Bunsen flame. This kind of glass is made heat-proof by adding boron oxide to the raw materials to make borosilicate.

Optical fibre is hair-thin and flexible

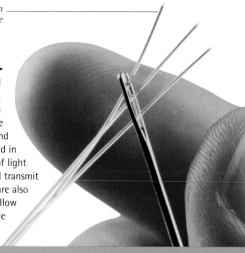

OPTICAL FIBRES ▶
Molten glass can be pulled into extremely thin tubes called optical fibres. A beam of light is reflected down the tube, even as it bends around corners. Optical fibres are used in telephone cables. Pulses of light pass down the tube, and transmit information. Optical fibres are also used in endoscopes that allow doctors to see right inside our bodies.

FIND OUT MORE ▶▶ New Materials 58–59 • Lenses 115 • Telecommunications 146

CERAMICS

JOSIAH WEDGWOOD
English, 1730-1795

This master potter and industrialist introduced many kinds of coloured pottery. He is best known for his Jasper Ware, with classical designs in white on blue or green. He also invented a pyrometer to measure kiln temperature.

The word ceramic comes from an ancient Greek word for "burned earth". Ceramics are made by firing (heating) clay (fine particles of earth) in an oven called a kiln or furnace. China, bricks, and tiles are made from ceramics. Over the past few decades, **ADVANCED CERAMICS** have been developed with superior or additional properties to traditional ceramics.

MAKING CERAMICS

◄ THROWING
Ceramics can be thrown (made) by shaping a lump of wet clay on a wheel (a turning plate). The potter places the clay on the centre of the wheel, then skilfully raises it into shape by hand.

◄ FIRING
The pot is fired in a kiln. The first firing, called the bisque firing, hardens the clay. A coating called a glaze is painted onto the pot, and the pot is fired again. Glaze waterproofs the pot.

◄ DECORATING
Once glazed, a decal (relief) is pre-soaked and smoothed onto the pot. The pot is fired again to stick the decal to the pot permanently. Pots can also be painted with enamel and then fired.

CONCRETE ARCHITECTURE ►
The spectacular Guggenheim Museum in New York, USA, was built of concrete in 1959. Concrete is still the main material used for buildings today. It contains cement, a ceramic made by crushing and heating clay, chalk, and sand. After drying to a powder, cement is mixed with water, sand, and gravel to create concrete. It sets (hardens) to form an extremely strong material.

▲ PORCELAIN
Porcelain has the finest texture of all ceramics. It is made from a white clay called kaolin, fired at very high temperatures. Most ceramics let water through until they are glazed, but porcelain is naturally water-resistant already. It is valued for its glassy smoothness and translucency.

ADVANCED CERAMICS

Bio-ceramics now replace teeth and bones. They are one example of advanced ceramics. Each type is made from a particular component of pure clay. It is heated at a specific temperature, sometimes in a specific gas environment, such as nitrogen. This changes the ceramic's chemical structure and properties.

ceramics

SPACE SHUTTLE CERAMICS ►
One type of advanced ceramic, called shuttle ceramic tiles, has been created to withstand temperatures up to 1,280°C (2,336°F). Space shuttles are covered in 30,000 of these lightweight tiles. They protect the shuttle from heat when it re-enters Earth's atmosphere from space.

FIND OUT MORE ►► New Materials 60–61 • Space Travel 190–191 • Synthetic Fabrics 56 • Telecommunications 146

SYNTHETIC FABRICS

Synthetic fabrics, such as nylon and polyester, are produced entirely from chemicals. Natural fabrics, such as cotton, silk, and wool are made of fibres from plants and animals. Synthetic fabrics are useful because they have very different or enhanced (improved) properties in comparison to natural materials. Plastic raincoats, for example, are waterproof, and stretchy Lycra® keeps its original shape.

▲ FASTSKIN™
A bodysuit made of Fastskin™ helps swimmers move through water faster than when wearing a traditional swimsuit. Fastskin™ is a stretchy fabric made of polyester and Lycra®. The bodysuit is made from several panels of Fastskin™, which hug the swimmer's body to make it as streamlined as possible.

Fabric has many V-shaped ridges

Water spirals along V-shaped ridge

Spiralling water collects and flows smoothly over the material

FASTSKIN™ TECHNOLOGY ▶
The developers of Fastskin™ looked to nature for inspiration. They found that sharks have V-shaped ridges – called denticles – on their skin, which channel water to pass over the skin very efficiently. Fastskin™ has similar built-in ridges. The ridges help to reduce the drag of the water and push the swimmer through the water.

▼ KEVLAR® GLOVE
These sharp, metal strips can be handled safely because the glove is made of a fabric called Kevlar®. Fibres of Kevlar® are made from long, complex molecular chains with strong bonds between the chains. This strongly bonded structure makes Kevlar® fabric, light, flexible, and five times stronger than steel. This makes Kevlar® ideal for bulletproof vests.

◀ MAKING SYNTHETIC FABRICS
The starting point for most synthetic fabrics is a liquid made from the products of coal, oil, or natural gas. The liquid is forced through the fine holes of a nozzle, called a spinneret. As the liquid emerges from the holes, it is cooled so that it solidifies to form tiny threads. These threads are woven together to make fabric.

synthetics

NYLON STRANDS ▶
The world's first synthetic fabric, nylon, was developed in 1938. Long chains of molecules, called polyamide, are made by heating a polymer solution to 260°C (500°F). The liquid is forced through a spinneret and the strands are treated in a cooling bath. The strands are woven together to make fabric for clothes and parachutes.

FIND OUT MORE ▶▶ Chemical Industry 50–51 • Elasticity 69 • Plastics 52–53

COMPOSITES

A composite is a combination of two or more different materials. The new material combines the best properties, such as strength and lightness, of each of the individual materials. There are examples of composites all around us. Boats, bikes, tennis rackets, even dental fillings are all made of composites. Most composites are synthetic materials, but they also occur in nature.

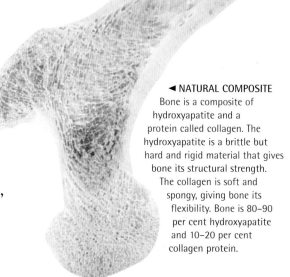

◄ NATURAL COMPOSITE
Bone is a composite of hydroxyapatite and a protein called collagen. The hydroxyapatite is a brittle but hard and rigid material that gives bone its structural strength. The collagen is soft and spongy, giving bone its flexibility. Bone is 80–90 per cent hydroxyapatite and 10–20 per cent collagen protein.

FIBREGLASS ►
This highly magnified macro-photograph of fibreglass shows how thin glass fibres are woven together. The woven fibres are laid in a mould and embedded with a plastic called a resin. The resulting material has the strength of glass and the flexibility of resin.

LIGHTWEIGHT GLIDER ▼
Fibreglass is the perfect material for a glider because it is incredibly light, yet strong. Its ability to be moulded easily also makes it ideal for the hulls (body) of lightweight sailing boats. The hulls are moulded from one piece of fibreglass and do not have any seams, so water will not leak into the boat.

Fibreglass tailplane is moulded from one piece of material

composites

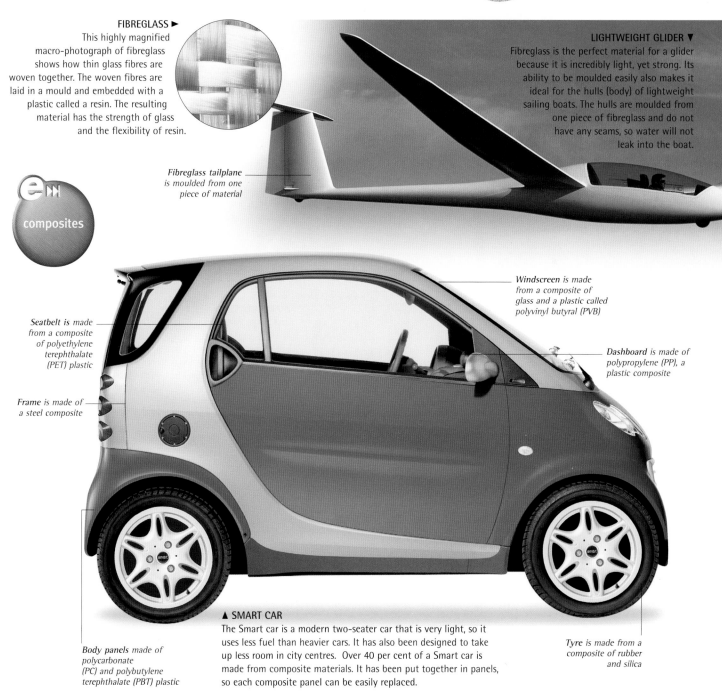

Windscreen is made from a composite of glass and a plastic called polyvinyl butyral (PVB)

Seatbelt is made from a composite of polyethylene terephthalate (PET) plastic

Dashboard is made of polypropylene (PP), a plastic composite

Frame is made of a steel composite

▲ SMART CAR
The Smart car is a modern two-seater car that is very light, so it uses less fuel than heavier cars. It has also been designed to take up less room in city centres. Over 40 per cent of a Smart car is made from composite materials. It has been put together in panels, so each composite panel can be easily replaced.

Body panels made of polycarbonate (PC) and polybutylene terephthalate (PBT) plastic

Tyre is made from a composite of rubber and silica

FIND OUT MORE ▶▶ Boats 95 • Glass 54 • Plastics 52–53 • Solids 12–13

NEW MATERIALS

Materials scientists develop and test new materials that do certain jobs better than existing materials, or are easier or cheaper to make. Scientists use their knowledge of how molecules form to combine atoms in new ways. They also apply heat and pressure to existing materials to create materials with new properties. They can even create materials with **SMART** properties that respond to their environment.

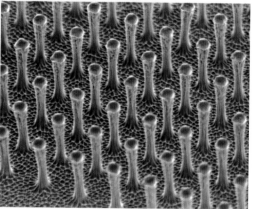

Hand insulated from naked flame by aerogel

Aerogel light enough to float on flame

Flame must be over 1,200°C (2,200°F) to melt aerogel

▲ ALL EXTREME FABRIC
In freezing conditions, climbers need clothes that are extremely warm and lightweight, so they can move easily. Aerogel is a new, super-light insulating material used to line extreme weather coats. It is made from silicon dioxide, the same material that glass is made from. However, aerogel is 99 per cent air, so it is 1,000 times less dense than glass.

new materials

▲ AEROGEL FOAM
One of the lightest substances on Earth, aerogel can even float on air in its pure form. It also has amazing insulation properties and can protect skin from the heat of a blowtorch. Aerogel is made by mixing a silicon compound with other chemicals to make a wet gel. The gel is dried at a high temperature and pressure.

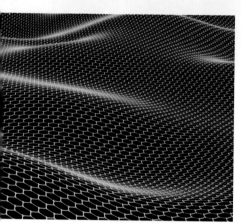

GRAPHENE
One of the world's most exciting new materials, graphene was discovered only in 2004. A flat sheet of carbon only one atom thick, one possible use could be to make super-fast transistors (computer switches) and highly efficient new batteries.

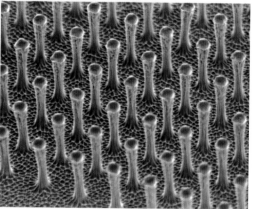

BIOMIMETIC MATERIAL
Biomimetics is a way of mimicking clever ideas used by animals and plants. This adhesive tape is covered in tiny plastic bars that stick like the hairs on a gecko's feet (spatulae), which let it walk upside down on a ceiling.

ECO-FRIENDLY PLASTIC
Plastics cause environmental problems because they can take 500 years to break down in nature. Biodegradable plastics are designed to rot away harmlessly in months. They are perfect for food packaging and growing crops.

CONDUCTIVE FABRICS ▶
This prototype computer is made from a fabric that conducts electricity. The fabric, called ElekTex™, contains fibres that have a thin coating of silver or copper – these metals are good electrical conductors. Microchips woven into the fabric translate the electrical impulses from the fibres into digital data, which can be viewed on the computer's screen.

Computer can be folded and rolled without losing its conductive properties

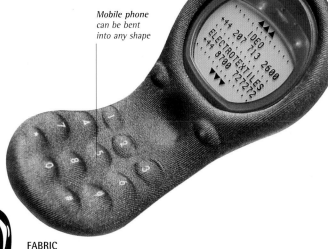

Mobile phone can be bent into any shape

FABRIC MOBILE PHONE ▲
Prototype mobile phones made from ElekTex™ are lightweight and water-resistant. They can also be folded up and crumpled without breaking. These properties make them a lot more hard-wearing than traditional mobile phones.

SMART MATERIALS

A smart material senses a change to its environment and responds by altering in some way. Each type of smart material has a different property that changes, such as stiffness, colour, shape, or conductivity (ability to conduct electricity). For instance, a piezoelectric material gives off a small electric current when bent, and is used in car passenger airbags. If the car slows down suddenly, the piezoelectric bends and sends out an electric charge, which blows up the airbag.

Tiny capsules contain monomer liquid

Black catalyst helps monomer molecules in liquid join together to form plastic

Plastic joint may develop tiny cracks through constant use

▲ PLASTIC HIP JOINT
Self-healing plastics could provide a breakthrough in surgery. If an artificial bone joint such as a metal hip wears out, it is difficult to replace it. However, a joint made of self-healing plastic would be able to repair itself, just like a bone can. If the artificial joint cracked, it could be almost good as new again within a few days.

▲ SELF-HEALING PLASTIC
Plastic is made of thousands of small molecules called monomers linked together to form polymers. A new, self-healing plastic contains tiny capsules filled with liquid monomer. If the plastic cracks, the capsules burst and release the liquid into the crack. A black catalyst in the plastic makes the liquid monomer molecules link up to create polymers and form new plastic that repairs the break.

FIND OUT MORE ▶▶ Electricity **126–127** • Molecules **28–29** • Plastics **52–53** • Synthetic Fabrics **56**

RECYCLING

Many things we normally throw away can be recycled (used again), including paper, glass, metals, plastics, and **BIODEGRADABLE** waste such as vegetable peelings and cut grass. Recycling saves natural resources, such as trees and crude oil. It can also save energy as it often takes less energy to make a product from recycled materials than it does to make the product from new materials. For example, 93 per cent more energy is needed to extract aluminium from ore than to recycle it.

Only magnetic metals are picked up by the magnet

▲ ELECTROMAGNET

Scrap steel and iron are magnetic metals that are sorted for recycling by an electromagnet. This electrically powered magnet is hung from a crane with three strong chains. The magnetic metals stick to the magnet and the non-magnetic materials are left behind. Cars contain a lot of steel and iron, so scrap cars are one of the main sources of recycled metal.

▲ ROLLING STEEL

Scrap steel is flattened or shredded, then melted in a furnace. The molten steel is poured into moulds to make slabs of steel called billets. Once solidified, the billets are reheated and rolled into thin sheets.

▲ SHEET STEEL

Steel sheets are used to make a range of products, such as food cans and car parts. Steel is 100 per cent recyclable. This means that recycled steel is exactly the same as the steel in the orginal material.

LANDFILL SITES ▶

Most rubbish is dumped in big pits called landfill sites. The pits are lined with clay to prevent poisons leaking into the surrounding soil and polluting water supplies. Pipes are inserted into the pit to collect and remove poisonous methane gas. Unless we recycle more, we will run out of places to put landfill sites.

Bulldozer squashes the rubbish so it takes up less space

◄ DISUSED PLANES
This aircraft graveyard in the desert in Arizona, USA, holds thousands of planes that cannot be used or work any longer. It is also a huge warehouse of spare parts that can be used again in other aircraft. Some parts of the planes can be melted down and recycled into new aluminium products, such as fizzy drinks cans.

Aluminium may once have been part of a plane

B52 bombers laid out in neat rows for recycling

Bookshelf made from recycled PET plastic

recycling

Sorting station at paper recycling plant

Recycled newspaper is slightly grey as it contains inks from the orginal paper

RECYCLING PAPER ►
Before recycling, paper is sorted into different grades. It is then mashed with water and chemicals to form a pulp. The pulp is cleaned (to remove staples, glue, or ink) and sprayed onto flat screens. When dry, the paper is used to make new products, such as newspapers.

RECYCLED PLASTIC ►
Plastics such as polyethylene terephthalate (PET), used in fizzy drinks bottles, can be recycled. This is because they are a kind of plastic called a thermoplastic. When heated, the plastic melts and can be moulded into a new shape. Only thermoplastics are recyclable. Thermosetting plastics burn rather than melt when heated.

Fizzy drinks bottle made from PET plastic

BIODEGRADABLE

Materials from living things are usually biodegradable. They break down into simpler substances, often with the help of micro-organisms. Leaves biodegrade into compost and carbon dioxide, both of which recycle in our environment. Most plastics are not biodegradable. They are so different from natural materials that micro-organsims cannot digest them.

GARDEN COMPOST ►
Making compost is a good way of recycling biodegradable materials that you would otherwise throw away. Vegetable peelings, sawdust, and grass cuttings can all be layered in a large container. Over a few months, micro-organisms will break down the biodegradable waste into compost. This rich, dark material can be scattered over the soil to provide plants with extra nutrients.

Fungus spores decomposing melon on compost heap

FIND OUT MORE ►► Fungi **282–283** • Groundwater **233** • Metals **34–35** • Plastics **52–53** • Pollution **250**

FORCES & ENERGY

FORCES

From the movements of the planets to the energy produced inside atoms, everything that happens in the Universe is ultimately caused by forces. A force is a push or pull that can make an object move or **TURN** around. The bigger the force, the more movement it can produce. When two or more forces act together on an object, their effects are **COMBINED**. Sometimes the forces add together to make a larger force, and sometimes they cancel each other out.

PULLING FORCE

▲ TUG O' WAR
In this game, two teams tug on a rope until one or other is pulled over the white line between them. Often there is no movement at all because the forces produced by the two teams are balanced (equal and opposite) and cancel each other out. The winning team is the one that produces the greater force on the rope.

TURNING FORCES

If an object is fixed at one point and can rotate around it, that point is called a pivot. If a force acts on the object, the object turns around the pivot. The turning force is called a torque and the effect it produces is called a moment. The bigger the force, the greater the moment. The moment also increases if the force acts at a greater distance from the pivot.

◄ WHEELBARROW
A wheelbarrow is free to pivot around the large wheel at the front. When the worker lifts the handles, the force causes the entire wheelbarrow to swing upwards and turn around the wheel. The long body and handles of a wheelbarrow increase the turning effect and make it easier to tip out a heavy load.

NEWTONS

SALTER
12
NEWTONS

LOAD

Forces are measured in units called newtons (N), named after English scientist Sir Isaac Newton. The size of a force can be measured using a device called a force meter or newtonmeter. As the load pulls on the hook, it stretches a spring to give a reading on the scale. On Earth, the force of gravity on 1 kg (2.2 lb) is 9.8 newtons.

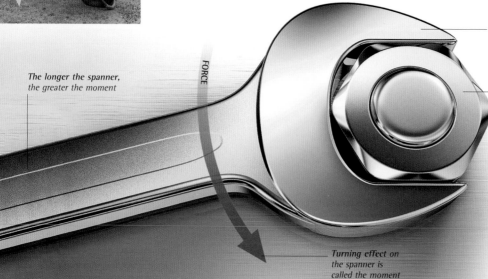

The longer the spanner, the greater the moment

FORCE

Spanner exerts a strong turning force on the nut

Nut is the pivot (turning point)

e ►►
forces

Turning effect on the spanner is called the moment

◄ INCREASING MOMENTS
It is easier to unscrew a nut with a spanner than with your fingers, because the spanner's long handle increases the turning effect or moment of the force. The size of a moment is equal to the force used times the distance from the pivot on which it acts. If you use a spanner twice as long, you double the moment, and the nut is twice as easy to turn.

▲ USING FORCE TO MOVE AN OBJECT
When a force acts on an object, it can produce movement. The football stays still on the ground until the player kicks it. Then it moves off in the direction in which it is kicked. The harder the kick, the more force is applied. The greater the force, the faster the ball flies through the air.

▲ USING FORCE TO CHANGE DIRECTION
As the football shoots up and into the goal, the goalkeeper's hand reaches out and pulls downwards on the ball. The force of the goalkeeper's hand cancels out the movement of the football and saves the goal. The faster the ball moves, the harder the goalkeeper has to work to stop it.

▲ BALANCED FORCES
Forces can change the shape of objects. The football boot pushing down is met by the balancing force of the ground, and the ball is squashed out of shape (compressed). The ball is made of stretchy (elastic) material, so it returns to its original shape when the force is removed.

PULLING FORCE

COMBINED FORCES

When forces act in the same direction, they combine to make a bigger force. When they act in opposite directions, they can cancel one another out. If the forces acting on an object balance, the object does not move, but may change shape. If the forces combine to make an overall force in one direction, the object moves in that direction.

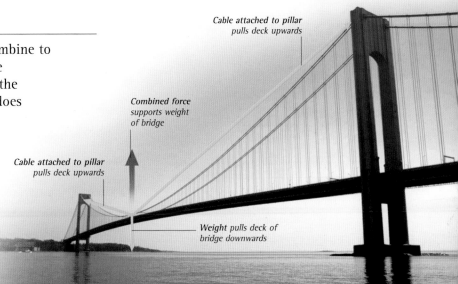

Cable attached to pillar
pulls deck upwards

Combined force
supports weight
of bridge

Cable attached to pillar
pulls deck upwards

Weight pulls deck of
bridge downwards

SUPPORTING A BRIDGE ►
A suspension bridge has to support the weight of its own deck, plus the weight of the vehicles that go across it. The deck of the bridge hangs from huge steel cables suspended over giant pillars. The cables and pillars are arranged so that there is no overall force in any direction. A bridge stays up because the forces on it are balanced and cancel one another out.

FIND OUT MORE ▶▶ Atoms 24–25 • Dynamics 66–67 • Elasticity 69 • Energy 76–77 • Solar System 172–173

DYNAMICS

Dynamics is the study of how objects move when forces act on them. Normally objects stay still or move along at a steady pace. They resist changes in their motion because of their **INERTIA**. Once they start moving, they tend to carry on doing so because of their **MOMENTUM**. Most types of everyday movement can be explained by just three simple **LAWS OF MOTION**. These were originally worked out by English physicist Sir Isaac Newton.

dynamics

POOL BALLS AT BREAK ▶
When the white cue ball hits a pack of coloured pool balls at high speed, it has lots of momentum – force due to its weight and speed. The cue ball hits the other balls so hard that it bounces back. During the collision, it slows down and loses some of its momentum. The pack balls gain momentum from the cue ball, and fly off in all directions.

Stationary pack balls have inertia, which is overcome by the force of the cue ball

LAWS OF MOTION

Newton's three laws of motion (often called Newton's laws) explain how forces make objects move. When the forces that are acting on an object are balanced, there is no change in the way it moves. When the forces are unbalanced, there is an overall force in one direction. This changes the object's speed or the direction in which it is moving. Physicists call a change in speed or direction an acceleration.

▲ NEWTON'S 1ST LAW
An object will stay still or move along at a steady pace unless a force acts on it. For example, a rocket on a launchpad remains in place because there is no force acting on it to make it move.

▲ NEWTON'S 2ND LAW
When a force acts on an object, it makes the object change speed or move in a different direction. When the rocket's engines fire, the force they produce lifts the rocket up off the launchpad and into the air.

▲ NEWTON'S 3RD LAW
When a force acts on an object, the object pulls or pushes back. This reaction is equal to the original force but in the opposite direction. As the hot gases shoot down from the engines, an equal force pushes the rocket up.

SIR ISAAC NEWTON
English, 1642–1727

Newton's three laws of motion enabled him to produce a complete theory of gravity, the force that dominates our Universe, and to explain why the Moon circles round Earth. Newton also made major discoveries about optics (the theory of light) and explained how white light is composed of many colours.

DIRECTION OF
CUE BALL

CUE BALL REBOUNDING

CUE BALL APPLIES
FORCE AT IMPACT

*First ball to be struck
by cue ball gains
momentum from it*

*The total momentum
of the pack balls and
the cue ball is equal to
the momentum of the
cue ball before
the collision.*

*Each pack ball
flies off in a
different direction
according to the
forces acting
upon it*

INERTIA

Newton's first law explains that objects remain where they are or move along at a steady speed unless a force acts on them. This idea is known as inertia. The greater the weight (or mass) of an object, the more inertia it has. Heavy objects are harder to move than light ones because they have more inertia. Inertia also makes it harder to stop heavy things once they are moving.

CRASH-TEST DUMMIES ▶
As a car accelerates, passengers are thrown backwards; when a car brakes or crashes, passengers are thrown forwards. In both cases, this is because the inertia caused by their mass resists the change in movement. During crash-tests, dummies that weigh the same as a human body are used to help test safety belts and airbags.

MOMENTUM

Moving objects carry on moving because they have momentum. The momentum of a moving object increases with its mass and its speed. The heavier the object and the faster it is moving, the greater its momentum and the harder it is to stop. If a truck and a car are travelling at the same speed, it takes more force to stop the truck because its greater mass gives it more momentum.

▲ COMPARING MOMENTUM
A foal is smaller and has less mass than a horse. When a foal and a horse gallop along together at the same speed, the horse has more momentum because of its greater mass. This means that it is easier for the foal to start moving, stop moving, and change direction than the horse. The momentum of a moving object is equal to its mass times its velocity.

FIND OUT MORE ▶▶ Energy **76–77** • Forces **64–65** • Gravity **72** • Motion **70–71**

Bobsleigh team keep their heads down to minimize air resistance

FRICTION

If you kick a ball across a playground, it bounces and rolls on the ground's rough surface and soon comes to a halt. What slows it down is friction, which is the force between a moving object and whatever it touches. Cars travel faster if they are **STREAMLINED** to reduce a type of friction called air resistance. Friction can sometimes be helpful. Without friction between the tyres and the road, cars would not have enough grip to go around corners.

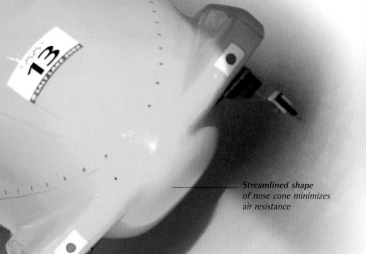

◄ BOBSLEIGH ON ICE
It is easy to start this bobsleigh moving because there is very little friction between its highly polished runners and the ice underneath them. As the bobsleigh moves along, the pressure of its runners melts the ice slightly. Some of the ice turns to water. This reduces friction and helps the bobsleigh travel faster.

Streamlined shape of nose cone minimizes air resistance

Bobsleigh runners are made of thin metal to move over the ice with minimum friction

LUBRICATING MACHINERY ▲
Slippery substances such as oil reduce the friction between two surfaces. This is known as lubrication. Machinery has to be lubricated to prevent its moving parts from wearing out due to friction. Most machines are oiled or greased when they are made and are lubricated from time to time as they are used.

STREAMLINING

When objects move, the air around them generates a type of friction called air resistance, or drag, that slows them down. Fast-moving objects such as cars, trains, and aeroplanes are all streamlined – designed with curved and sloping surfaces to cut through the air and reduce drag. This helps them to move faster and use less fuel. Boats can be streamlined too, to reduce water resistance.

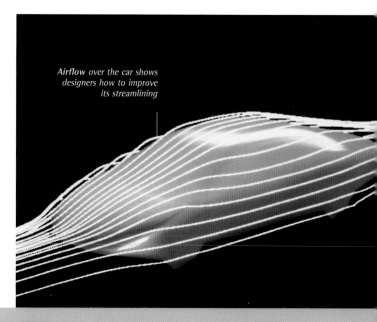

Airflow over the car shows designers how to improve its streamlining

friction

CAR IN WIND TUNNEL ►
When a new car is designed, the engineers make a model of its body and place it in a testing chamber called a wind tunnel, which blows air past the car. The engineers spray white smoke into the tunnel and watch how it flows around the car body. They can see where friction is greatest, and how to improve the car's streamlining by changing its shape.

FIND OUT MORE ▶▶ Boats **95** • Flight **96** • Forces **64–65** • Machines **88–91**

ELASTICITY

Forces make things move, but they can also stretch things, squeeze them, and change their shape. A rubber ball changes shape when you use force to squeeze it, but it returns to its original shape when you stop squeezing. Materials that do this have elasticity. They are made up of particles called molecules that can stretch apart. Other materials, such as modelling clay, change shape easily when a force is applied, but they do not return to their original shape when the force is no longer applied. These materials have **PLASTICITY**.

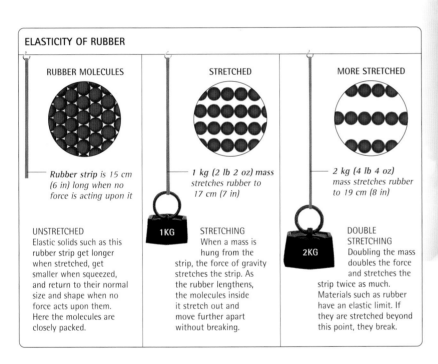

ELASTICITY OF RUBBER

RUBBER MOLECULES	STRETCHED	MORE STRETCHED
Rubber strip is 15 cm (6 in) long when no force is acting upon it	*1 kg (2 lb 2 oz) mass stretches rubber to 17 cm (7 in)*	*2 kg (4 lb 4 oz) mass stretches rubber to 19 cm (8 in)*

UNSTRETCHED
Elastic solids such as this rubber strip get longer when stretched, get smaller when squeezed, and return to their normal size and shape when no force acts upon them. Here the molecules are closely packed.

1KG

STRETCHING
When a mass is hung from the strip, the force of gravity stretches the strip. As the rubber lengthens, the molecules inside it stretch out and move further apart without breaking.

2KG

DOUBLE STRETCHING
Doubling the mass doubles the force and stretches the strip twice as much. Materials such as rubber have an elastic limit. If they are stretched beyond this point, they break.

TRAMPOLINING ▲
A trampoline is made of stretchy rubber fastened to a metal frame by metal springs. When you land on a trampoline, you stretch the rubber and the springs. Both rubber and springs are elastic. As they return to their original shape, they pull back upwards and push you into the air.

PLASTICITY

Materials have plasticity when they are easily moulded into shape and do not return to their original shape when the moulding force is removed. When we talk about plastics, we usually mean various colourful materials that have been made out of chemicals produced from oil. In fact, the word plastic applies to any material that can be easily moulded into different shapes. Even metals can be plastic because, if heated, they soften and can be shaped.

elasticity

MANUFACTURING PLASTIC OBJECTS ►
These spoons are made from a chemical called a polymer. When the polymer is hot, it is a runny liquid made up of molecules that slide past one another easily. It is said to be plastic because it can be shaped easily. The polymer is poured into a spoon-shaped mould. As the polymer cools, it hardens and sets, and the spoons take their final shape.

FIND OUT MORE ►► Forces 64–65 • Gravity 72 • Molecules 28–29 • Plastics 52–53

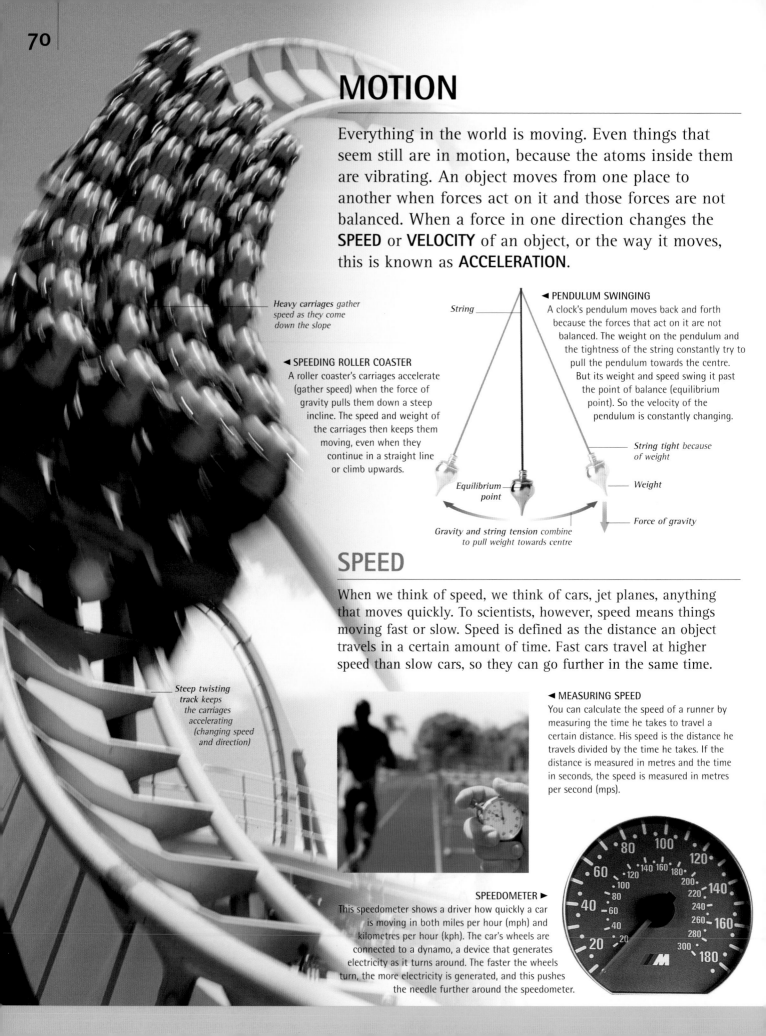

MOTION

Everything in the world is moving. Even things that seem still are in motion, because the atoms inside them are vibrating. An object moves from one place to another when forces act on it and those forces are not balanced. When a force in one direction changes the **SPEED** or **VELOCITY** of an object, or the way it moves, this is known as **ACCELERATION**.

Heavy carriages gather speed as they come down the slope

◄ SPEEDING ROLLER COASTER
A roller coaster's carriages accelerate (gather speed) when the force of gravity pulls them down a steep incline. The speed and weight of the carriages then keeps them moving, even when they continue in a straight line or climb upwards.

◄ PENDULUM SWINGING
A clock's pendulum moves back and forth because the forces that act on it are not balanced. The weight on the pendulum and the tightness of the string constantly try to pull the pendulum towards the centre. But its weight and speed swing it past the point of balance (equilibrium point). So the velocity of the pendulum is constantly changing.

String

String tight because of weight

Weight

Equilibrium point

Force of gravity

Gravity and string tension combine to pull weight towards centre

Steep twisting track keeps the carriages accelerating (changing speed and direction)

SPEED

When we think of speed, we think of cars, jet planes, anything that moves quickly. To scientists, however, speed means things moving fast or slow. Speed is defined as the distance an object travels in a certain amount of time. Fast cars travel at higher speed than slow cars, so they can go further in the same time.

◄ MEASURING SPEED
You can calculate the speed of a runner by measuring the time he takes to travel a certain distance. His speed is the distance he travels divided by the time he takes. If the distance is measured in metres and the time in seconds, the speed is measured in metres per second (mps).

SPEEDOMETER ►
This speedometer shows a driver how quickly a car is moving in both miles per hour (mph) and kilometres per hour (kph). The car's wheels are connected to a dynamo, a device that generates electricity as it turns around. The faster the wheels turn, the more electricity is generated, and this pushes the needle further around the speedometer.

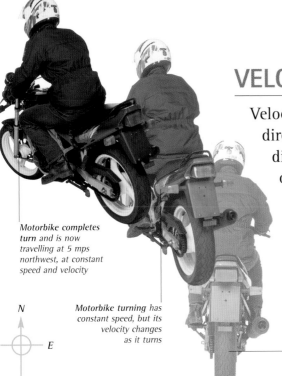

Motorbike completes turn and is now travelling at 5 mps northwest, at constant speed and velocity

N
W — E
S

Motorbike turning has constant speed, but its velocity changes as it turns

VELOCITY

Velocity is the speed of an object moving in a particular direction. Two cars driving at the same speed have different velocities if one of them goes north and the other goes south. Velocity is measured in metres per second (mps), which divides the distance travelled by the time taken, in a specific direction.

◀ **CHANGING VELOCITY**
When moving objects change speed, they change velocity. As they change direction, they also change velocity, even if their speed stays the same. When the bike goes faster or slower, the force that makes it change velocity is the engine or brakes. When it turns, the rider provides the force by turning the handlebars.

motion

Motorbike travelling at a velocity of 5 mps north has constant speed and velocity

ACCELERATION

When we talk of things accelerating, we usually mean they are speeding up. In science, however, acceleration means any change in an object's velocity, whether it goes faster, slower, or changes direction. According to Newton's second law of motion, a force is always needed to produce an acceleration. The bigger the force, the faster the change in velocity.

0 SECONDS   0.1   0.2   0.3   0.4

SPRINT START ▲
It takes time for a runner to reach his top speed. To begin with, he travels slowly. As he gets into his stride, he travels greater and greater distances in the same amount of time. He moves in a straight line, but his speed and velocity are constantly increasing. He is accelerating.

Force towards centre pulls ball around in a circle

If the force is removed, the ball carries on in the same direction

◀ **CIRCULAR MOTION**
An object that moves in a circle, such as this ball swinging on a string, constantly changes direction. Even when it turns at a steady speed, its velocity is always changing. It takes a force to make it accelerate like this. When an object moves in a circle, the force that constantly pulls it towards the centre and stops it flying off in a straight line is called centripetal force.

String under tension pulls ball towards centre

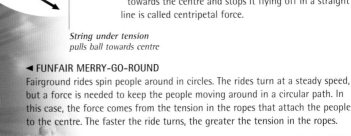

◀ **FUNFAIR MERRY-GO-ROUND**
Fairground rides spin people around in circles. The rides turn at a steady speed, but a force is needed to keep the people moving around in a circular path. In this case, the force comes from the tension in the ropes that attach the people to the centre. The faster the ride turns, the greater the tension in the ropes.

FIND OUT MORE ▸▸ Atoms **24–25** • Dynamics **66–67** • Forces **64–65** • Gravity **72**

GRAVITY

Gravity is the force that makes things fall to the ground on Earth and holds the planets in their orbits (paths) around the Sun. The force of gravity acts over immense distances between objects in the Universe and holds them all together. The gravitational force between objects increases with their **MASS**. It also increases the closer they are. The gravity between objects on Earth is usually too small to notice.

Centre of gravity is still above the car's axis, or base

Axis – when centre of gravity moves beyond axis, car tips over

CENTRE OF GRAVITY ▶
On Earth, objects have a point, often near their centre, which is called their centre of gravity. The lower it is, the more stable they are. Cars are designed with their heavy engines near to the ground, to keep their centre of gravity low. This means they can corner at speed without tipping over.

◀ ZERO GRAVITY
These astronauts are training for the lack of gravity they will find in space. Their specially modified aeroplane climbs high above Earth, then dives steeply back. As the aeroplane zooms downwards, everything falls together and gravity seems to disappear. This state is known as zero gravity. The astronauts now become weightless and float about.

▲ FALLING FORCE
Which falls faster, a ball or a feather? In Earth's atmosphere, the ball reaches the ground first because air resistance slows the feather down. In a vacuum, there is no air and therefore no air resistance. The feather and the pool ball fall at the same rate because gravity pulls them with exactly the same amount of force.

MASS

The mass of an object is the amount of matter it contains. The greater the mass of an object, the more matter it contains, and the more it pulls on other objects with the force of gravity. The mass of an object does not vary unless the amount of matter inside it changes for some reason. Mass is measured in kilograms (kg).

gravity

▲ MASS VERSUS WEIGHT
The weight of this astronaut is the effect of gravity acting on the mass of his body. The Moon has a mass roughly one-sixth that of Earth, so its gravity is one-sixth as strong. On the Moon his mass is the same as on Earth, but he weighs only one-sixth as much.

FIND OUT MORE ▶▶ Dynamics 66–67 • Forces 64–65 • Friction 68 • Solar System 172–173 • Space Travel 190–191

RELATIVITY

Einstein realized that the speed of light is always the same. He then calculated that an object travelling near this speed acts strangely: it shrinks in length, increases in mass, and time slows down. He also calculated that mass alters space. So small objects do not travel in straight lines near a large object – instead they follow the distortions in space made by it. Centuries after gravity was identified as a force, Einstein's theory of relativity explained why it works the way it does.

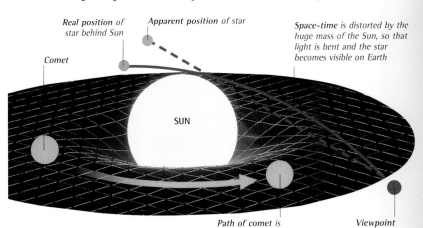

Real position of star behind Sun

Apparent position of star

Comet

Space-time is distorted by the huge mass of the Sun, so that light is bent and the star becomes visible on Earth

SUN

Path of comet is distorted by gravity

Viewpoint on Earth

▲ EINSTEIN'S GRAVITY

In traditional physics, gravity attracts one mass to another. This explains why a comet follows a curved path around the Sun. Einstein's general theory of relativity explains gravity differently. Masses warp space and time a bit like heavy balls resting on a sheet of rubber. The bigger the mass, the more distortion, and the greater the pull of gravity. In 1921 Einstein was proved correct when the light from a star was shown to be bent by the warping effect of the Sun's mass.

Path of light beam as seen from Earth

B

Path of light beam flashing from one rocket to the other, as seen from the rockets

A

Behind the traditional clockface lies a radio with a link to a central atomic clock

Digital display shows highly accurate time

▲ ACCURATE TIMEKEEPING

The effects of relativity are only detectable when things travel at very high speeds. To detect them, scientists need accurate clocks that use atoms to tell the time. Atoms of the element caesium vibrate at a precise rate. Atomic clocks measure time by counting these vibrations. Clocks such as the one above use a radio link to a central atomic clock to relay the precise time.

ALBERT EINSTEIN
German, 1879-1955

When Albert Einstein was expelled from school, no one imagined he would become one of the most brilliant physicists of the 20th century. His theory of relativity was so strange that people refused to believe it at first. It was widely accepted only after he won the Nobel Prize for Physics in 1921.

▲ RELATIVITY EXPLAINED

Strange things happen when objects, such as these two rockets, travel at near the speed of light (300,000 kps or 186,000 mps). A beam of light flashing between them would seem to observers on the rockets to be line A. To observers on Earth not travelling so fast, the beam is seen as line B. Speed equals distance over time, and since the speed of light is constant, and the distance it travels is longer when viewed from Earth, the only possible explanation is that time is passing faster on Earth than on the rockets.

FIND OUT MORE ▸▸ Atoms 24–25 • Black Holes 169 • Gravity 72 • Motion 70–71 • Periodic Table 26–27 • Sun 170–171

PRESSURE

When you press or push something, the force you apply is called pressure. Pressure is measured as the force you use divided by the area over which you use it. If you use a bigger force, or if you use the same force over a smaller area, you increase the pressure. We experience **AIR PRESSURE** all the time because of the weight of air pressing in on our bodies. **WATER PRESSURE** increases as you go deeper in the ocean.

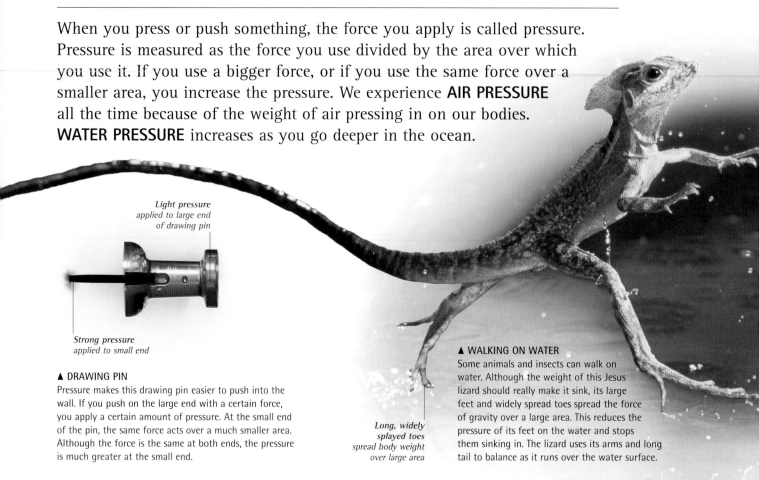

Light pressure applied to large end of drawing pin

Strong pressure applied to small end

▲ DRAWING PIN
Pressure makes this drawing pin easier to push into the wall. If you push on the large end with a certain force, you apply a certain amount of pressure. At the small end of the pin, the same force acts over a much smaller area. Although the force is the same at both ends, the pressure is much greater at the small end.

Long, widely splayed toes spread body weight over large area

▲ WALKING ON WATER
Some animals and insects can walk on water. Although the weight of this Jesus lizard should really make it sink, its large feet and widely spread toes spread the force of gravity over a large area. This reduces the pressure of its feet on the water and stops them sinking in. The lizard uses its arms and long tail to balance as it runs over the water surface.

AIR PRESSURE

The gases in Earth's atmosphere are made up of tiny molecules that are constantly crashing into your body and trying to press it inwards. This pressing force is called air pressure. It is greatest at ground level where there are most air molecules. At greater heights above Earth, there are fewer air molecules and the air pressure is much less. It is possible to compress (squeeze) air, and this is used to inflate vehicle tyres and to power machines such as pneumatic drills.

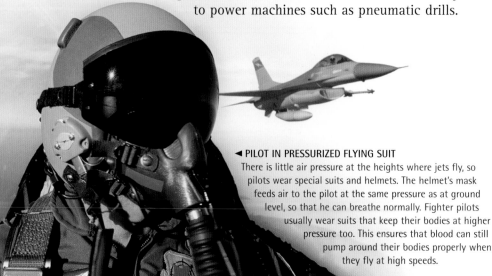

◄ PILOT IN PRESSURIZED FLYING SUIT
There is little air pressure at the heights where jets fly, so pilots wear special suits and helmets. The helmet's mask feeds air to the pilot at the same pressure as at ground level, so that he can breathe normally. Fighter pilots usually wear suits that keep their bodies at higher pressure too. This ensures that blood can still pump around their bodies properly when they fly at high speeds.

▲ AIR PRESSURE IN TYRES
Heavy construction machines have large tyres for two reasons. The compressed air in the tyre helps to absorb bumps, so the ride is much smoother than it would be with a solid wheel. Large tyres also help to spread the weight of the machine over a much bigger area. This reduces the pressure on the ground and stops the machine sinking into the mud.

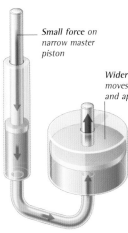

Small force on narrow master piston

Wider slave piston moves short distance and applies large force

▲ HOW HYDRAULICS WORK

Liquid pressure is used to carry force through pipes. The small force pushing down does not compress the liquid but moves through the liquid to push another piston a small distance upwards. The wider area of this piston increases the force applied.

pressure

WATER PRESSURE

Water behaves differently from air when it is under pressure. It cannot be compressed (squeezed). This makes it useful for transmitting force in machines, using a system called hydraulics. Water is also heavier than air, and an increase in water pressure affects humans more than a drop in air pressure. Even with a snorkel or other breathing apparatus, it feels much harder to breathe underwater. The water above you presses down from all sides on your body, so your lungs find it harder to expand to take in air. The deeper you go, the more water there is above you and the greater the pressure on your body.

◄ HYDRAULICS IN ACTION

Hydraulic pipes move the arms of this aerial platform up and down. The engine pushes hydraulic fluid into the rams, which fills them and extends them upwards. Hydraulics are an effective way of transferring force from the engine to other parts of the machine. Hydraulics are used in vehicle brakes, in jacks that lift cars, and in factory machines.

Hydraulic ram powered by pressure of hydraulic fluid

Hydraulic pipes supplying hydraulic fluid to ram

DEEP-SEA DIVER'S NEWT-SUIT ►

Water pressure increases rapidly the deeper you go in the ocean, and divers need to wear special suits so that they can breathe properly. This newt-suit enables a diver to go down to depths as great as 300 m (1,000 ft). It has its own air supply, bendy joints, so the diver can move his arms and legs, and a built-in radio so that he can talk to colleagues on the surface.

CHANGING AIR AND WATER PRESSURE

The higher we go, the less air there is in the atmosphere above us. The deeper in the sea we go, the more water there is pressing down on us.

20,000 m (65,600 ft) HIGH

At this height, air pressure is less than one-tenth that at sea level.

AIRLINERS 11,000 m (36,000 ft)

Aircraft cabins are pressurized to allow us to breathe as easily as at sea level. Oxygen is also supplied in case of emergency, as there is less air at this height.

MOUNTAIN TOPS 7,500 m (24,600 ft)

At this height, climbers often use breathing apparatus to give them more oxygen.

SEA LEVEL

The human body is ideally adapted to deal with the air pressure at sea level.

120 m (400 ft) DEEP

Divers cannot go any deeper than this without special suits to protect them from the pressure of the water.

SUBMERSIBLES 6,500 m (21,300 ft)

Underwater craft such as submarines have strong, double-skinned hulls to withstand water pressure. The world's deepest-diving crewed submersible can dive to 6,500 m (21,300 ft).

10,000 m (32,800 ft) DEEP

At this depth, the pressure of water is 1,000 times greater than it is at sea level.

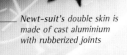

Newt-suit's double skin is made of cast aluminium with rubberized joints

FIND OUT MORE ►► Atmosphere **234–235** • Forces **64–65** • Gases **15** • Liquids **14** • Machines **88–89** • Water **40–41**

ENERGY

Scientists define energy as the ability to do work. Energy makes things happen. The energy in sunlight makes plants grow, the energy in food enables us to move and helps us to keep warm, and the energy in fuel powers engines. Energy comes in many different forms and can be converted from one form into another. The main types include **POTENTIAL ENERGY**, **KINETIC ENERGY**, and **CHEMICAL ENERGY**.

POTENTIAL ENERGY

Energy that is stored up ready to be used in the future is called potential energy, because it has the potential (or ability) to do something useful later on. An object usually has potential energy because a force has moved it to a different position or changed it in some other way. When an object releases its stored potential energy, this energy is converted into energy of a different form.

◄ ELECTRICAL POTENTIAL ENERGY
When thunderclouds move through the sky, they build up a large amount of electricity inside themselves. This is known as static electricity, which is a store of energy. When a cloud builds up more static electricity than it can store, some of the electricity flows from the cloud to Earth in a bolt of lightning.

The elasticity of the bow stores energy, which is released and transferred to the arrow

Arrow

GRAVITY PULLS SNOW DOWNWARDS

◄ ELASTIC POTENTIAL ENERGY
This type of potential energy powers bows and catapults. It takes effort to stretch a piece of elastic or rubber because the forces between its molecules try to resist being pulled apart. As the elastic stretches, the molecules move away from one another and gain potential energy. The energy stored in stretched elastic can also be used to power such things as toy cars and model aeroplanes.

▲ GRAVITATIONAL POTENTIAL ENERGY
A snowdrift on top of a mountain has a huge amount of potential energy. This is known as gravitational potential energy because it is gravity that is constantly trying to pull the snow down the mountain to the bottom. When an avalanche occurs, the snow gathers speed and its stored potential energy is turned into kinetic energy (the energy of movement).

KINETIC ENERGY

Moving objects have a type of energy called kinetic energy. The more kinetic energy something has, the faster it moves. When objects slow down, their kinetic energy is converted into another type of energy, such as heat or sound. Objects at rest have no kinetic energy. Kinetic energy is often produced when objects release their potential energy.

*The **kinetic energy** of the moving hammer is transferred to the nail*

*The **nail** is driven hard into the wood*

energy

◄ HAMMER STRIKING NAIL
A moving hammer has a lot of kinetic energy. As it strikes the nail, it slows down and loses its kinetic energy. The energy does not disappear, however. Some of it goes to split the wood to make way for the nail, some passes into the wood as heat energy, and some is converted into sound.

FOOD AS CHEMICAL ENERGY

When humans or other animals eat food, they use its stored energy to keep warm, maintain and repair their bodies, and move about. Different types of food store different amounts of energy. The amount of energy a food contains is measured in kilocalories (called Calories for short).

TYPE OF ANIMAL	DAILY CALORIES FROM FOOD
Elephant	40,000
Panda	20,000
Man	2,600 (moderate activity)
Woman	2,300 (moderate activity)
Child (7-10 yrs)	2,000
Mouse	20

CHEMICAL ENERGY

This is the energy involved in chemical reactions, when elements join together into compounds. This energy is stored inside the compounds as chemical potential energy. The stored energy can be released by further chemical reactions. The food we eat stores energy that is released by digestion. Energy can also be released by burning the chemicals in a process called combustion. Fuels are chemical compounds that release heat energy by combustion.

▲ CHARCOAL FIRE
Fuels such as charcoal are hydrocarbons, chemical compounds made mainly from hydrogen and carbon. When a fuel burns in air, the hydrocarbons break up into simpler compounds. The chemical potential energy they contain is then released as heat energy. Light energy is produced at the same time and this is what makes a fire glow as it burns.

FIND OUT MORE ►► Chemical Reactions **30–31** • Elasticity **69** • Gravity **72** • Heat **80–81** • Molecules **28–29** • Work **78–79**

WORK

Scientists use the word work to describe the energy needed to do a task, by making a force move through a distance. The amount of work done is equal to the energy used and both are measured in **JOULES** (J). It takes energy to lift a weight a certain distance, because you have to do work against the force of gravity. **POWERFUL** machines can do lots of work in a short time. **EFFICIENT** machines waste relatively little energy when doing work.

TUGBOAT TOWING LOGS ▶
Although logs float, they are heavy. They also drag in the water. The tugboat has to use a great deal of force to overcome water resistance and move the logs. The work it does is to pull the logs a certain distance through the water.

Logs pick up energy to move from the movement of the tugboat

▲ LEAFCUTTER ANT
This tiny ant can lift and carry many times its own body weight. The ant does work by lifting the leaf up in the air against the force of gravity. It carries the leaf sideways to reduce air resistance. This greatly reduces the amount of work it has to do.

EFFICIENCY

Efficiency is a measure of how much of its energy a machine converts into useful work. No machine ever converts all its energy into work: some energy is always wasted in the process. Car engines convert fuel into the energy they need in order to move, but get hot as they do so. This heat does not help the car to move, so a car is relatively inefficient, compared to other machines.

◀ EFFICIENT MACHINE
Bicycles are efficient machines. They allow riders to convert muscle power into movement with little wasted energy. Racing cyclists wear aerodynamic clothing. Less energy is wasted overcoming air resistance, so more energy is used to move the bicycle.

Tugboat is not 100% efficient, as it wastes some energy moving water, and some as heat and sound

HOW EFFICIENT ARE OUR MACHINES?	
Bicycle	90%
Power station's steam turbine	35%
Human body	24%
Car's petrol engine	20–25%
Electric light bulb	5%

POWER

Some machines can do work more quickly than others, and these are said to be more powerful. Power is the amount of work that something can do in a certain amount of time. Cars with bigger engines can go faster, which means they cover more distance in the same time. This means faster cars do work more quickly than slower cars, so they are more powerful machines.

Large bucket to lift huge loads, powered by hydraulic arms

POWERFUL DIGGER ▶
This digger does work by using a force to move heavy loads over a distance. The bigger the bucket at the front of the digger, the greater the load it can move in one go. Diggers with large buckets can do more work in the same time as diggers with small buckets. That makes diggers with large buckets more powerful machines.

JOULES

The amount of work done when a force acts over a distance equals the size of the force (measured in newtons) times the distance through which it moves (measured in metres). The work done is measured in joules, named after English physicist James Prescott Joule (1818–1889). An amount of work takes the same amount of energy to do it, so energy is also measured in joules.

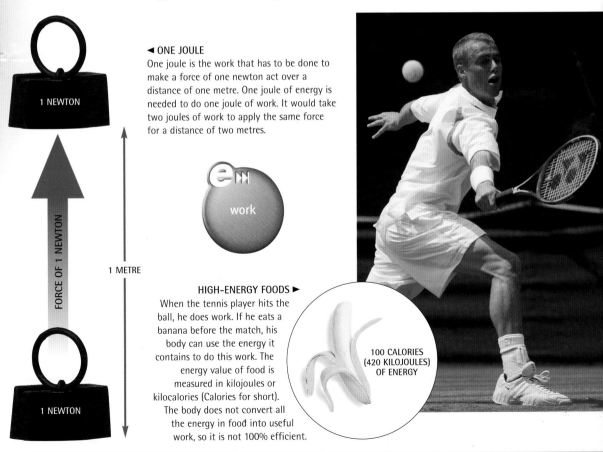

1 NEWTON

FORCE OF 1 NEWTON

1 METRE

1 NEWTON

◀ ONE JOULE
One joule is the work that has to be done to make a force of one newton act over a distance of one metre. One joule of energy is needed to do one joule of work. It would take two joules of work to apply the same force for a distance of two metres.

work

HIGH-ENERGY FOODS ▶
When the tennis player hits the ball, he does work. If he eats a banana before the match, his body can use the energy it contains to do this work. The energy value of food is measured in kilojoules or kilocalories (Calories for short). The body does not convert all the energy in food into useful work, so it is not 100% efficient.

100 CALORIES (420 KILOJOULES) OF ENERGY

FIND OUT MORE ▶▶ Energy 76–77 • Engines 92 • Forces 64–65 • Friction 68 • Gravity 72 • Machines 88–91

HEAT

Metal heated in a furnace shows that it is hot by glowing red and sending out sparks – but there is also some heat in ice and snow. Heat is the energy of movement, or kinetic energy, stored inside every object, hot and cold alike. Heat energy makes the particles (atoms and molecules) inside the object move about. **TEMPERATURE** is how hot or cold an object is, depending on its heat energy. Temperature is measured with a **THERMOMETER**.

◀ MOLTEN METAL
When iron is heated in a furnace, it glows red-hot and then melts at a temperature of 1,535°C (2,795°F). At this temperature, its particles move about with lots of kinetic energy. This view shows the particles at this temperature in vigorous motion. At higher temperatures they move even faster, and the iron in the furnace starts bubbling.

PARTICLES OF HOT IRON

▲ ICEBERG
Ice is cold, but it still contains some heat energy. An iceberg is made up of particles of water, held in a rigid crystal structure. They still vibrate slightly. If the iceberg cooled down so that its particles stopped moving altogether, it would be at the lowest possible temperature that can ever, in theory, be reached. This is absolute zero.

PARTICLES OF ICE

▼ TEMPERATURE SCALES
Temperature is measured in degrees Celsius or Fahrenheit (°C or °F) or on the absolute temperature scale, in units called Kelvins (K). The Celsius (also called Centigrade) scale runs from freezing point (0°C) to boiling point (100°C).

Absolute zero is the lowest possible temperature and equals -273°C, -460°F, zero K

Antarctica experiences the lowest temperature on Earth: -89°C, -128°F, 184K

Body temperature for a healthy human is 37°C, 98.6°F, 310K

Water boils at 100°C, 212°F, 373K

TEMPERATURE

Temperature is a measure of how hot or cold something is. Things that have high temperature are hotter than things that have lower temperature, because they have more heat energy inside them. Any object can transfer heat energy to a colder object. As it does so, it cools down and its temperature falls. The colder object warms up and its temperature rises.

SPACE SHUTTLE ▶
When the Space Shuttle re-enters Earth's atmosphere, it is travelling extremely fast. Air resistance heats the body of the Shuttle up to temperatures of around 3,500°C (6,332°F). This is hot enough to melt most materials, but the Shuttle is covered in special ceramic tiles to withstand this heat.

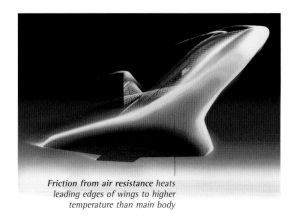

Friction from air resistance heats leading edges of wings to higher temperature than main body

Roof is well insulated, allowing little heat to escape

Curtained window shows up orange because heat is lost through the glass

Uncurtained window shows up yellow because more heat is lost through the glass

▲ THERMAL IMAGE OF HOUSE
A house at night is usually warmer than the cool air outside, so heat tends to flow out from the inside. This image from a heat-sensitive camera shows the hotter parts of the house as yellow and orange, and the cooler parts as pink and violet. A lot of heat is being lost through the windows and doors, and some heat is escaping through the chimney.

heat

THERMOMETER

This is a device that measures how hot or cold something is on a temperature scale. When things get hotter, their heat energy makes them expand or get bigger. This is how a thermometer measures temperature. As the liquid inside expands, it creeps up a tube, which is marked with a scale and numbers that show the temperature.

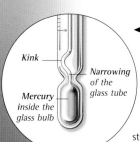

◀ MEDICAL THERMOMETER
This type of thermometer is designed to measure the temperature of the human body. The temperature of our bodies cannot change much, so the thermometer has only a short temperature range, from 32°C to 42°C (90–108°F). This means the marks on the temperature scale can be quite far apart, which makes the thermometer easier to read accurately.

Kink

Narrowing of the glass tube

Mercury inside the glass bulb

◀ MERCURY IN BULB
The thermometer contains a small amount of liquid mercury in a glass bulb at the bottom. To take someone's temperature, the glass bulb is placed inside their mouth. As the mercury is warmed by the person's body, it expands up the tube, and climbs the temperature scale. A kink in the tube stops the mercury falling back too quickly, so the temperature can be read and recorded.

Bar moves towards or away from strip

Contact is made, and the air conditioner is turned on

Screw moves bar to control temperature

Brass

Iron

As the strip heats up, the brass expands more than the iron and the strip bends

Electric current to bar

Electric current to air conditioner

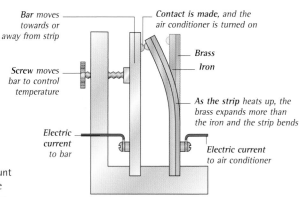

▲ THERMOSTAT
A thermostat switches an air-conditioning unit on and off to keep a room at a constant temperature. As the room heats up, the brass strip inside the thermostat expands more than the iron strip attached to it. The strip bends inwards, completes an electrical circuit, and switches on the air-conditioning unit.

Paper burns at 184°C, 363°F, 457K

Natural gas burns at 660°C, 1,220°F, 933K

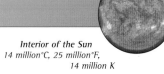

Interior of the Sun 14 million°C, 25 million°F, 14 million K

FIND OUT MORE ▶▶ Atoms **24–25** • Circuits **128–129** • Energy **76–77** • Heat Transfer **82–83** • Skin **351**

HEAT TRANSFER

Heat energy can be transferred from one place to another by three main processes. In **CONVECTION**, heat energy is carried by the movement of particles of matter. In **CONDUCTION**, heat is transferred by particles vibrating. In **RADIATION**, heat is carried directly by electromagnetic waves. When a hot object touches a cool object, heat moves from the hot one to the cool one. When objects transfer heat, they cool down to a lower temperature unless the heat energy they lose is constantly replaced.

▲ EVAPORATION
Another process by which heat is transferred is called evaporation. When a dog sticks out its tongue and breathes hard (pants), the moisture on the tongue turns into water vapour — it evaporates. Heat energy is needed to turn a liquid into a gas, so heat is removed from the dog's tongue in the process. This helps to cool the dog down. People cool themselves down by sweating through pores (tiny holes) in their skin, which removes heat from their bodies in the same way.

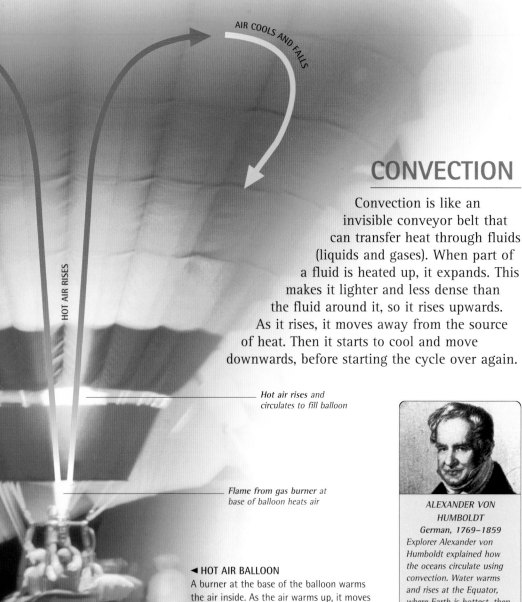

AIR COOLS AND FALLS

HOT AIR RISES

CONVECTION

Convection is like an invisible conveyor belt that can transfer heat through fluids (liquids and gases). When part of a fluid is heated up, it expands. This makes it lighter and less dense than the fluid around it, so it rises upwards. As it rises, it moves away from the source of heat. Then it starts to cool and move downwards, before starting the cycle over again.

Hot air rises and circulates to fill balloon

Flame from gas burner at base of balloon heats air

◄ HOT AIR BALLOON
A burner at the base of the balloon warms the air inside. As the air warms up, it moves upwards, cools, and moves round in a circular pattern known as a convection current. When the balloon is full of hot air, it lifts off the ground because the hot air inside it is less dense and lighter than the cold air outside it.

ALEXANDER VON HUMBOLDT
German, 1769–1859
Explorer Alexander von Humboldt explained how the oceans circulate using convection. Water warms and rises at the Equator, where Earth is hottest, then flows along the surface before cooling and sinking at the poles. He gave his name to the Humboldt Current, which travels up the South American coast.

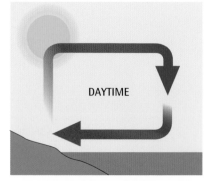

DAYTIME

NIGHTTIME

▲ SEA BREEZES
During the day, sunlight warms the land more quickly than the sea. Warm air rises from the land by convection, moves out to sea, and cools, creating a circular current. This is why, at ground level, sea breezes blow from sea to land during the day. At night, the land cools more quickly than the sea. Warm air rises from the sea, the convection current reverses, and the breezes blow from land to sea.

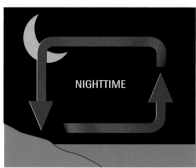

CONDUCTION

Heat travels through solids by conduction. If one end of a metal bar is heated, heat energy moves rapidly along the bar. The hot particles do not move along the bar, but vibrate and pass energy to their neighbours. Materials that conduct electricity are also good conductors of heat. Metals conduct heat well, but wood, plastics, and glass conduct heat only poorly.

▲ ALUMINIUM POT
Cooking pots are made from metal, often aluminium. This metal is a good conductor of heat, so it rapidly transfers heat energy from the stove to the food. The handles of cooking pots are often made of wood or plastic. These materials do not conduct heat very well and are called insulators.

RADIATION

All the light and heat energy we receive on Earth comes from the Sun, and travels through space in invisible electromagnetic waves known as radiation. Space is vast and empty, so heat energy cannot travel from the Sun by conduction or convection. Hot objects on Earth, such as fires and radiators, also radiate heat.

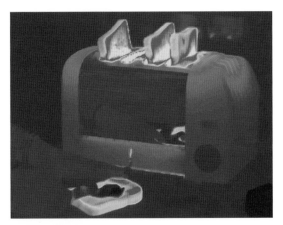

▲ TOASTER RADIATING HEAT
Inside a toaster, electricity heats metal wires so that they glow red-hot. The bread does not touch the wires, but is toasted by the heat radiation they give off. Surfaces inside the toaster are made of reflective metal to maximize the radiation. This photo taken with a heat-sensitive camera shows the hottest parts as red and yellow, and the coolest as blue and green.

▼ HOT METAL BAR
When the end of an iron bar is heated, the end glows red, then orange, then yellow, and finally white as the temperature increases. Heat energy flows along the bar by conduction. The hottest part of this iron bar is the yellow tip closest to the fire. The orange and red parts of the bar are also very hot.

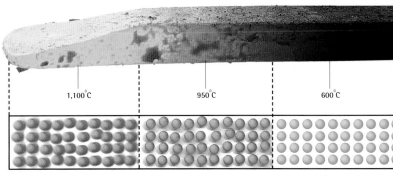

| 1,100°C | 950°C | 600°C |

▲ HOW ATOMS OF IRON CONDUCT HEAT
When an iron bar is heated, its atoms start to move about more vigorously. They move more because they have more energy, which they get from the fire. As the atoms jiggle about, they cause nearby atoms to move more vigorously as well. In this way, heat energy is transferred through the whole bar.

e ►► heat transfer

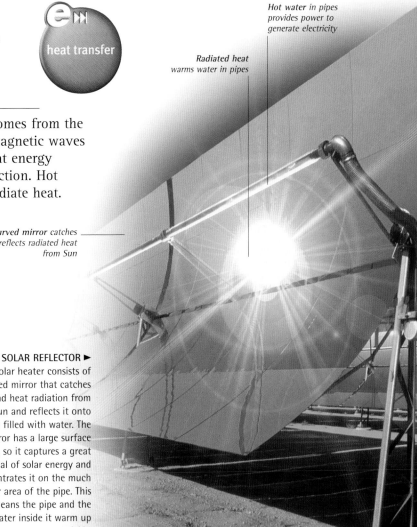

Hot water in pipes provides power to generate electricity

Radiated heat warms water in pipes

Curved mirror catches and reflects radiated heat from Sun

SOLAR REFLECTOR ►
This solar heater consists of a curved mirror that catches light and heat radiation from the Sun and reflects it onto a pipe filled with water. The mirror has a large surface area, so it captures a great deal of solar energy and concentrates it on the much smaller area of the pipe. This means the pipe and the water inside it warm up very quickly.

FIND OUT MORE ►► Conductors **130** • Energy Waves **98–99** • Heat **80–81** • Oceans **228–229** • Sun **170–171**

RADIOACTIVITY

The atoms of some chemical elements are unstable. They try to rearrange themselves to make more stable atoms. In the process, they give off radiation particles or tiny bursts of radiation. This process is called radioactivity. Although radioactivity can be harmful to people, it can also be important to us in everyday life. It is used to make nuclear energy and preserve food, and it also plays a vital role in the treatment of cancer.

▲ DANGER: RADIATION
Some types of radioactivity are harmful, because they damage or destroy the tissues of the human body. If people receive large doses of radioactivity, they can become ill with radiation sickness, which often causes cancer. Radiation sickness can also affect people's ability to have children.

| ALPHA | BETA | GAMMA |

▲ TYPES OF RADIOACTIVITY
The three types of radiation are alpha and beta particles, and gamma radiation, named after the Greek letters above. An alpha particle is two protons joined to two neutrons. A beta particle is an electron. Gamma radiation is high-energy electromagnetic radiation.

ALPHA DECAY

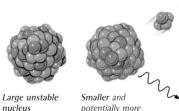

Alpha particle (two neutrons and two protons) is released

Gamma ray is also released

Large unstable nucleus　　*Smaller and potentially more stable nucleus*

An alpha particle is made when the nucleus (central part) of a large, unstable atom rearranges itself, or decays, to make a smaller, more stable atom. The new and smaller atom has two protons and two neutrons fewer than the original atom. These join together to make the alpha particle that is given off. Some energy is also released as a gamma ray. This is high-energy and high-frequency radiation, travelling at the speed of light.

BETA DECAY

Electron (beta particle) is ejected

Gamma ray is also released

Unstable nucleus　　*New nucleus has one more proton and one less neutron*

Beta decay is quite different from alpha decay. One of the neutrons in the nucleus of the unstable atom changes into a proton and an electron. The proton joins onto the nucleus, but the electron is ejected from the atom at high speed. This fast-moving electron is called a beta particle. Some energy is also released as a gamma ray.

Radiotherapy machine directs radiation at patient

radioactivity

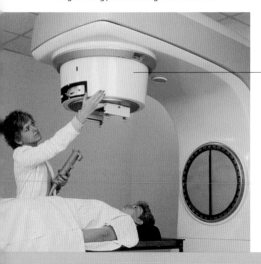

◄ RADIOTHERAPY
Radioactivity can cause cancer if it harms healthy cells in the human body. It can also help to cure cancer if it is used to destroy unhealthy cells. In radiotherapy, a powerful machine fires carefully targeted beams of radiation at tumours (cancer cells) in the patient's body. The radioactivity destroys the cells and helps to improve the patient's chances of survival.

FIND OUT MORE ►► Atoms 24–25 • Disease 370–371 • Elements 22–23 • Energy 76–77 • Energy Waves 98–99

NUCLEAR ENERGY

Atoms are small but can release lots of energy. When an unstable atom changes into a more stable one, it gives off radioactivity. It also gives off some of the potential energy locked inside the nucleus of the atom. Some atoms can be made to produce a constant supply of nuclear energy in a process called a chain reaction. Nuclear energy makes possible the destructive power of nuclear bombs, but it also generates much of the world's electricity.

◄ **NUCLEAR EXPLOSION**
When a nuclear bomb detonates, it starts a runaway chain reaction and releases enormous amounts of energy very quickly. A lump of radioactive plutonium the size of a tennis ball can produce as much energy as tens of thousands of tons of powerful explosives.

Mushroom-shaped cloud of smoke and gases caused by nuclear explosion

nuclear energy

Smoke and flames produced by intense heat of explosion

LISE MEITNER
Austro-Swedish, 1878–1968

Physicist Lise Meitner was one of the first to explain the process of nuclear fission. She also predicted the idea of the nuclear chain reaction before anyone had managed to make it work. She supported development of nuclear power, but opposed the production of nuclear bombs.

NUCLEAR FISSION

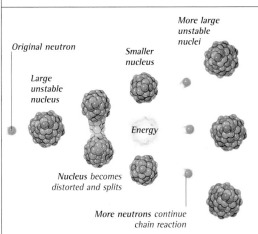

Original neutron

Large unstable nucleus

Smaller nucleus

More large unstable nuclei

Energy

Nucleus becomes distorted and splits

More neutrons continue chain reaction

In nuclear fission (splitting), large atoms break into smaller ones and give off energy. When a neutron is fired at the nucleus of a large atom, the atom becomes unstable and splits into two smaller atoms. Energy is produced and some neutrons are given off too. They collide with more large unstable nuclei of the original material and continue the chain reaction.

NUCLEAR FUSION

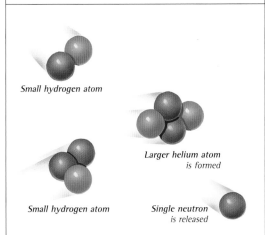

Small hydrogen atom

Larger helium atom is formed

Small hydrogen atom

Single neutron is released

In nuclear fusion (joining), massive energy is given off when small atoms fuse together to make larger atoms. A neutron is released at the same time. Stars like the Sun make their energy when nuclear fusion happens inside them at extremely high temperatures and pressures. Scientists are hoping that nuclear power stations will one day use fusion to provide Earth with a clean and inexpensive source of energy.

FIND OUT MORE ▶▶ Atoms **24–25** • Elements **22–23** • Energy **76–77** • Energy Sources **86–87** • Hydrogen **38**

ENERGY SOURCES

Everything we do takes energy, which we get from many different sources. Most of the energy on Earth originally came as light and heat from the Sun. It has been stored in fuels such as coal and oil, formed from the fossilized remains of plants or animals over millions of years. Supplies of these fossil fuels are limited. This is why we are now turning to supplies of **RENEWABLE ENERGY** that never run out. Another alternative is **GEOTHERMAL ENERGY**, produced deep inside the Earth.

▲ NODDING DONKEY OIL PUMP
Crude oil can be brought to the surface by a pump like this at the head of an oil well. Fuels such as heating oil, petrol, and gas are obtained from crude oil. This is made when sea organisms called plankton die and decay, and, over millions of years, pressure turns them into carbon-rich oil. Most crude oil forms under the sea or underground near the coast.

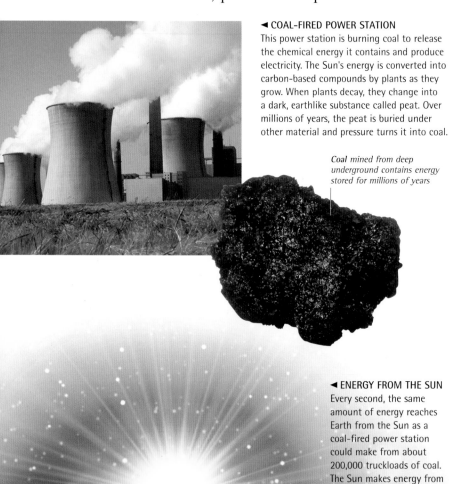

◄ COAL-FIRED POWER STATION
This power station is burning coal to release the chemical energy it contains and produce electricity. The Sun's energy is converted into carbon-based compounds by plants as they grow. When plants decay, they change into a dark, earthlike substance called peat. Over millions of years, the peat is buried under other material and pressure turns it into coal.

Coal mined from deep underground contains energy stored for millions of years

◄ ENERGY FROM THE SUN
Every second, the same amount of energy reaches Earth from the Sun as a coal-fired power station could make from about 200,000 truckloads of coal. The Sun makes energy from nuclear reactions deep inside it. In some ways it is like a giant nuclear power station.

INSIDE A NUCLEAR POWER STATION

Nuclear reactions take place in the fission reactor of a nuclear power station. Coolant (cold water) is pumped around the reactor and is turned to steam by the heat generated by the reactions. The steam drives an electricity-generating machine called a turbine.

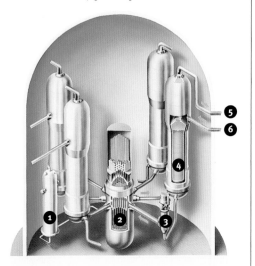

1 Water pressurizer keeps the coolant pressure high so that the water cannot boil

2 Reactor core containing fuel and control rods

3 Coolant pump circulates water around reactor

4 Hot water produces steam in the steam generator

5 Steam goes to drive the turbine

6 Steam is condensed back to water and returned to the steam generator

RENEWABLE ENERGY

Long after fossil fuels have run out, the tides will still be turning, the wind will still be blowing, and the Sun will still be shining. Ocean, wind, and solar power is called renewable energy because it never runs out. Using renewable energy is better for the environment. Unlike fossil fuels, it produces no harmful pollution and does not add to the problem of global warming.

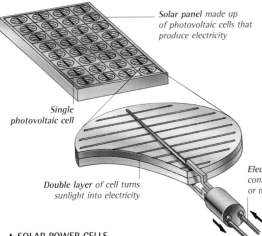

Solar panel made up of photovoltaic cells that produce electricity

Single photovoltaic cell

Double layer of cell turns sunlight into electricity

Electric wires connect to battery or machine

WIND FARM ▶
When the blades of these giant wind turbines turn, they gain some of the wind's energy. The spinning blades turn a generator and produce an electric current. Together, the many turbines in this wind farm can generate enough electricity for a small community.

▲ SOLAR POWER CELLS
A solar panel can be made up of many small solar photovoltaic cells. A photovoltaic cell is an electronic device that converts light into electricity. When sunlight falls on the cell, it makes electrons move from one layer of the cell to the other. The movement of electrons makes electricity flow through the cell, and out through wires to be used or collected in a battery.

TIDAL POWER GENERATOR ▶
As the tides ebb and flow, they make water move back and forth in rivers that end in estuaries at the coast. A tidal power generator is a type of bridge that blocks the mouth of an estuary so the tide has to move through it. Each time the water flows in or out, it turns a turbine inside the power generator and produces electricity.

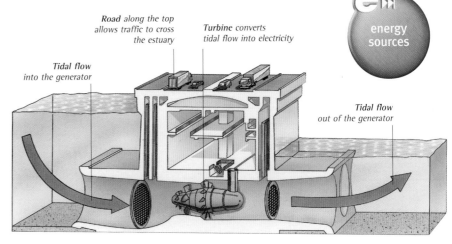

Road along the top allows traffic to cross the estuary

Turbine converts tidal flow into electricity

Tidal flow into the generator

Tidal flow out of the generator

energy sources

GEOTHERMAL ENERGY

This form of energy is not generated by the Sun. It is made by the nuclear reactions taking place all the time deep inside the Earth. These make heat energy in the Earth's core, and the heat moves around inside the Earth by convection. Volcanoes and hot geysers release geothermal energy at the Earth's surface.

GEOTHERMAL PLANT ▶
A geothermal energy plant takes its power from the Earth's heat. It works by pumping cold water down a hole drilled into the Earth. The Earth's geothermal energy heats up the water and it returns to the surface as hot water and steam. The hot water can be pumped to homes and factories nearby. The steam is used to drive a turbine and make electricity.

FIND OUT MORE ▶▶ Chemical Industry **50–51** • Earth's Resources **248–249** • Generators **137** • Nuclear Energy **85** • Sun **170–171**

MACHINES

In science, a machine is any device that changes a force into a bigger or smaller force, or alters the direction in which a force acts. Machines come in all shapes and sizes. Large machines such as cranes, bulldozers, and tipper trucks are based on smaller, simpler machines called **LEVERS**, **WHEELS**, **PULLEYS**, **SCREWS**, and **GEARS**. Simple tools such as a spade, a knife, a drawing-pin, and a nutcracker are also machines.

LEVERS

Most levers are force multipliers. They reduce the effort needed to work against a force called the load. They magnify a small force into a larger force. When a force acts on an object that is fixed at one point, the object turns around this pivot point. The further away the force is from the pivot point, the easier it is to turn the object. That is how levers make work easier.

Front of tipping body moves through a long arc as it dumps its load

Tipper chassis is built for strength to carry large loads

TIPPER TRUCK ▶
The body of this tipper truck pivots just above and behind the rear wheels. Hydraulic rams push upwards to lift the body. The main weight of the load is in the truck body, between the rams' effort and the pivot, so the body acts as a giant class two lever. This reduces the effort needed to raise the body and tip out the load.

TYPES OF LEVERS

Levers can work in three ways. Class one and class two levers turn the effort into a larger force to work against the load. Class three levers work in the opposite way, to reduce the force and increase the control of it over a greater distance.

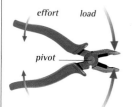

CLASS ONE LEVER
Pliers reduce the effort needed to grip something tight. The load and the effort are on opposite sides of the pivot. The load is greater than the effort.

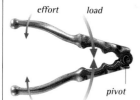

CLASS TWO LEVER
A nutcracker reduces the effort needed to crack a nut. The effort is applied further away from the pivot than the load. The load is greater than the effort.

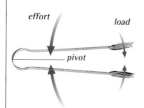

CLASS THREE LEVER
Sugar tongs and tweezers reduce the force you apply and increase your control of it. The effort is applied closer to the pivot than the load, and is greater than the load.

WHEELS

A wheel and the axle it turns around combine as a machine that works like a lever. The distance between the rim of the wheel and the axle multiplies either speed and distance or force. If the effort is applied to the axle, the rim of the wheel turns further, and so faster, than the axle, but with less force. If the effort is applied to the rim of the wheel, the axle turns with more force but not so far or fast.

Light turning force
on steering wheel

Driveshaft
carries force
from engine
to **rear axle**

Rear axle carries driving
force from driveshaft to
rear wheels

Steering column
is axle from
steering wheel

Road wheels
multiply speed
from rear axle

Strong turning force
on steering column

Steering rack transfers
force to turn front wheels

▲ WHEEL AND AXLE
A steering wheel multiplies force because the rim turns further than the steering column. It multiplies your effort and turns the car's wheels with more force than you actually apply. The road wheels multiply speed. The car's engine turns the driveshaft and rear axle at a certain speed. The axle turns the large road wheels further and therefore faster.

machines

Hydraulic rams
apply effort
upward to tip
out the load

785

Load of
building
materials is
tipped out of
body of truck

SEE OVER ▶▶ Pulleys, Screws, Gears **90–91**

PULLEYS

A pulley is a rope looped around one or more wheels to make a heavy load easier to lift. The more ropes and wheels are used, the less force is needed to lift the load, but the further the rope has to be pulled. Pulleys make it easier to lift things using less force, but the same amount of work has to be done whether or not a pulley is used. Cranes lift huge weights using large pulleys.

◄ GIANT CRANE

Instead of using ropes, this large crane uses strong steel cables to support its massive loads. Several pulleys and cables work together to reduce the effort needed to lift a heavy load. The weight of the load is shared by the cables in the pulleys, so less force is needed to lift it.

TYPES OF PULLEY

Pulleys vary in usefulness, depending on the number of wheels and ropes they have. A simple pulley changes only the direction of a force. Doubling the wheels and ropes halves the force needed to lift a given weight, but the rope must be pulled twice as far.

Newtonmeter shows 10 newtons of force to lift weight

Newtonmeter shows 5 newtons of force to lift same weight

SIMPLE PULLEY

A simple pulley has one wheel and one rope. It does not reduce the effort needed to lift a load, just the direction of the force. Using this simple pulley, it takes 10 newtons of force to lift a weight of 10 newtons.

DOUBLE PULLEY

This pulley has two wheels and its rope is looped into a double length. The double pulley reduces by half the effort needed to lift the load, needing only 5 newtons, but the rope has to be pulled twice as far.

KEY

1. Six wheels on the crane's top pulley.
2. Cable between the pulleys hangs in six loops.
3. Six wheels on the bottom pulley.
4. Heavy load is lifted slowly by the cable running through the pulleys.

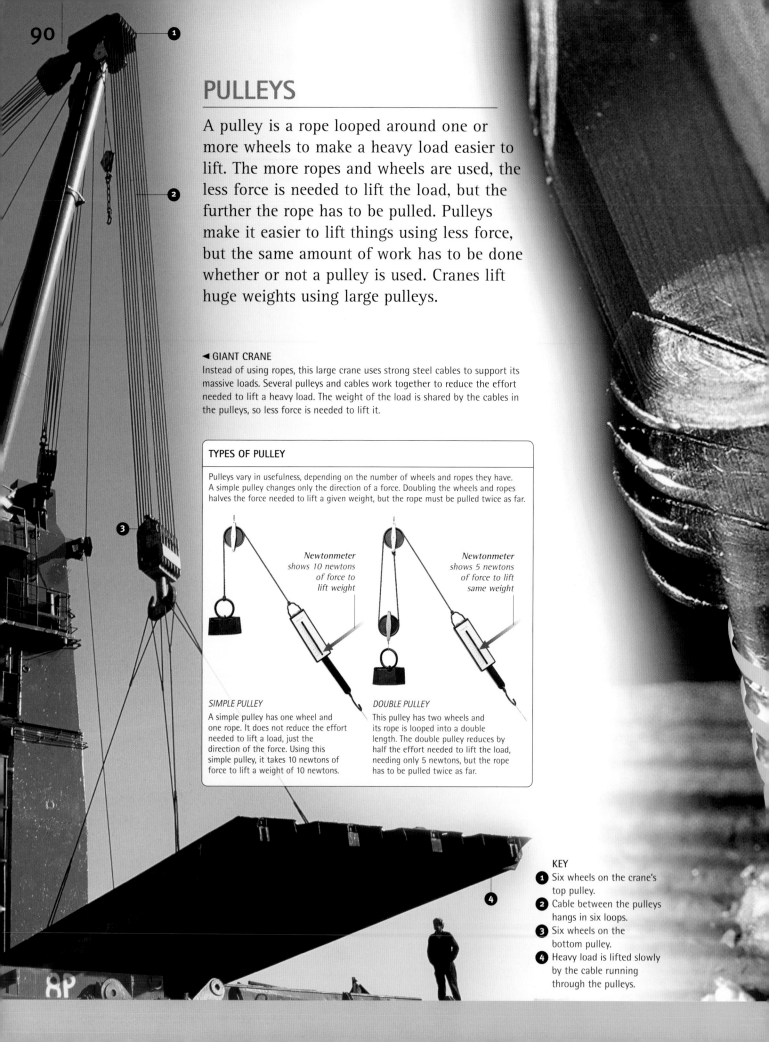

SCREWS

A simple screw that holds pieces of wood together is also a type of machine. The spiral thread of a screw is designed to reduce the effort needed to drive it into a piece of wood. Turning a screw is like pushing something up a spiral ramp instead of trying to lift it straight up. It reduces the force needed, but that force has to be used over a greater distance and for a longer time.

Screwdriver turns with a light downward force

◄ DRIVING SCREW INTO WOOD
When you turn the head of a screw, the long spiral groove down its side pulls the screw into the wood. Although you turn the screw head many times with a screwdriver, the screw moves forward into the wood only a short distance. However, the screw bites into the wood with a lot of force.

Many turns of the screw drive it with great force a short distance into the wood

MOUNTAIN ROAD ▲
An inclined plane or ramp makes it easier to move something upwards. Increasing the distance moved even more makes it even easier. It would take a very powerful engine to drive a car straight up the side of this hill. If the car drives up the long, winding road, it takes much longer to reach the top but the engine does not have to use as much force. In this way, the road is like the spiral thread of a screw that has been unwound.

GEARS

Gears are pairs of wheels with teeth around their edges that mesh and turn together. Gears are machines because they multiply turning force or speed. If one gear wheel drives another that has more teeth, the wheel with more teeth turns more slowly but with greater force than the other. If a gear wheel drives another with fewer teeth, the wheel with fewer teeth turns with less force but faster.

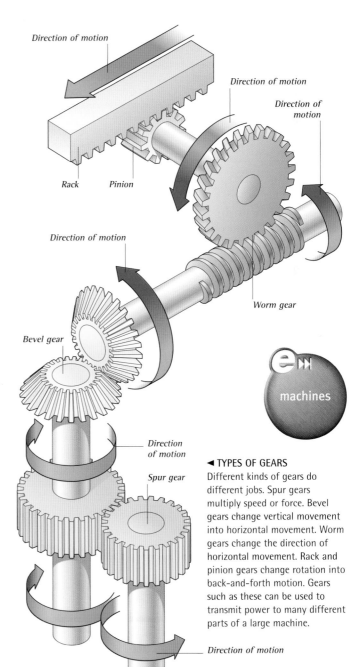

Direction of motion

Direction of motion

Direction of motion

Rack   Pinion

Direction of motion

Worm gear

Bevel gear

Direction of motion

Spur gear

Direction of motion

machines

◄ TYPES OF GEARS
Different kinds of gears do different jobs. Spur gears multiply speed or force. Bevel gears change vertical movement into horizontal movement. Worm gears change the direction of horizontal movement. Rack and pinion gears change rotation into back-and-forth motion. Gears such as these can be used to transmit power to many different parts of a large machine.

FIND OUT MORE ▶▶ Forces **64–65** • Machines **88–89** • Motion **70–71** • Road Vehicles **93** • Work **78–79**

ENGINES

Many modern machines, from motorbikes to jet aircraft, are powered by engines. An engine is a machine that turns fuel into movement. The fuel is burned to generate heat energy. The heat is then converted into mechanical power. In a car or motorbike engine, the power comes from pistons and cylinders. In a jet aircraft, power comes from hot gases rushing past a spinning wheel called a **TURBINE**. Engines also produce various waste gases, which cause pollution.

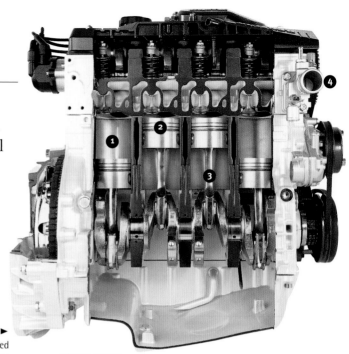

engines

INSIDE A CAR ENGINE ▶
A car engine gets energy from burning fuel inside closed chambers called cylinders. When the fuel burns, it makes hot gases that move the pistons downwards. As the pistons move, they turn a rod called a crankshaft that makes the car's wheels rotate. This engine has four cylinders. Each one provides power at a slightly different time to keep the crankshaft turning continuously.

KEY TO PARTS
❶ Cylinder where fuel is burned to produce energy.
❷ Piston compresses fuel - spark plug ignites it.
❸ Piston rod moves up and down to turn crankshaft below.
❹ Outlet to pipe to remove exhaust gases.

THE FOUR-STROKE CYCLE

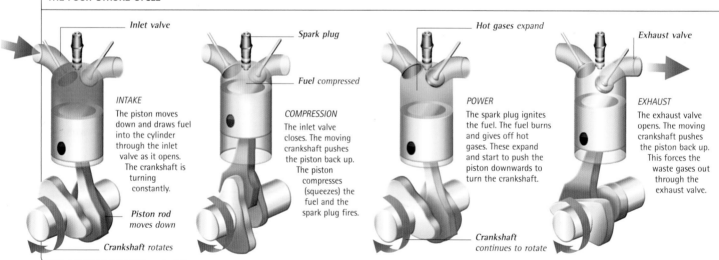

Inlet valve

INTAKE
The piston moves down and draws fuel into the cylinder through the inlet valve as it opens. The crankshaft is turning constantly.

Piston rod moves down

Crankshaft rotates

Spark plug

Fuel compressed

COMPRESSION
The inlet valve closes. The moving crankshaft pushes the piston back up. The piston compresses (squeezes) the fuel and the spark plug fires.

Hot gases expand

POWER
The spark plug ignites the fuel. The fuel burns and gives off hot gases. These expand and start to push the piston downwards to turn the crankshaft.

Crankshaft continues to rotate

Exhaust valve

EXHAUST
The exhaust valve opens. The moving crankshaft pushes the piston back up. This forces the waste gases out through the exhaust valve.

TURBINES

These are machines that extract and use the energy from a moving liquid or gas. Windmills and waterwheels were the very first turbines. Their sails and paddles took power from the movement of wind and water. Turbines are still important today, especially as they are used in power stations and jet engines.

Exhaust gases rush backwards, pushing plane forwards

Giant fan pulls air into the engine and compresses it

Air intake where cold air is sucked into engine

Blades of turbine that spins fan

Combustion chamber where air and fuel are burned

◀ TURBOJET ENGINE
An aeroplane's jet engine has one or more large fans at the front. These mix air with fuel and compress the mixture. In the combustion chamber, the fuel ignites, burns, and produces hot gases. As the gases expand, they turn a turbine that spins the fans. The force of the hot gases rushing backwards out of the engine propels the aircraft forwards.

FIND OUT MORE ▸▸ Aircraft **97** • Dynamics **66–67** • Energy **76–77** • Energy Sources **86–87** • Pollution **250**

ROAD VEHICLES

Vehicles use engines of different kinds to move people and cargo from place to place. Most cars and motorbikes have petrol engines, but vans and trucks use larger diesel engines. A diesel engine produces more power than a petrol engine by compressing the air and fuel much more. Petrol and diesel engines produce large amounts of pollution. **ELECTRIC CARS** are less polluting.

Heavy load of liquid or gas is carried in tanker body

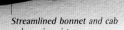

Big wheels help to spread heavy load

Streamlined bonnet and cab reduce air resistance

◄ INNER-CITY COMMUTER BIKE
This motorbike's compact petrol engine, under the driver's seat, powers the rear wheel. Gears increase the speed of the bike along a straight and the power of the bike when it climbs uphill. The handlebars are levers that help to turn the front wheel to steer. The roof provides weather and crash protection.

▲ DIESEL TANKER TRUCK
Trucks have big diesel engines that produce more power than a car engine, but they also use more fuel and produce more pollution. A truck is heavier and moves with more momentum (force) than a car travelling at the same speed. This is why a truck needs much more powerful brakes than a car and takes a longer distance to come to a stop.

ELECTRIC CARS

Electric cars use batteries or fuel cells instead of engines and petrol. Batteries have to be charged up every so often, from the mains or from an engine, and the car then runs until the batteries are flat. Fuel cells work in a different way. Like an engine, a fuel cell takes in a steady supply of fuel, usually hydrogen gas. Like a battery, it produces a constant stream of electricity that powers an electric motor.

Solar panels convert the Sun's rays into power to run the electric motor

▲ SOLAR-POWERED CAR
The world's fastest solar-powered car, Nuna II, has a top speed of 160 kph (100 mph). It is built in plastic and covered in solar panels. These convert the Sun's energy into electricity and store it in batteries, so the car can also drive in the shade. The body, solar panels, and batteries were originally developed for spacecraft.

Petrol engine used on open roads

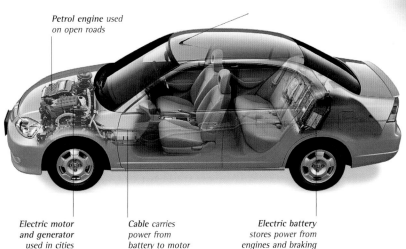

Electric motor and generator used in cities

Cable carries power from battery to motor

Electric battery stores power from engines and braking

◄ HYBRID PETROL/ELECTRIC CAR
Petrol engines are good for driving at constant, higher speeds on open roads. Electric motors are good for stop-start driving in city centres. They have lower top speeds than petrol engines. Hybrid cars have both a petrol engine and an electric motor. The car automatically switches between the two to suit varying traffic conditions.

vehicles

FIND OUT MORE ►► Electric Motors 136 • Electricity 126–127 • Energy 76–77 • Engines 92 • Work 78–79

FLOATING

When an object such as a boat or an airship rests in a fluid (a liquid or gas), it has to displace (push aside) some of the fluid to make room for itself. The object's weight pulls it downwards. But the pressure of the fluid all around the object tries to push it upwards with a force called upthrust. The object **SINKS** if the upthrust is less than its weight, but floats if the upthrust is equal to, or more than its weight.

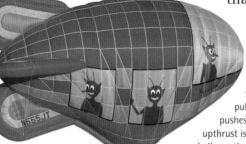

◄ FLOATING AIRSHIP

Hot air is less dense than cooler air, so the hot air in a balloon weighs less than the same volume of cool air. The weight of the airship pulls it downwards, but the air around the balloon pushes it upwards with a force called upthrust. If the upthrust is equal to or greater than the total weight of the balloon, the basket, and the hot air, the airship floats.

FLOATING SHIP ►

When a boat floats, it displaces some of the water underneath it. As the weight of the boat pushes down on the water, the water pushes up on the boat with an upward force called buoyancy. The larger the boat, the greater the buoyancy. The boat floats if the buoyancy is as great as or greater than the weight of the boat.

floating

ARCHIMEDES
Greek, 287–212 BC

Archimedes is best known for realizing that a floating object displaces its own weight in a fluid. Legend has it that he worked this out in his bath. As he stepped into the bathtub, water splashed over the side. He found that the weight of this water equalled his body weight. This idea is known as Archimedes' Principle.

SINKING

Not everything will float. A block of wood will float on water, but a lump of iron exactly the same size will sink. This is because a piece of wood of a certain size weighs less than the same volume of water, so wood floats on water. However, iron is much heavier than either wood or water. A block of iron weighs more than the same volume of water. This is why iron sinks in water.

◄ PLIMSOLL LINE

The more cargo a ship carries, the deeper it sits in the water. Ships also displace varying amounts of water according to the saltiness and temperature of the water. This varies from ocean to ocean around the world. Large ships have a mark called a Plimsoll line painted on their sides. This shows how much weight they can safely carry in different parts of the world.

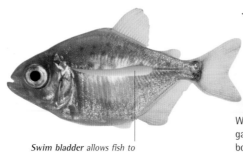

Swim bladder allows fish to float or sink

◄ SWIM BLADDER

Some fish can raise or lower themselves in water using their swim bladder. This is an organ inside their body that they can fill with gas to make their bodies lighter, so they rise towards the surface. When they reduce the amount of gas in the swim bladder, their bodies become heavier and sink.

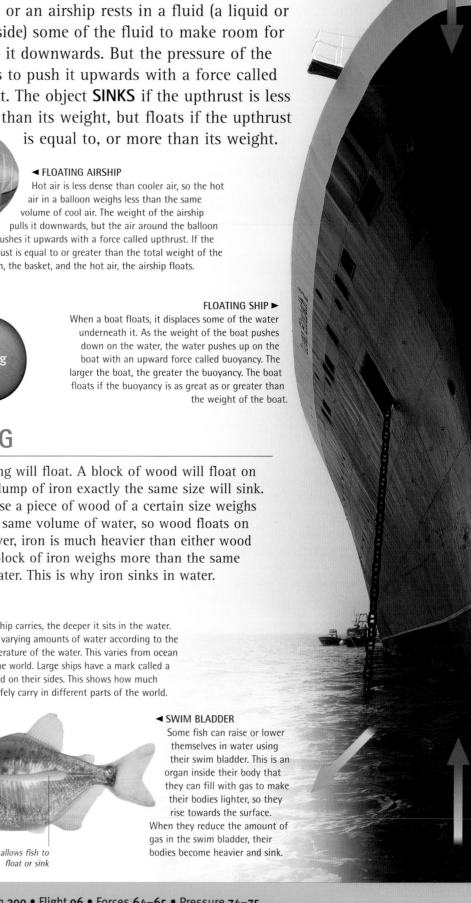

FIND OUT MORE ►► Dynamics **66–67** • Fish **300** • Flight **96** • Forces **64–65** • Pressure **74–75**

Weight of boat pushes downward because of force of gravity

BOATS

Although boats float, their weight makes them settle a little way into the water. This means they create some resistance or drag when they move through the water. The bow (front) of a boat is V-shaped and curved. This raises the boat up as it goes faster, and helps reduce drag. Boats are powered by sails, oars, or engines that turn propellers at the rear. The propellers push water backwards, and this backward thrust moves the boat forwards.

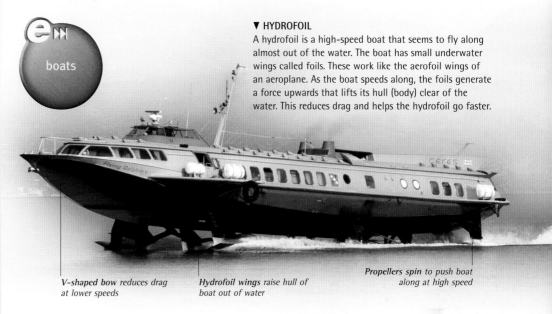

▼ HYDROFOIL
A hydrofoil is a high-speed boat that seems to fly along almost out of the water. The boat has small underwater wings called foils. These work like the aerofoil wings of an aeroplane. As the boat speeds along, the foils generate a force upwards that lifts its hull (body) clear of the water. This reduces drag and helps the hydrofoil go faster.

V-shaped bow reduces drag at lower speeds

Hydrofoil wings raise hull of boat out of water

Propellers spin to push boat along at high speed

SUBMARINES

Submarines can float on the sea, sink just beneath the surface, or dive to the seabed. They dive or surface using tanks that work like a fish's swim bladder. When the tanks are filled with water, the submarine dives. When they are filled with air, it surfaces. A submarine can select its level in the sea by changing the mixture of air and water in its tanks.

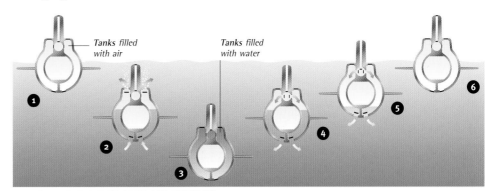

Tanks filled with air

Tanks filled with water

HOW A SUBMARINE DIVES AND SURFACES
1 The tanks are filled with air. The submarine floats on the surface of the sea.
2 The tanks are opened. Sea water enters and pushes the air out. The submarine begins to dive.
3 When the tanks are full of water, the submarine sinks to the seabed.
4 Compressed air is pumped into the tanks. Water is pushed out. The submarine begins to rise.
5 When the mixture of air and water is exactly right, the submarine floats beneath the surface.
6 The tanks are filled with air from the surface. The submarine floats on the surface again.

Displaced water pushed aside by bulk of ship below the surface

Buoyancy pushes upwards on ship and makes it float

FIND OUT MORE ▶▶ Aircraft **97** • Floating **94** • Friction **68** • Pressure **74–75**

FLIGHT

When something flies, it overcomes the force of gravity and moves through the air. Birds and aeroplanes fly using curved aerofoil wings that produce an upward force called lift. When birds flap their wings, they generate lift and move their bodies forwards at the same time. Aeroplanes generate lift with their aerofoil wings, but need engines to move them forwards.

LIFT

THRUST

DRAG

GRAVITY

HOW TO CONTROL FLIGHT

Pilots control and steer an aeroplane using the ailerons, rudder, and elevators. These are swivelling flaps built into the wings and the tail of the aeroplane.

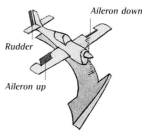

Aileron down

Rudder

Aileron up

AILERON
The pilot can bank (roll) the aeroplane by using the ailerons. For example, he turns to the right by tilting the right aileron up and the left aileron down. This increases lift on the left wing, reduces lift on the right wing, and makes the plane bank and turn to the right.

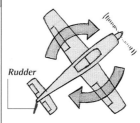

Rudder

RUDDER
The rudder is a vertical flap on the rear edge of the tailfin. The pilot can swivel it from side to side to help turn the aeroplane to the left or to the right without banking.

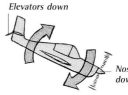

Elevators down

Nose down

ELEVATORS
The elevators are horizontal flaps at the back of the tailfin. The pilot can tilt them up or down to raise or lower the nose of the aeroplane, to climb or dive.

◀ FORCES OF FLIGHT
Four forces act on an aeroplane as it flies. The engine produces a force called thrust that pushes the plane forwards, while air resistance (drag) pulls in the opposite direction. As the plane moves forwards, the aerofoil wing creates lift. To stay in the air, the plane must move fast enough so that the lift is at least equal to its weight, caused by the force of gravity.

AEROFOIL ▶
An aerofoil wing generates lift because of its curved shape. As the wing moves forwards, air has to travel faster over the curved top of the wing to keep up with the air moving underneath it. This lowers the air pressure above the wing and creates an upward force that overcomes the aeroplane's weight. An aerofoil also creates drag that pulls the aeroplane backwards.

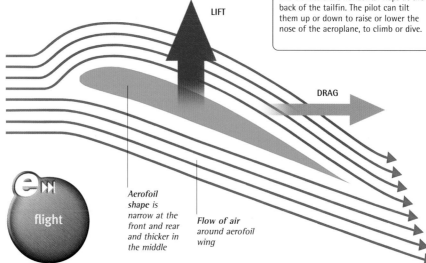

LIFT

DRAG

flight

Aerofoil shape is narrow at the front and rear and thicker in the middle

Flow of air around aerofoil wing

FIND OUT MORE ▶▶ Aircraft 97 • Birds 303 • Forces 64–65 • Friction 68 • Gravity 72 • Pressure 74–75

AIRCRAFT

Both aeroplanes and **HELICOPTERS** are aircraft. These machines use engines and aerofoil wings to lift off the ground and move through the air. Aeroplanes use either conventional engines with propellers or jet engines. Jets burn lots of fuel to generate huge forward thrust and go very fast. The faster an aeroplane moves, the more lift its wings produce.

Tailfin with rudder, which controls left and right movement

Tailplane with elevators, which control up and down movement

aircraft

◄ **AIRLINER BEING BUILT**
Airliners can carry hundreds of passengers and huge amounts of cargo, so they are extremely heavy. They need to have very wide wings to generate enough lift to overcome the force of gravity and get them into the air. The huge wings also contain fuel tanks. The liquid fuel is piped directly to the jet engines under the wings.

Ailerons are tilting flaps at the back of each wing

Wing containing fuel tanks

Powerful jet engine forces hot gases backwards to move plane forwards

Escape hatches provide emergency exits

Propeller is a spinning aerofoil that generates thrust

Slim aerofoil shape of the propeller blade cuts through the air

WOODEN PROPELLER ▲
A propeller is a twisted aerofoil, driven by an engine, that spins around at high speed. As a propeller turns, it generates a backward draft of air that moves the aeroplane forwards. Aircraft propellers spin faster than ships' propellers. This is because aeroplanes need to move forwards more quickly to generate the lift that keeps them in the air.

HELICOPTERS

This type of aircraft generates lift and thrust using a huge overhead propeller or rotor. The rotor has several blades shaped like aerofoils. As the blades spin, they generate lift that overcomes the helicopter's weight and lifts it into the air. The pilot can move a helicopter forwards, backwards, or from side to side by tilting the rotor blades slightly as they spin around.

Cockpit houses the aeroplane's control systems

▼ **HOVERING HELICOPTER**
When a helicopter hovers above the ground without moving, the lift from its rotors is exactly equal and opposite to its weight. Although a normal aeroplane can fly along at a steady height, it cannot hover. It must move forwards all the time to generate the lift that keeps it flying.

LIFT

Spinning rotor

Tail rotor stops helicopter from spinning round

Nose cone is bullet-shaped to reduce drag

GRAVITY

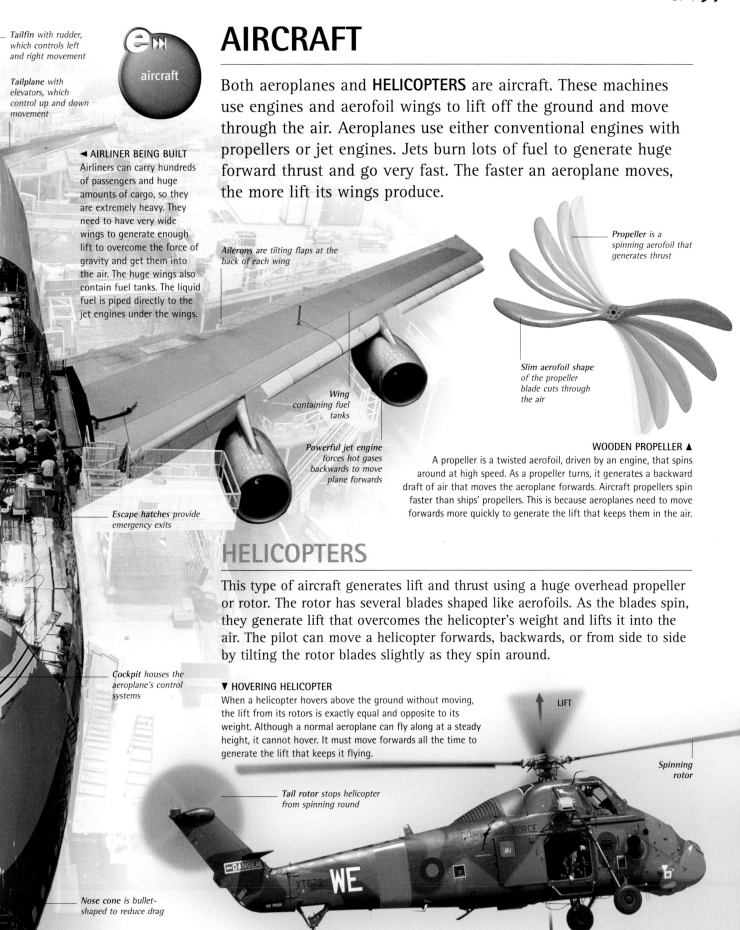

FIND OUT MORE ▸▸ Dynamics 66–67 • Engines 92 • Gravity 72 • Pressure 74–75 • Work 78–79

ENERGY WAVES

Many different kinds of energy travel in waves. Sound waves carry noises through the air to our ears. **SEISMIC WAVES** travel inside the Earth and cause earthquakes. Light, heat, radio, and similar types of energy are carried by a variety of waves in the **ELECTROMAGNETIC SPECTRUM**. Some energy waves need a medium, such as water or air, through which to travel. The medium moves back and forth as waves carry energy through it, but it does not actually travel along with the wave.

OCEAN WAVES ▶
When an ocean wave crashes against the shore, it releases a large amount of energy. Ocean waves are transverse waves that carry huge amounts of energy across the surface of the sea as they move up and down. A wave 3 m (10ft) high carries enough energy to power around 1,000 lightbulbs in every 1 m (3 ft) of its length.

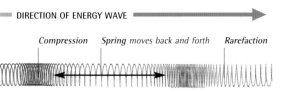

DIRECTION OF ENERGY WAVE

Compression   Spring *moves back and forth*   Rarefaction

▲ LONGITUDINAL WAVE
Suppose you fix a slinky spring at one end and push the other end back and forth. Some parts of the spring, called compressions, are squeezed together. Other parts of the spring, called rarefactions, are stretched out. The compressions and rarefactions travel down the spring carrying energy. This type of wave is called a longitudinal or compression wave.

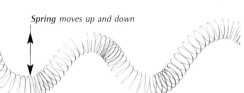

DIRECTION OF ENERGY WAVE

Spring *moves up and down*

▲ TRANSVERSE WAVE
Suppose you fix the slinky at one end and move it up and down. Energy travels along the spring's length in S-shaped waves, so the forward direction in which the energy moves is at right angles (or transverse) to the up-and-down direction of the movement of the spring. This is called a transverse wave.

HEINRICH HERTZ
German, 1857-1894

In 1887, physicist Heinrich Hertz became the first person to prove that waves carry electromagnetic energy between two places. This extremely important finding eventually led to the development of radio and television. Hertz did not live to see these inventions, however. He died in 1894, aged only 36.

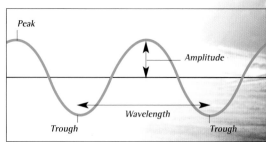

Peak

Amplitude

Wavelength

Trough   Trough

▲ MEASURING WAVES
Waves have three important measurements. The amplitude is the height of a peak or trough. The wavelength is the distance between any two peaks or troughs. The frequency is the number of waves that pass by in one second. Amplitude and wavelength are measured in metres. Frequency is measured in hertz (Hz). One Hz is equal to one wave passing by each second.

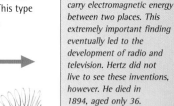

ELECTROMAGNETIC SPECTRUM

Radios, televisions, mobile phones, and radar use signals made up of electromagnetic waves. These are waves that carry energy as electricity and magnetism at the speed of light. Light we can see is also an electromagnetic wave, but other types of electromagnetic wave are invisible. The various types of electromagnetic wave have different frequencies and wavelengths. Together, they make up the electromagnetic spectrum.

energy waves

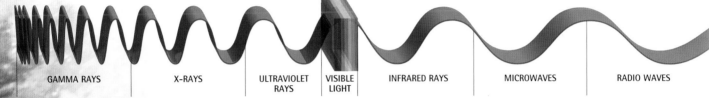

| GAMMA RAYS | X-RAYS | ULTRAVIOLET RAYS | VISIBLE LIGHT | INFRARED RAYS | MICROWAVES | RADIO WAVES |

▲ **GAMMA RAYS**
These are produced by radioactivity. They have a short wavelength and a high frequency and carry large amounts of energy. They are very harmful and can cause cancer in humans and animals.

▲ **X-RAYS**
X-rays are high-energy waves that pass through flesh but not bone. In medicine, X-ray photographs are used to check people's bones for damage. In high doses, X-rays can harm people.

▲ **ULTRAVIOLET RAYS**
These invisible waves are slightly shorter than visible violet light and carry more energy. We wear sunglasses and sunblock to prevent damage to our eyes and skin by ultraviolet rays.

▲ **INFRARED RAYS**
Infrared rays are slightly longer waves than visible red light. Although we cannot see infrared, we can feel it as heat. When heat energy is transferred by radiation, it is carried by waves of infrared.

▲ **RADAR**
Radar is a way of locating aeroplanes and ships using a type of radio waves called microwaves. These have much longer wavelengths than visible light. Cooking is another use for microwaves.

▲ **RADIO WAVES**
Radio waves are the longest in the spectrum. They carry radio and TV signals around Earth. Radio waves from outer space are picked up by radio telescopes and used in studies of the universe.

SEISMIC WAVES

When the energy stored in rock deep inside the Earth is suddenly released, it travels up to the Earth's surface in huge seismic shock waves. The waves move along weaknesses in the rock known as faults. As they do so, they produce violent shaking of the ground and an earthquake. The largest earthquakes can release as much energy as a small atomic bomb.

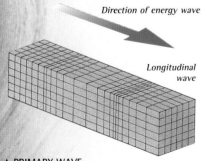

Direction of energy wave

Longitudinal wave

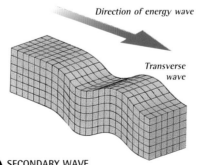

Direction of energy wave

Transverse wave

▲ **PRIMARY WAVE**
Some seismic waves are longitudinal or compression waves called primary or P-waves. They cause damage by pushing and pulling things back and forth in the same direction that the wave travels. P-waves move extremely quickly through the Earth's interior at a speed of about 25,000 kph (15,500 mph).

▲ **SECONDARY WAVE**
Following the P-waves are transverse seismic waves known as secondary or S-waves. These shake rocks up and down or from side to side as they move forwards, which causes a twisting or shearing motion. S-waves also travel through the Earth's interior, but at about half the speed of P-waves.

EARTHQUAKE ▲
Earthquakes kill around 10,000 people every year worldwide. This one happened in Mexico City in 1985 and was one of the biggest ever recorded. Many cities now have buildings that absorb the energy in seismic waves. They may wobble, but they do not collapse.

FIND OUT MORE ▶▶ Earthquakes **210–111** • Heat **80–81** • Light **110–111** • Radio **143** • Radioactivity **84** • Sound **100–101**

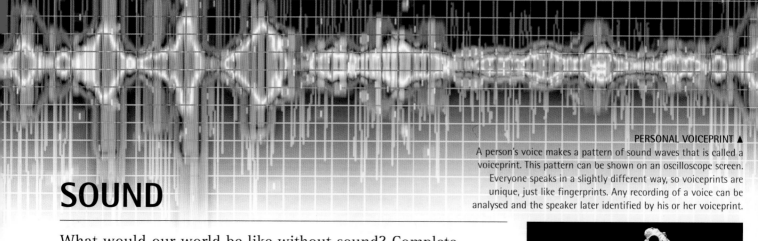

PERSONAL VOICEPRINT ▲
A person's voice makes a pattern of sound waves that is called a voiceprint. This pattern can be shown on an oscilloscope screen. Everyone speaks in a slightly different way, so voiceprints are unique, just like fingerprints. Any recording of a voice can be analysed and the speaker later identified by his or her voiceprint.

SOUND

What would our world be like without sound? Complete silence might seem peaceful, but there would be no speech, music, or birdsong. Sound is a type of energy that objects produce when they vibrate (move back and forth). The energy travels at high **SPEED** through air, water, or another substance, in a pattern of sound waves. When the sound waves reach us, they make the eardrums of our inner ears vibrate. Our brains recognize these vibrations as sounds made by different things.

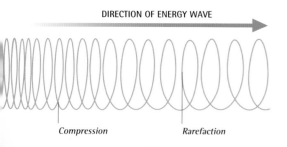

DIRECTION OF ENERGY WAVE

Compression   *Rarefaction*

◄ SOUND WAVE
Sound energy travels through air in waves. When an alarm clock rings, nearby air molecules vibrate. The air molecules jiggle about and make neighbouring molecules vibrate too, starting a wave of energy that travels out from the clock. The wave of sound travels in a pattern of compressions (where the air molecules are squeezed together) and rarefactions (where the air molecules are stretched apart).

▲ CONVERSATION IN SPACE
It is impossible for astronauts to talk to one another in space as they would on Earth, no matter how loud they shout. There is no air in space, so there is nothing for sound waves to travel through. In this totally silent place, astronauts have to communicate by radio, using microphones and headsets in their space helmets.

▼ WAVE MAKER
When an alarm clock goes off, vibrations from the bell inside the clock create sound waves that travel through the air and quickly reach our ears. If there were no air, there would be nothing to carry the sound waves and we would not be able to hear the clock. Sound always needs to travel through some kind of medium, such as air, water, wood, or metal.

Alarm clock creates sound waves that travel out in all directions

Large outer ear funnels distant sounds towards eardrum, just inside the hare's head

SENSITIVE EARS ►
Animals such as this hare have large outer ears that help them to detect passing sound waves. The ears can be swivelled towards the source of the sound. Inside the head, another part of the ear converts the sound waves into a form that the brain can understand. Without ears, the sounds of the world would be lost to us.

SPEED OF SOUND

sound

Sound travels faster through some substances than through others. When passing through dry air at a temperature of 0°C (32°F), sound travels at a speed of 1,190 kph (740 mph). It travels faster than that in warmer air, and more slowly in colder air. Sound moves about four times faster in water than in air. Dense, heavy substances (made up of molecules closely packed together) allow sound to pass through more quickly than lighter substances do.

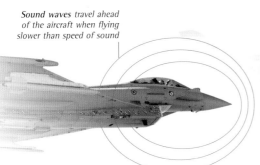

Sound waves travel ahead of the aircraft when flying slower than speed of sound

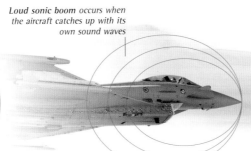

Loud sonic boom occurs when the aircraft catches up with its own sound waves

Supersonic speed enables the aircraft to travel ahead of its sound waves

SLOWER THAN SOUND ▲
An aircraft engine sends out sound waves in all directions. When an aircraft travels at a subsonic speed (slower than the speed of sound), sound waves travel ahead of the plane. If you look up after hearing an aircraft travelling at such a speed, the sound it makes appears to be coming from the aircraft itself, just as you would expect.

AT THE SPEED OF SOUND ▲
When an aircraft reaches the speed of sound (sometimes called the sound barrier), it catches up with its own sound waves. The sound waves are squeezed up in front of it and this produces a loud bang, called a sonic boom. People on the ground hear a sonic boom as a very loud, thundery noise that seems to sweep past them.

FASTER THAN SOUND ▲
Travelling at a supersonic speed (faster than the speed of sound), an aircraft surges ahead of its own sound waves. When you look up into the sky as an aircraft flying at a supersonic speed passes overhead, the noise it makes seems to be coming from some distance behind it. The sound waves only reach you after the plane has passed by.

WHALE SOUNDS ▶
Whales communicate with one another by making eerie, low-frequency (deep-pitched) moaning noises. Sound waves can carry great distances in water, and the sounds that whales make can travel hundreds or even thousands of miles across entire oceans. In contrast, dolphins communicate over shorter distances by exchanging clicking noises of a higher frequency.

FIND OUT MORE ▶▶ Energy Waves **98–99** • Hearing **347** • Loudness **102** • Pitch **103** • Sound Reproduction **108–109**

DECIBEL SCALE

SHUTTLE TAKE-OFF
150–190 DB

CIRCULAR SAW
100 DB

TRAFFIC
70–90 DB

BIRDS
30–50 DB

FALLING LEAVES
10 DB

LOUDNESS

Some sounds are so loud they are painful to our ears; others are so quiet they may be hard to hear at all. Things that vibrate a lot, such as car engines, can make a tremendous noise; they sound louder because the sound waves they generate carry more energy. The amount of energy carried by a sound wave is called its intensity. Sound waves of higher intensity are louder to our ears. The loudness of a sound is measured in decibels (dB).

ROCK MUSIC ▶

Very loud, amplified rock music can sometimes make your ears hurt. It can reach an intensity of 120–140 decibels, which is loud enough to cause temporary or permanent damage to your hearing. A sound of 140 dB is 100,000 billion times louder than the sound of falling leaves. Sounds begin to cause pain in the ears at around 120–140 dB.

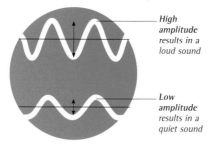

High amplitude results in a loud sound

Low amplitude results in a quiet sound

loudness

▲ AMPLITUDE

When something vibrates and produces a sound, the sound waves coming from it move up and down as they travel. Loud sounds are carried by waves that have a higher amplitude (height between peak and trough) than quiet sounds. The bigger the amplitude of a sound wave, the louder it sounds to our ears.

◀ NOISE CANCELLING

Pilots wear special headphones that reduce the roar of an aeroplane engine to a quiet hum. Each earpiece has a built-in microphone that samples the unwanted noise many times each second. Electronics inside the earpiece produce sound waves exactly opposite in shape. When these sound waves are added to the noise, they cancel it out, protecting the pilot's hearing.

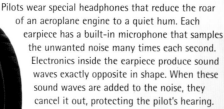

Peak

Engine noise is cancelled out

Trough

◀ DECIBEL SCALE

Loudness is measured on the decibel scale. The quietest sound our ears can detect measures 0 dB, but even gently falling leaves make a sound that is 10 times more intense than that. Traffic on a busy road produces sounds of around 70–90 dB, around a million times more intense than the measurement for falling leaves. At up to 190 dB, the blast of a launching rocket is loud enough to damage people's hearing permanently.

FIND OUT MORE ▶▶ Energy Waves **98–99** • Musical Sound **104–105** • Sound **100–101**

PITCH

Sound can be low-pitched, like the rumble of a large truck, or high-pitched, like a whistle. The pitch of a sound depends on the frequency of its wave. Piano keys produce notes that increase in pitch from the left side of the keyboard to the right. We can hear sounds of different pitches, but there are some sounds we cannot hear. Our ears cannot detect very low-pitched noises, known as infrasound, or very high-pitched noises, called **ULTRASOUND**.

HEARING ABILITY ▶
Bats can hear frequencies up to 120,000 Hz. Other animals cannot hear such high-pitched sounds. Mice can hear frequencies up to 100,000 Hz, dogs up to 35,000 Hz, and cats up to 25,000 Hz. Humans hear sounds only up to about 17,000 Hz, but children can usually hear higher-frequency sounds than adults.

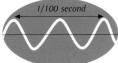

High-frequency wave makes a high-pitched sound — 1/100 second

Low-frequency wave makes a low-pitched sound — 1/100 second

▲ FREQUENCY
Pitch and frequency are not the same thing. Objects that vibrate slowly produce low-frequency sound that we hear as low-pitched. Things that vibrate more quickly make sounds of a higher frequency that our ears hear as more high-pitched.

pitch

Behind car the sound waves become longer and the sound becomes lower

Sound deepens when the car passes by

Ahead of car the sound waves are shorter and the car's sound is higher

DOPPLER EFFECT ▶
A racing car coming towards you bunches up the sound waves made by its engine. This makes them travel slightly faster. When the car passes by, the sound waves spread out and sound lower in pitch. This change in pitch is called the Doppler effect.

ULTRASOUND

Sound waves with a frequency of 20,000 Hz or more are known as ultrasound. Humans cannot hear ultrasound, but bats, dogs, porpoises, and many other animals can easily detect it. Caused by things vibrating extremely quickly, ultrasound has many uses, from toothbrushes that clean your teeth with sound to submarine navigation.

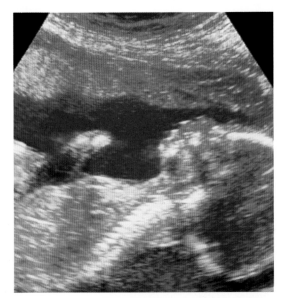

ULTRASOUND SCANNING ▶
Ultrasound allows doctors to check the health of a baby while it is still in the womb. High-frequency sound waves are sent into the mother's body, where some reflect off the baby and bounce back to a receiver. A computer uses the reflected sound waves to create a scan (picture) of the growing baby.

FIND OUT MORE ▶▶ Energy Waves 98–99 • Hearing 347 • Musical Sound 104–105 • Senses 316–317

MUSICAL SOUND

Music is one of the glories of sound. When a musician plays a note of a certain pitch, the musical instrument vibrates or **RESONATES** and produces a complex pattern of sound waves made up of many different frequencies. The most noticeable sound wave is called the fundamental, but there are other waves with higher frequencies, called harmonics. Notes from a flute sound more pure than those from a saxophone because they contain fewer harmonics. Musical instruments often make very quiet sounds, but some are designed to **AMPLIFY** the sounds they make so we can hear them more easily.

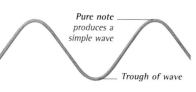

Pure note produces a simple wave

Trough of wave

◄ TUNING FORK
Hitting the two metal prongs of a tuning fork causes them to vibrate at a precise frequency. As they vibrate, they make the air around them vibrate, too. This produces sound waves in the form of a single, pure note. If you stand the base of the vibrating fork on a table, the table vibrates as well. This amplifies the note by making louder sound waves.

Complex note produces a complex wave

Trough of wave

◄ WOODWIND INSTRUMENT
When you play any form of pipe instrument, such as a flute, the air inside vibrates in complex patterns. Sound waves come out and you hear them as musical notes. A long flute can make a long sound wave and a low-pitched note. A short piccolo makes shorter sound waves and higher notes. By blocking holes in a pipe with your fingers or by pressing keys, you can play notes of different pitch.

Flute makes deeper sounds when more holes in the pipe are blocked

Spiky sound wave created by a note played on a violin

◄ STRING INSTRUMENT
A violin makes musical sounds when its strings vibrate. If you pluck a violin string and watch it closely, you can see it vibrating very quickly. The vibrations begin with the strings, but quickly make the large wooden body of the instrument vibrate as well. The vibrating body amplifies the sound greatly.

Treble clef tells players that the scale is in the mid range, not made up of deep, bass notes

Hollow, wooden body amplifies sounds made by the strings

Each note is a sound of a particular pitch named by a letter

Staff consists of lines and spaces that correspond to particular notes in the scale

B C D E F

C D E F G A

▲ MUSICAL SCALE
People compose music using sounds of different pitch. When musical sounds are arranged from low pitch to high pitch, they make a scale that can be written down on a staff. Each note on a scale is a sound of a different pitch. Different scales can be made by choosing different notes or by changing the way the pitch increases from one note to the next.

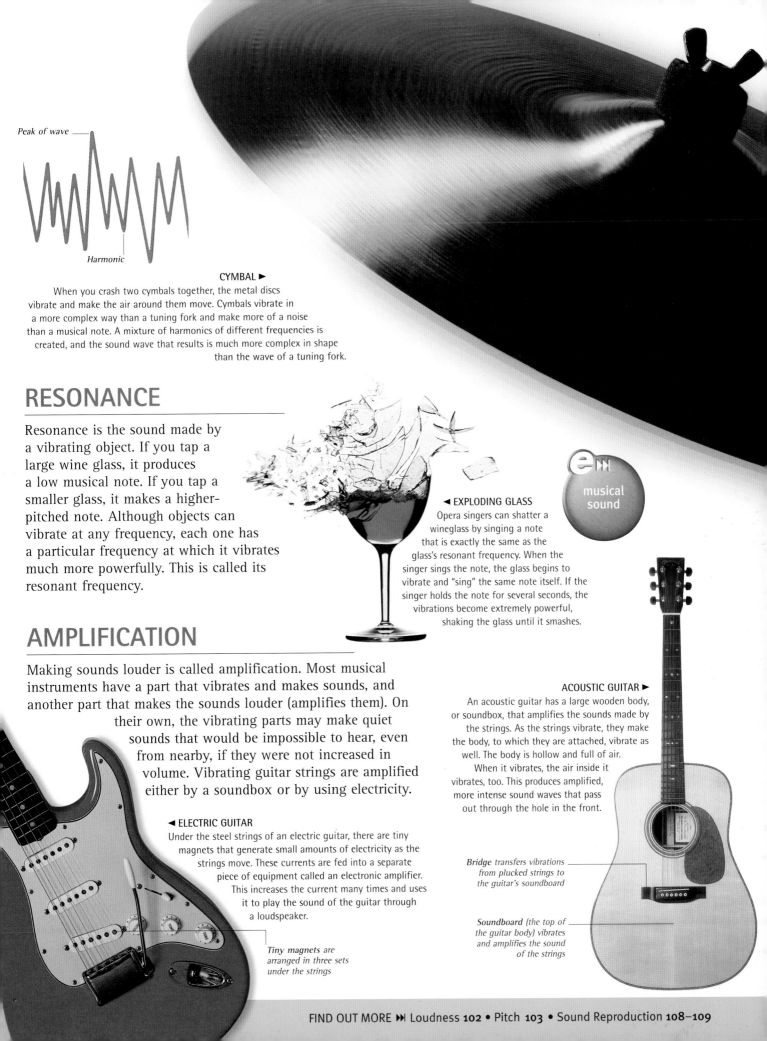

Peak of wave

Harmonic

CYMBAL ►
When you crash two cymbals together, the metal discs vibrate and make the air around them move. Cymbals vibrate in a more complex way than a tuning fork and make more of a noise than a musical note. A mixture of harmonics of different frequencies is created, and the sound wave that results is much more complex in shape than the wave of a tuning fork.

RESONANCE

Resonance is the sound made by a vibrating object. If you tap a large wine glass, it produces a low musical note. If you tap a smaller glass, it makes a higher-pitched note. Although objects can vibrate at any frequency, each one has a particular frequency at which it vibrates much more powerfully. This is called its resonant frequency.

◄ EXPLODING GLASS
Opera singers can shatter a wineglass by singing a note that is exactly the same as the glass's resonant frequency. When the singer sings the note, the glass begins to vibrate and "sing" the same note itself. If the singer holds the note for several seconds, the vibrations become extremely powerful, shaking the glass until it smashes.

musical sound

AMPLIFICATION

Making sounds louder is called amplification. Most musical instruments have a part that vibrates and makes sounds, and another part that makes the sounds louder (amplifies them). On their own, the vibrating parts may make quiet sounds that would be impossible to hear, even from nearby, if they were not increased in volume. Vibrating guitar strings are amplified either by a soundbox or by using electricity.

◄ ELECTRIC GUITAR
Under the steel strings of an electric guitar, there are tiny magnets that generate small amounts of electricity as the strings move. These currents are fed into a separate piece of equipment called an electronic amplifier. This increases the current many times and uses it to play the sound of the guitar through a loudspeaker.

Tiny magnets are arranged in three sets under the strings

ACOUSTIC GUITAR ►
An acoustic guitar has a large wooden body, or soundbox, that amplifies the sounds made by the strings. As the strings vibrate, they make the body, to which they are attached, vibrate as well. The body is hollow and full of air. When it vibrates, the air inside it vibrates, too. This produces amplified, more intense sound waves that pass out through the hole in the front.

Bridge transfers vibrations from plucked strings to the guitar's soundboard

Soundboard (the top of the guitar body) vibrates and amplifies the sound of the strings

FIND OUT MORE ►► Loudness **102** • Pitch **103** • Sound Reproduction **108–109**

ACOUSTICS

The science of how sound behaves, especially when it travels through our everyday world, is called acoustics. Sound waves normally travel in straight lines directly outwards from their source, but they do not always travel in that way. An object standing in the path of a sound wave can affect its movement. When a sound wave hits a hard object, the sound reflects back towards the source in the form of an **ECHO**. When soft objects get in the way, they can **ABSORB** the sound and stop it from travelling any further. Scientists use sound reflection and absorption to investigate places that they cannot visit, such as the depths of oceans and the interior of the Earth.

▲ HOLLYWOOD BOWL
The Hollywood Bowl is a famous, open-air amphitheatre in California, USA. An amphitheatre is a bowl-shaped place that reflects sound naturally and evenly into the landscape around it. The Hollywood Bowl was carved into the side of a mountain at Bolton Canyon in the 1920s and can seat 20,000 people.

Curved ceiling reflects sound to seats in all parts of the auditorium

Curved panel directs the sound of the orchestra to listeners opposite

Rear seats receive sound reflected from the ceiling

Padded seats absorb sound when raised

Hard baffle board reflects sound to the side seats

Gap in ceiling absorbs sound where reflection is not needed

Seated people will absorb sound with their bodies and prevent echo

Direct sound travels straight from the orchestra to the audience

▲ CONCERT HALL
Music must sound clear in an auditorium, no matter where people are sitting in the audience. It should sound the same whether the hall is full or nearly empty. The curved shapes in modern concert halls are designed to help distribute the sound evenly to every seat in the auditorium.

Orchestra pit is located centrally, so sound can flow evenly in all directions

Curved surface prevents sound from being reflected back and forth across the hall in echoes

ECHOES

If you shout at a distant wall, you can hear your voice return as a reflected sound wave, or echo. When the reflected sound wave has to travel some distance, it takes time to return and you hear it separately from the original sound. Sound waves that reflect off nearer objects return almost instantly. Our brains blend these waves with the original sound and we hear no echo.

Dolphin's head amplifies echoes

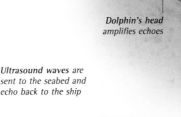

Ultrasound waves are sent to the seabed and echo back to the ship

▲ EXPLORING WITH SONAR
The depth of the ocean can be measured using SONAR (SOund Navigation And Ranging). A loudspeaker under the ship sends down a beam of high-frequency ultrasound. Echoes of the sound waves are detected by hydrophone (underwater microphone) as they bounce back up.

▲ ECHOLOCATION
Like many other sea creatures, dolphins use sound to find their way around, locate their companions, and discover sources of food. The clicking sounds they make are reflected back from the seabed and objects around them and are picked up by the dolphins' long, bony heads. Using echoes to find things is known as echolocation.

Shipwreck stands out from the blues of the surrounding seabed

◄ WRECK-FINDING WITH RADAR
This shipwreck, lying deep on the ocean floor, was found by a submarine using radio-wave echoes (radar). The submarine scanned the seabed by sending out beams of radio waves. Some of these reflected back from the wreck and were picked up by a detector. A computer made an image of the wreck using the reflected waves.

ABSORPTION

Hard objects reflect sounds, but soft materials absorb sounds and silence them. When sound waves reach a soft material, their energy is soaked up and they travel no further. Things that absorb sound can be useful for reducing noise. Trees are sometimes planted by motorways so that their leaves will reduce the sound of traffic. Walls can be padded with soft materials to stop sound from travelling through them.

acoustics

ANECHOIC CHAMBER ►
Engineers test loudspeakers and audio equipment in specially designed laboratories called anechoic chambers. The walls and ceiling are covered by spikes of soft foam that absorb sound and stop any echoes and reverberations (very fast echoes). Sounds made inside an anechoic chamber sound very dull or "dead", which is why the chambers are also called "dead rooms".

FIND OUT MORE ▶▶ Communication 318–319 • Pitch 103 • Sound 100–101 • Sound Reproduction 108–109

FROM RECORDING TO PLAYBACK

◄ MICROPHONE
Inside a recording studio, a microphone is turning the sound energy of this singer's voice into electrical energy. She holds it close up to cut out background sounds. A wire carries the pulses of electricity to sound recording equipment elsewhere in the studio.

◄ MIXING DESK
Lots of different singers and instruments can appear on a record, and each one has to be recorded by a separate microphone. The knobs on this mixing desk control the signals from the different microphones. Each knob can make a singer or player louder or quieter in the mix.

◄ COMPACT DISK (CD)
The final version of the recording is put on sale on CD. The music is recorded in the surface of the plastic disk as a series of tiny bumps (seen as red and yellow in this highly magnified photograph). The bumps are covered by fine metal film and a layer of plastic.

◄ CD PLAYER
Every CD player contains a laser that reads the series of bumps on the CD surface as a long string of numbers. The CD's shiny metal film reflects back the light of the laser. The numbers are converted back to the same pulses of electricity that originally made the bumps in the CD.

◄ LOUDSPEAKER
The pulses of electricity are fed through an amplifier into a loudspeaker, which works in the opposite way to a microphone. It turns electrical energy back into sound by using electricity to make the air around it vibrate. In this way, the music is reproduced exactly.

SOUND REPRODUCTION

Most sounds happen only briefly and are then lost to us. Fortunately, there are two ways in which we can record sounds and later reproduce them (play them back). One way is to convert and store sound using other forms of energy, such as electricity and magnetism. The other, **DIGITAL** way involves converting and storing sound in the form of numbers.

Diaphragm vibrates differently to high and low-frequency sounds

Wire coil vibrates up and down along with the diaphragm

Magnet interacts with the coil to produce electrical signals

▲ MICROPHONE
A microphone changes sound waves into tiny bursts of electricity. Inside it, sound waves cause a flexible disc (diaphragm) to vibrate. The up-and-down movement of a wire coil fixed to the diaphragm interacts with a magnet to produce a varying electrical current that can be stored and played back.

▼ VINYL RECORD
Before CDs and digital music became popular, music was recorded on flat discs made of a special plastic called vinyl. Discs containing up to an hour of sound were known as long-playing (or LP) records. The sound was stored in tiny bumps in a long spiral groove on the LP's surface. Both sides of the disc were used.

▲ CASSETTE TAPE
Now obsolete, audio cassettes were popular in the 1980s and 1990s. The reel of tape inside stored sound as a pattern of magnetic pulses laid out along its length. When the tape was used in a cassette player, the magnetic pulses were turned back into electricity and sound.

Polydor

▼ MAKING A MASTER COMPACT DISK (CD)

When a piece of music is ready for transfer to CD, a blank master CD is set up to spin round at very high speed in front of a laser that switches on and off very quickly. Each time the laser switches on, it burns a tiny bump onto the CD's plastic surface. The pattern of bumps is a coded version of the music stored on the CD. Copies of the master CD have the same surface bumps, which are read by CD players.

Motorized wheel spins the CD at high speed

Laser beam burns bumps onto the disk surface

DIGITAL SOUND

Sounds can be stored in digital form by using electronics to turn them into patterns of numbers. A CD stores music on its surface as a pattern of bumps. The bumps represent a coded pattern of numbers that the CD player turns back into sound waves. Digitally recorded sounds are not affected by background noises. Digital sounds are easily edited and mixed with the aid of computerized equipment.

ORIGINAL SOUND WAVE

SOUND WAVE MEASURED BY FEW SAMPLES

◄ SOUND WAVE

A sound wave is stored digitally by a process called sampling. The amplitude (height) of the wave is measured every so often and stored as a number. When all these numbers are written together, they make a longer number that represents the entire wave. The more times the wave is measured, or sampled, the better it sounds during playback.

sound reproduction

SOUND WAVE MEASURED BY MANY SAMPLES

Track details are displayed on the player's screen

Modern MP3 players can often play videos too

▲ MP3

Many people now listen to music stored in digital files called MP3s. These files can either be downloaded using an ordinary computer, or "ripped" (extracted) from CDs by using a program. They are easily transported onto a portable MP3 player, and can then be heard by using headphones or by plugging into a speaker system.

FIND OUT MORE ▶▶ Computers 148–149 • Digital Electronics 140–141 • Lasers 112 • Musical Sound 104–105

LIGHT

Light makes the world seem bright and colourful to our eyes. Light is a type of electromagnetic radiation that carries energy from a **SOURCE** (something that makes light) at the very high speed of 300,000 kps (186,000 miles per second, or 670 million mph). Light rays travel from their source in straight lines. Although they can pass through some objects, they bounce off others or pass around them to make **SHADOWS**.

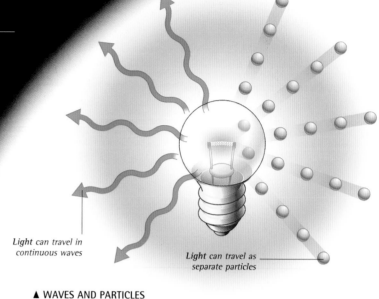

Light can travel in continuous waves

Light can travel as separate particles

▲ WAVES AND PARTICLES

Sometimes light seems to behave as though it carries energy in waves. Other times it seems to carry energy in particles or packets, called photons, fired off in quick succession from the source. Scientists argued for many years over whether light was really a wave or a particle. Now they agree that light can behave as either a wave or a particle, depending on the situation.

LIGHTHOUSE ►

The powerful beam from a lighthouse illustrates that light travels in straight lines. Under normal circumstances, light never bends or goes round corners but travels in a perfectly straight path, making what is known as a light ray. Nothing can travel faster than light. The beam from a lighthouse travels its full length in a tiny fraction of a second.

e ▸▸
light

◄ SOAP BUBBLE

When light shines on a soap bubble, some of the rays reflect back from its outer surface. Others travel through the thin soap film and bounce back from its inner surface. The two sorts of reflected rays are slightly out of step because they travel different distances. They interfere with one another and produce colourful swirling patterns on the bubble's surface.

TRANSMISSION OF LIGHT

Some objects transmit light better than others. Transparent objects, such as glass, let virtually all light rays pass straight through them. When you look at a glass of orange juice, you can see the juice inside very clearly. You can also see other things through the glass.

Translucent objects, such as plastic, allow only part of the light through. A plastic bottle lets some light rays pass through it. It is possible to see the orange juice inside the bottle, but you cannot see anything behind the bottle.

Opaque objects, such as metal, reflect all the light falling on them and allow none to pass through. When you look at a can of orange juice, all you can see is the can. It is impossible to tell, just from looking, whether or not the can has any orange juice in it.

TRANSPARENT

Glass contents are clearly visible

Glass allows all light to pass through it

TRANSLUCENT

Bottle contents are visible but appear milky

Plastic allows some light to pass through it

OPAQUE

Can contents are not visible

Metal allows no light to pass through it

LIGHT SOURCES

Things that give off light are called light sources. When we see something, light rays have travelled from a source of light into our eyes. Some objects appear bright to us because they give off energy as light rays; these objects are said to be luminous or light-emitting. Other objects do not make light themselves, but appear bright because they reflect the light from a light source.

◄ SUNLIGHT
The Sun shines because it produces energy deep in its core. The energy is made when atoms join together in nuclear fusion reactions. The Sun fires off the energy into space in all directions in the form of electromagnetic radiation. Some of the radiation travels to Earth as the light and heat we know as sunlight. The Sun is a luminous light source because it makes energy inside itself.

BIOLUMINESCENCE ►
Some sea organisms can make their own light. This ability is called bioluminescence, which means making light biologically. Transparent polychaete worms such as this one make yellow light inside their bodies. In their dark seawater habitat they can glow or flash to scare off predators. Other bioluminescent sea creatures include shrimps, squid, and starfish.

▲ MOONLIGHT
The Moon shines much less brightly than the Sun. Unlike the Sun, the Moon does generate its own energy, so it produces no light of its own. We can see the Moon only because its grey-white surface reflects sunlight towards Earth. If the Earth passes between the Sun and the Moon, the Moon seems to disappear from the sky. This is called a lunar eclipse.

SHADOWS

Shadows are made by blocking light. Light rays travel from a source in straight lines. If an opaque object gets in the way, it stops some of the light rays travelling through it, and an area of darkness appears behind the object. The dark area is called a shadow. The size and shape of a shadow depend on the position and size of the light source compared to the object.

YOUR CHANGING SHADOW ►
When you stand with the Sun behind you, the light rays that hit your body are blocked and create a shadow on the ground in front of you. When the Sun is high in the sky at midday, your shadow is quite short. Later on, when the Sun is lower, your shadow is much longer.

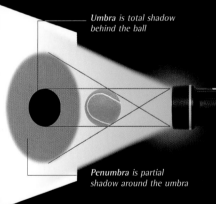

Umbra is total shadow behind the ball

Penumbra is partial shadow around the umbra

▲ UMBRA AND PENUMBRA
Shadows are not totally black. If you look closely at a shadow, you will see a dark area in the centre and a lighter area around it. The central dark area, called the umbra, occurs where rays of light from the source are totally blocked. The outer area, called the penumbra, is lighter because some rays do get through.

FIND OUT MORE ►► Energy Waves 98–99 • Nuclear Energy 85 • Reflection 113 • Refraction 114 • Sun 170–171

LASERS

Some beams of light are powerful enough to cut through metal. Others are precise enough to use for delicate surgery on people's bodies. These remarkable forms of light are made by lasers. Laser stands for Light Amplification by Stimulated Emission of Radiation. A laser is a device that concentrates light rays so they all travel exactly in step. Laser rays are much more powerful and precise than other light rays.

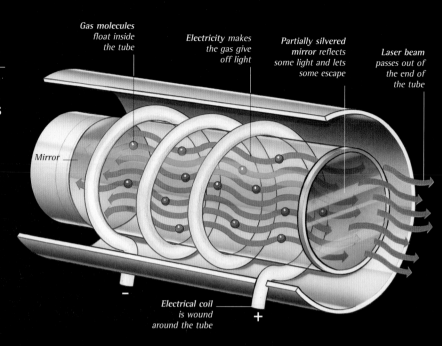

Gas molecules float inside the tube

Electricity makes the gas give off light

Partially silvered mirror reflects some light and lets some escape

Laser beam passes out of the end of the tube

Mirror

Electrical coil is wound around the tube

▲ LASER CUTTING
This machine is using a carbon dioxide laser to cut metal. A laser of this type makes its beam by passing electricity through carbon dioxide gas. A computer controls the laser cutting with great accuracy. Clothing manufacturers often use lasers, as a single laser beam can cut through hundreds of thicknesses of clothing material at once.

▲ HOLOGRAM
A three-dimensional photograph that seems to hover inside a piece of plastic or glass is called a hologram. Although the hologram appears to be a solid object, it is actually an image that was stored in the plastic or glass by scanning a laser beam over the object.

INSIDE A LASER ▲
A laser makes light by passing electricity through a gas. This makes the gas emit (give out) light waves at a precise wavelength. The light waves bounce back and forth along a tube between two mirrors. This encourages the gas to give out more light exactly in step with the original light waves. It also amplifies (makes brighter) the beam of light.

lasers

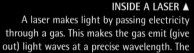

LASER LIGHT SHOW ▶
Some lasers are so powerful that they can shine great distances into the sky. They are often used at rock concerts or to provide spectacular light shows in the open air. This light show is held regularly near the pyramids at Giza in Egypt. Several powerful, computer-controlled lasers produce strong beams of red light that reach up and sweep through the air.

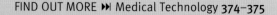

FIND OUT MORE ▸▸ Medical Technology **374–375**

REFLECTION

Reflections are usually caused by shiny things, such as **MIRRORS**, that show a reversed image of whatever is placed in front of them. The image seems to be as far behind the mirror as the object is in front of it. Not only mirrors make reflections, however. Most objects reflect some of the light that falls on them. In daytime we see familiar objects like grass, trees, and the sky only because they reflect light from the Sun into our eyes.

▲ REGULAR REFLECTION
When light rays bounce off a completely smooth surface, such as a still pool of water, a mirror, or even something like a shop window, we are able to see a very clear reflection on the surface. Every ray of light is reflected perfectly from the surface and bounces back in a regular way. The reflected image is very clear and sharp.

▲ IRREGULAR REFLECTION
A rough surface, such as this rippling pond, causes light rays to bounce off it in many different directions. It may still be possible to make out an image on the surface, or, if it is very rough, the image is very broken up. Most objects reflect light in this irregular way. Although we can see them, we cannot see any images reflected in their surfaces.

reflection

MIRRORS

A mirror is a very smooth, highly polished piece of metal or plastic that reflects virtually all the light that falls onto it. The reflection appears to be behind the mirror and may look bigger, smaller, or the same size as the thing it is reflecting, depending on the mirror's shape. We use mirrors when checking our appearance or driving. They also play an important part in telescopes, microscopes, cameras, and other optical (light-based) instruments.

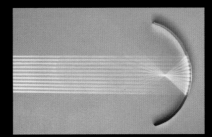

▲ CONCAVE MIRROR
A concave mirror curves or bends inwards and makes an object look bigger and nearer than it actually is. It works by making light rays seem to come from a point in front of the mirror, which is closer to our eyes. Concave mirrors are important in such things as bicycle reflectors and reflecting telescopes.

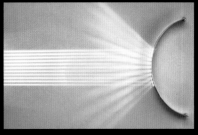

▲ CONVEX MIRROR
A convex mirror curves or bends outwards and makes an object look smaller and further away than it actually is. It makes light rays seem to come from a point behind the mirror, further from our eyes. Things look smaller, but convex mirrors are helpful because they can show a wider picture or field of view.

◄ SHAVING MIRROR
This man is shaving with the help of a concave mirror. Its curved surface makes the man's face seem closer to him than it really is. The reflected image he sees is magnified and he can easily see what he is doing. The mirror's drawback is that less of the man's face fits into the mirror than in a flat mirror of the same size.

CAR WING MIRROR ►
Drivers use mirrors to see traffic coming up behind them. It is important for drivers to see as much of the road behind as they can, so wing mirrors and rear-view mirrors are convex. A drawback is that they make vehicles on the road behind look smaller and further away than they would in a flat mirror of the same size. Drivers must remember that the vehicles are nearer than they appear.

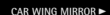

FIND OUT MORE ►► Cameras **118–119** • Light **110–111** • Microscopes **116** • Telescopes **117**

REFRACTION

Light rays usually travel in straight lines, but when they pass from one material to another they can be forced to bend (change direction and continue on a new straight path). The bending is called refraction. It happens because light travels at different speeds in different materials. If light rays travel through air and enter a more dense material, such as water, they slow down and bend into the more dense material. Light rays moving into a less dense material, such as from water to air, speed up and bend outwards.

PUZZLE FOR THE EYE ▼
If you stand a straw in a glass of water, the top and the bottom of the straw no longer seem to fit together. This trick of the light is caused by refraction. Light bends outwards when it travels from water to air, so the eye sees the bottom of the straw (in the water) as deeper than the top of the straw (in the air).

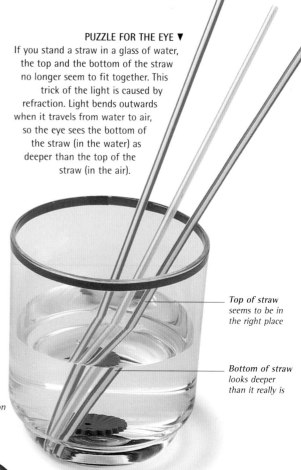

Top of straw seems to be in the right place

Bottom of straw looks deeper than it really is

PATH OF LIGHT ▶
Light rays bend or refract if they enter a glass block at an angle. When they pass from air into glass, they bend inwards and slow down. They travel in a straight line through the glass at an angle to their original direction. As they pass out from the glass into air, they bend outwards and speed up again.

Light rays travel in a straight line towards the glass block

Rays change direction inwards as they enter the more dense material

Rays change direction outwards as they leave the block

refraction

▼ REFRACTION IN HEAT HAZE
On hot days, the surface of the Earth is warmer than the sky above it. This means that air close to the ground is generally much warmer than the air higher up. Hot air rising from the ground can bend and distort the light rays passing through it. This gives a very hazy appearance to objects, such as this giraffe, as they move on the horizon.

Light rays from giraffe are refracted by rising hot air

MIRAGE

People who travel through hot deserts often think they can see water or trees on the ground ahead of them, when really there is nothing there. This trick of the light is called a mirage. Layers of warm and cold air bend or refract light rays coming from distant objects – perhaps real trees over the horizon. Our eyes are fooled into thinking the light rays come from objects on the ground instead of from the sky.

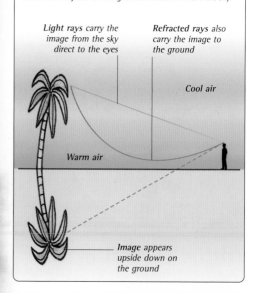

Light rays carry the image from the sky direct to the eyes

Refracted rays also carry the image to the ground

Cool air

Warm air

Image appears upside down on the ground

FIND OUT MORE ▶▶ Light **110–111** • Reflection **113**

LENSES

A lens is a piece of transparent plastic or glass that can make things seem to change size. It works by bending light rays so they appear to come from a slightly different place. Some lenses make things look nearer and bigger. Others make things look smaller and further away. Without their spectacle lenses, many people would be unable to see clearly, read books, or drive safely.

FRESNEL LENS ►

A lighthouse must send a long beam of light far out to sea. To do that a very large and heavy lens would normally be needed. Instead, lighthouses use a specially shaped Fresnel lens. It has steps like a staircase, each of which helps to bend the light into a single, powerful beam. A Fresnel lens can be made from glass or lightweight plastic.

Stepped lens bends light into a powerful beam

Rotating base sweeps the beam far across the sea

CONTACT LENS ►

A contact lens is a tiny piece of plastic or glass that rests on the front of the eyeball. It bends light rays before they enter the eye in ways that help the wearer to see more clearly.

Lens is shaped to fit snugly on the eyeball

lenses

▼ CONCAVE AND CONVEX LENSES

The two main types of lens are called concave and convex. A concave lens is thin in the middle and thick at the edges, so it seems to "cave" inwards. It makes light rays bend outwards, or diverge. A convex lens works in the opposite way. It is thicker in the middle and thinner at the edges. Light rays passing through a convex lens bend inwards, or converge.

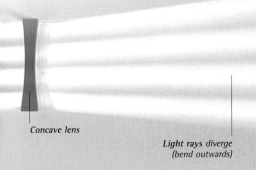

Concave lens

Light rays diverge (bend outwards)

Light rays converge (bend inwards)

Convex lens

SLIDE PROJECTOR ►

By shining a powerful beam of light through a transparent photographic slide, a projector can make a much larger image on a wall. The little image on the slide is shined through a concave lens, which spreads the light rays outwards. The further away the projector is from the wall, the bigger the image becomes.

MAGNIFYING GLASS ►

To magnify means to increase in apparent size. If you look closely at a magnifying glass, you can see that it is a large convex lens, thicker in the centre than at the edges. When you hold a magnifying glass over an object, it makes light rays from the object seem to come from a nearer point, causing it to look bigger than it really is.

FIND OUT MORE ►► Microscopes 116 • Refraction 114 • Telescopes 117

MOSQUITO (SEEN BY CAMERA ZOOM)

MOSQUITO (OPTICAL MICROSCOPE)

MOSQUITO (ELECTRON MICROSCOPE)

MOUTH HAIRS (ELECTRON MICROSCOPE)

MICROSCOPES

Some objects are so small that our eyes cannot see them. We cannot see atoms or molecules, or the cells of our bodies, or viruses that carry disease. A microscope uses lenses to make tiny things appear bigger so we can see them clearly. There are two main kinds of microscope. Optical microscopes create a magnified image using light from an object. Electron microscopes are much more powerful and use a beam of electrons instead of light.

Eyepiece moves to focus the image onto the eye

Lens increases the degree of magnification

Bigger lens can be swivelled round for greater magnification

Objective lens provides the first magnification of the object on the slide

Lens magnifies light from the objective lens

Glass slide holds in place the object to be viewed

Light rays enter the microscope

Mirror swivels to collect light rays and direct them towards the glass slide

microscopes

▲ MAGNIFICATION
The top picture shows a mosquito through a camera's zoom lens. The second picture down depicts a mosquito on the slide of an optical microscope. In the third picture down, an electron microscope provides greater three-dimensional detail. Finally, a powerful electron microscope has looked inside a mosquito's mouth to reveal the hairs growing there.

▲ OPTICAL MICROSCOPE
An optical microscope uses light. The object to be viewed is cut very thin so light will pass through it, then placed on a piece of glass called a slide. A mirror at the bottom gathers light and reflects it up through the slide. A system of lenses magnifies the object, making a bigger image that may be seen in the eyepiece at the top.

ELECTRON MICROSCOPE ▶
An electron microscope uses a beam of electrons instead of light. The object to be viewed is placed on a small stand in the middle. An electron gun, similar to the ones in TV sets, fires a beam of electrons down onto the object. As the electron beam scans (passes over) the object, a very detailed picture of the object appears on a TV screen.

FIND OUT MORE ▶▶ Atoms **24–25** • Lenses **115** • Light **110–111** • Reflection **113**

TELESCOPES

Just as our eyes cannot see small objects, so they cannot see things that are very far off. Even when things are millions of miles away, telescopes show them very clearly, The long tubes gather light rays from distant objects and make magnified images of them that seem nearer. Some telescopes use lenses to gather light, while others use mirrors.

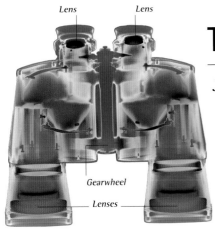

Lens   Lens

Gearwheel

Lenses

▲ X-RAY OF BINOCULARS
A pair of binoculars works like two small telescopes joined together, one for each eye. A gearwheel in the centre of the binoculars alters the distance between the lenses and brings the image in the binoculars into sharp focus.

BINOCULARS' VIEW OF MOON

AMATEUR TELESCOPE

30-CM (1-FT) LENS

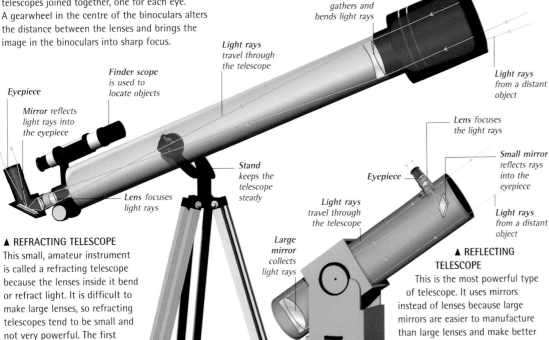

Large lens gathers and bends light rays

Light rays travel through the telescope

Light rays from a distant object

Finder scope is used to locate objects

Eyepiece

Mirror reflects light rays into the eyepiece

Stand keeps the telescope steady

Lens focuses light rays

Lens focuses the light rays

Small mirror reflects rays into the eyepiece

Eyepiece

Light rays travel through the telescope

Light rays from a distant object

Large mirror collects light rays

▲ REFRACTING TELESCOPE
This small, amateur instrument is called a refracting telescope because the lenses inside it bend or refract light. It is difficult to make large lenses, so refracting telescopes tend to be small and not very powerful. The first refracting telescope was built in 1608 by a Dutch scientist, Jan Lippershey, who lived from about 1570 to 1619.

▲ REFLECTING TELESCOPE
This is the most powerful type of telescope. It uses mirrors instead of lenses because large mirrors are easier to manufacture than large lenses and make better images. A large, central mirror collects light from a distant object and a smaller mirror reflects the light into the eyepiece.

1.2-M (4-FT) LENS

telescopes

GIANT TELESCOPE ▶
The biggest astronomical telescopes in the world are all of the reflecting type and use large mirrors to form images of the stars. This large reflecting telescope is part of the European Space Observatory, which is located on the top of a mountain at La Silla, in Chile. It has a mirror 3.6 m (12 ft) in diameter and is mounted inside a huge metal dome that protects it from the weather.

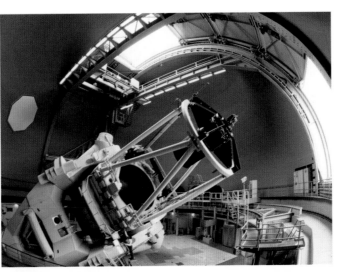

▲ LOOKING AT THE MOON
The top picture shows the Moon seen through a pair of binoculars. A small amateur telescope magnifies the Moon much more and reveals some of the "seas" as dark patches on its surface. A telescope with a 30-cm (1-ft) lens makes small craters on the Moon clearly visible. A 1.2-m (4-ft) lens is four times more powerful and provides much more detail.

FIND OUT MORE ▶▶ Moon **177** • Observatories **187** • Space Observatories **196–197**

◄ BITUMEN PHOTO

The oldest surviving photograph was taken by French physicist Joseph Niépce (1765–1833) in 1827. Instead of using photographic film, he used a piece of pewter metal covered with a tar-like substance called bitumen. Light had to enter his primitive camera for eight hours to take the photograph.

◄ DAGUERROTYPE

French painter Louis Daguerre (1787–1851) invented a much better method of taking photos in 1839. Known as a daguerrotype, it caught images on silver plates coated with a silver-based chemical that was sensitive to light. Taken in just a few minutes, they were clear and showed good detail.

◄ PHOTO ON PAPER

Also in 1839, British inventor William Henry Fox Talbot (1800–1877) took the first ever photograph on paper. Using a different process to Daguerre's, his cameras captured a reverse image called a negative. Then he used chemicals to turn this into a final image on paper, called a positive.

◄ COLOUR PHOTO

In 1861, the distinguished British physicist James Clerk Maxwell (1831–1879) became the first person to make a colour photograph. He took three photographs of this tartan ribbon in three separate colours, then added them together to make a single colour picture.

CAMERAS

A camera is a device that records pictures. It consists of a sealed box that catches the light rays given off by a source. A lens at the front of the camera brings the light rays to a focus and makes the picture seem closer or further away. Traditional cameras store pictures in a chemical form using **PHOTOGRAPHIC FILM**. Modern **DIGITAL** cameras store pictures electronically.

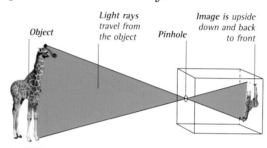

Object — *Light rays travel from the object* — *Pinhole* — *Image is upside down and back to front*

▲ PINHOLE CAMERA

The simplest camera is a small box with a tiny hole in its front wall and a piece of photographic film taped inside its back wall. Light rays cross as they travel between the object and the film, passing through the pinhole. All photographic images are small, upside down, and back to front when inside the camera.

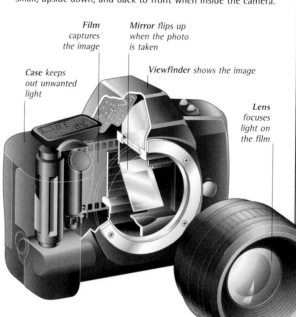

Film captures the image — *Mirror flips up when the photo is taken* — *Viewfinder shows the image* — *Case keeps out unwanted light* — *Lens focuses light on the film*

▲ FILM CAMERA

The lens on this camera captures light rays and focuses them. A mirror in the centre of the camera reflects light from the lens into the viewfinder. When a photo is taken, the mirror flips out of the way and light from the lens briefly travels through the camera to the film at the back. The film records an image of the view through the lens.

PHOTOGRAPHIC FILM

A film camera records light on a thin piece of transparent plastic coated with a light-sensitive emulsion. The emulsion consists of crystals of silver compounds in a jelly-like substance called gelatin. When light is allowed to briefly strike the film, it causes a chemical reaction in the emulsion and an image is formed.

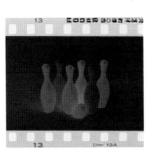

▲ BLACK AND WHITE NEGATIVE
The image taken by a camera, called a photographic negative, looks very different to what was photographed. It is a strange-looking version of the original scene in which dark and light areas are reversed. If you take a photograph of black ink on white paper, the negative shows white ink on black paper.

COLOUR NEGATIVE ▶
Colour films produce colour negatives, in which all the different colours in the image are replaced by their complementary, or opposite, colours. The dark colours appear as light areas and the light ones appear dark.

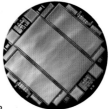

SLIDE ▶
Colour negatives can be used to make either paper prints or plastic slides like this one. A slide is just like the original colour negative but the colours have been reversed and appear normal.

DIGITAL PHOTOGRAPHY

An ordinary photograph is a piece of paper onto which a picture has been printed from a negative. A digital photograph is a computerized file in which a picture is made up of a string of numbers. A digital image can be loaded into a computer, edited, printed out, sent by email, or stored on a website.

◀ PIXELS
A digital photograph is made up of small squares called pixels. The more pixels a photograph has, the sharper it looks. Digital photographs have low or high resolution, depending on the amount of detail they show. The bottom of this flower is made up of large pixels and looks quite fuzzy; it has low resolution. The top of the flower is made up of tiny, almost invisible pixels and has high resolution.

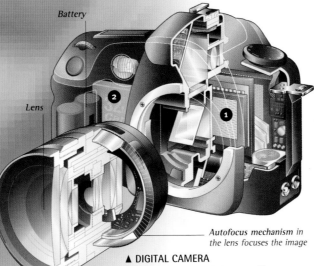

Battery

Lens

Autofocus mechanism in the lens focuses the image

▲ DIGITAL CAMERA
A digital camera is quite similar to a film camera and has similar components and controls. Instead of film, however, a digital camera has a light-sensitive sensor or microchip inside it called a charge-coupled device (CCD). This sensor turns light rays into a pattern of numbers, and the whole photograph is stored as one very long number.

❶ SENSOR
The CCD is made up of many tiny square segments arranged in a grid. Each segment measures the amount of light falling onto it and generates one pixel of the digital photo.

❷ MEMORY MICROCHIP
The CCD turns each pixel into a number that represents the colour and brightness of that part of the picture. All the pixel numbers are stored together on a memory microchip.

FIND OUT MORE ▶▶ Digital Electronics **140–141** • Lenses **115** • Microelectronics **142**

CINEMA

In a cinema movie, many still photographs are projected onto a screen in quick succession. Our eyes do not see them as separate still photographs, but blend them into a single moving image. Early movies had black and white pictures and little or no sound. Modern movies are colourful, have realistic sound, and use **DIGITAL EFFECTS**.

Cardboard disc is made to spin

Diner appears to feed himself rapidly with his spoon

cinema

AUGUSTE AND LOUIS LUMIÈRE
Cinema, as we call it today, was invented by the French brothers Auguste (1862–1954) and Louis (1862–1948) Lumière. They developed the first practical film projector, a machine they called the cinèmatographe, in 1895. Also in that year, they made the first ever motion picture and opened the first cinema to show movies.

◄ PHENAKISTOSCOPE
The first "movies" were little more than mechanical toys. This phenakistoscope, invented in 1832, has a series of still pictures printed around the surface of a large cardboard disc. Each picture depicts one stage of a continuous movement. When you spin the disc quite fast and stare at a single point, the pictures merge together and give the illusion of movement.

▲ MOVIE STORYBOARD
Filming is an expensive process, so it is usually planned in advance. After writers provide a story, artists sketch the scenes that need to be filmed on a storyboard. The director (who is responsible for the overall look of a movie) uses this to work out how to arrange cameras, lighting, and other equipment.

▲ MOVIE STUDIO
Making a movie is a huge team effort that can involve hundreds of people. In this shot from *The Matrix*, sound technicians, camera operators, special effects people, and lighting men are preparing to film the actor Keanu Reeves. The railway track enables the heavy camera to move smoothly across the studio

DIGITAL EFFECTS

Computers can be used when movies require dazzling special effects that would be impossible to create in real life. The effects are called digital because they are created with digital technology. Movies that need digital effects are turned into a series of digital photographs. Once in digital form, they can be edited, mixed with animation, and changed in other ways.

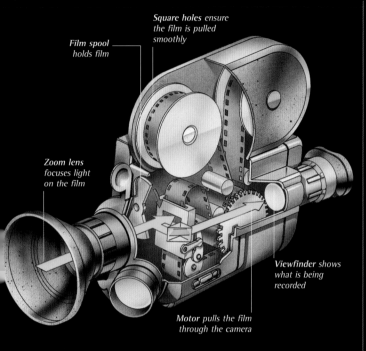

Film spool holds film

Square holes ensure the film is pulled smoothly

Zoom lens focuses light on the film

Viewfinder shows what is being recorded

Motor pulls the film through the camera

▲ MOVIE CAMERA

A movie camera works in much the same way as a still camera and has many of the same components. Instead of taking only one photograph, it takes 24 separate photographs each second. A motor inside the camera works a mechanism that pulls film past the lens from a large spool. Small, square holes punched along the edges of the film ensure that the film is pulled through steadily and at exactly the right speed.

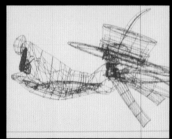

◄ WIRE-FRAME MODEL

Artists start with a wire-frame model, which is a simple drawing that shows a character's outline. A computer turns this into a long series of digits (numbers) that are altered to move the character.

FINAL ARTWORK ►

Computers are used to add colour and texture (the surface look) to the wire-frame images and to work out lighting and other special effects. The character shown is from Charles Russell's 1994 movie *The Mask*.

▲ SOUND STUDIO

Movie sound is usually recorded at the same time as the filming. The sound may need to be edited in a recording studio like this one. The sound editor's job is to make sure the sound is exactly in step with the pictures. He or she is also responsible for adding music (called the score) to the movie.

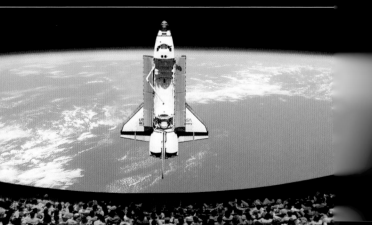

▲ IMAX CINEMA

Movie action can be very dramatic and exciting when seen on the gigantic screen of an IMAX cinema. The screen is so big that it completely fills your field of view (what you can see), and you easily forget the people and other things around you. That is why you feel so affected by the action on the screen.

FIND OUT MORE ▸▸ Cameras **118–119** • Digital Electronics **140–141** • Sound Reproduction **108–109**

RED
ORANGE
YELLOW
GREEN
BLUE
INDIGO
VIOLET

COLOUR

On a sunny day, the world seems light and colourful because our eyes are able to see differences in the wavelengths of light as different colours. Some animals cannot do this and live in a colourless world. Sunlight looks white or yellow to us, but is really a mixture of light of many different colours. Coloured light is one of the things that makes objects look different from one another. A tomato looks red because it reflects red light into our eyes, while an apple looks green because it reflects green light.

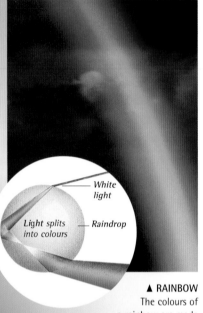

▲ RAINBOW
The colours of a rainbow are made when sunlight shines through raindrops. When a ray of sunlight enters a raindrop, the tiny drop of water splits up the white light into different colours. Although a rainbow usually looks semicircular from the ground, it appears as a complete circle if you look at it from an aeroplane.

White light
Light splits into colours
Raindrop

▲ SPLITTING LIGHT
When white light shines into a solid triangle of glass, called a prism, the glass in the prism refracts or changes the direction of the light as it passes through. Different colours of light are made by light of different wavelengths. The prism bends the shorter, blue wavelengths of light more than the longer, red wavelengths. This is how the prism splits white light into its spectrum of colours.

▲ SUNSET
The whole sky can look red at dawn or dusk when the Sun sits low on the horizon. At these times of day, sunlight reaches your area of Earth only after travelling through a thick layer of the atmosphere. Particles in the atmosphere scatter the blue part of sunlight away from Earth. The sunlight and sky seem to turn red because they are missing this blue light.

visible spectrum

◄ COLOUR SPECTRUM
White light is made of an infinite number of different colours, from violet at one end through to red at the other. This band of visible colours is known as the spectrum. Light at the blue end has a shorter wavelength and higher frequency than light at the red end. Most people can see only seven distinct colours in the spectrum: red, orange, yellow, green, blue, indigo, and violet.

BLUE MAGENTA RED
WHITE
CYAN YELLOW
GREEN

◄ WHITE LIGHT
Just as a prism can split white light into different colours, so lights of different colour can be added together to make white light. If three torches shine red, blue, and green light together, the colours combine to make white light. Yellow light appears where the red and green lights overlap. Magenta occurs where the red and blue lights meet. Cyan appears where the blue light meets the green.

SEEING COLOUR ►

Objects look coloured because they reflect or absorb the different colours in white light. A golf ball looks white because it reflects all the wavelengths of light that fall on it. A lemon absorbs all wavelengths of light except yellow, which it reflects into our eyes. A black helmet absorbs all wavelengths of light and reflects none, and so it looks dark to us.

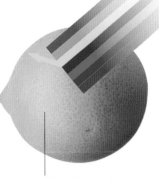

White golf ball reflects all colours in white light

Yellow lemon reflects yellow light and absorbs other colours

Black helmet absorbs all colours and reflects no colours

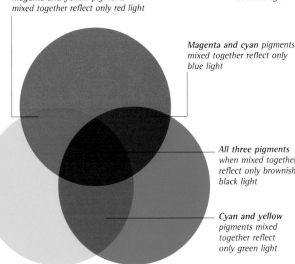

Magenta and yellow pigments mixed together reflect only red light

Magenta and cyan pigments mixed together reflect only blue light

All three pigments when mixed together reflect only brownish black light

Cyan and yellow pigments mixed together reflect only green light

▲ MIXING PIGMENTS

Coloured inks and paints (sometimes called pigments) mix in a completely different way to coloured lights. Each pigment reflects light of a different colour. When two coloured pigments are mixed together, the number of colours they can reflect is reduced. When three pigments are mixed, the mixture does not reflect any colours and appears a brownish black.

CYAN

MAGENTA

YELLOW

BLACK

▲ OVERPRINTING COLOURS

The colour pictures in this book were printed on the paper not once but four times. Each time the paper went through the printing press a further colour was added on top, or overprinted. Any colour should be printable by combining just three coloured inks: magenta, cyan, and yellow. A true black is difficult to make from these colours, however, so black ink is usually added.

▲ BLUE FILTER

Using a camera with a blue light filter totally transforms the colours of this vase of flowers. The white and yellow flowers turn pale blue because the blue filter stops weak red and green light from passing through. The purple flowers turn to darker blues because the filter traps their strong red light. The filter also blocks the green light of the leaves.

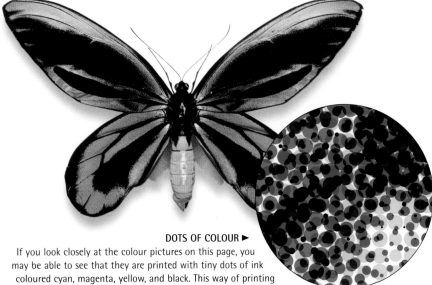

DOTS OF COLOUR ►

If you look closely at the colour pictures on this page, you may be able to see that they are printed with tiny dots of ink coloured cyan, magenta, yellow, and black. This way of printing colour is called colour separation. Any colour or shade of grey can be printed using dots of those four colours. Our eyes and brains blend the dots together and see natural-looking colours instead.

MAGNIFIED DOTS

FIND OUT MORE ►► Cameras **118–119** • Energy Waves **98–99** • Refraction **114**

ELECTRICITY & MAGNETISM

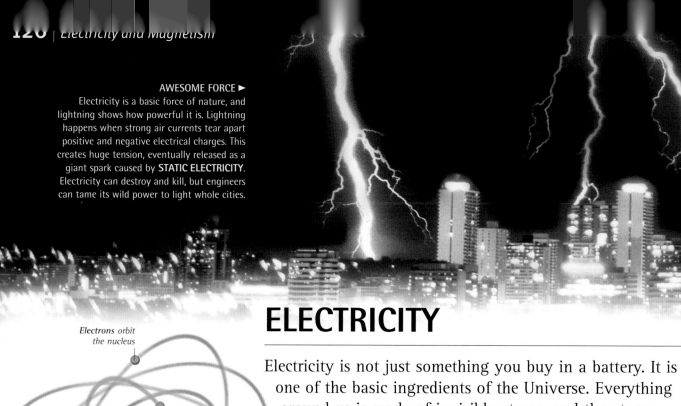

AWESOME FORCE ▶
Electricity is a basic force of nature, and lightning shows how powerful it is. Lightning happens when strong air currents tear apart positive and negative electrical charges. This creates huge tension, eventually released as a giant spark caused by **STATIC ELECTRICITY**. Electricity can destroy and kill, but engineers can tame its wild power to light whole cities.

ELECTRICITY

Electricity is not just something you buy in a battery. It is one of the basic ingredients of the Universe. Everything around us is made of invisible atoms, and the atoms contain particles that carry electric charge. Charge can be positive or negative. Particles with the same kind of charge repel each other, while opposite charges attract. When charges move, we get **CURRENT ELECTRICITY**, which drives much of the modern world.

Electrons orbit the nucleus

Nucleus (centre of the atom) is positively charged

Each electron is negatively charged

◀ INSIDE AN ATOM
Everything in the Universe is made of atoms, and atoms are held together by electricity. In an atom, negatively charged electrons swarm around a positively charged nucleus. A positive charge attracts a negative charge, so electrons rarely escape the pull of the nucleus. As the charges cancel each other out, the atom as a whole has no electric charge.

STATIC ELECTRICITY

We rarely notice the electricity all around us, because positive and negative charges usually balance. However, when objects touch, electrons can hop between them. This may leave each object with a static charge. A comb, for example, can strip electrons from hair, making the hair positively charged, crackly, and fly-away.

◀ ELECTROSTATIC INDUCTION
Charged objects are attracted to uncharged objects. This effect (electrostatic induction) is used in paint spraying. The object to be painted is connected to the ground so it stays uncharged. A spraygun charges the paint, and electrostatic induction pulls the paint onto the object so that every bit gets painted, even the back.

electricity

Metal ball at the centre of the plasma globe is charged with electricity

Electricity moves across the gas-filled globe to the glass wall

FANTASTIC PLASMA ▶
This plasma ball is an exciting demonstration of static electricity. The centre is charged to a very high voltage (electrical pressure), creating electrical stress in the low-pressure gas inside the ball. This tears the gas atoms apart to form particles that shift the charge to the outer wall. When the particles come together to form atoms again, they give out a bright light.

CURRENT ELECTRICITY

Static electricity depends on electrons not being able move around easily, so that charge builds up in one place. But in some materials – mostly metals – electrons can move freely to form an electric current. An electric current is measured by the amount of charge passing a fixed point each second. In most currents, the electrons move more slowly than a snail.

Electric charge is neutralized at the ball's surface

Light is given out as the particles come together

▲ ELECTRIC ACTION

At a rock concert, huge quantities of electricity are controlled by tiny electric currents in microphones to produce deafening sound. Electricity is also used to make lights blaze, and cameras turn the light into electrical signals to create giant images of the musicians above the stage. The whole show is run by electronic computers.

ELECTRIC HEAT ►

When electrons jostle their way through a metal, such as copper, they make the metal hot. The metal may even melt. This could be a disaster, but not when the process is used for joining metal parts by welding. In welding, a rod connected to a low-voltage supply of electricity is touched on to the metal parts that need joining. A brilliant electric arc forms as the tip of the rod is vaporized (turned into gas), and the parts join. Arc welding even works under water, to build and repair pipelines and oil rigs.

FIND OUT MORE ▶▶ Atoms **24–25** • Circuits **128–129** • Heat **80–81** • Light **110–111** • Metals **34–35**

CIRCUITS

An electrical circuit provides pathways along which current can flow to do work. Current is driven by a power source, such as a **BATTERY**. This produces an electrical pressure, known as voltage, which pushes electrons along the wires. Engineers classify circuits into two types. In a series circuit, the same current flows through all the components (such as light bulbs) in the circuit. In a parallel circuit, the same voltage is applied to all the components.

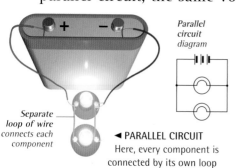

Parallel circuit diagram

Separate loop of wire connects each component

◀ PARALLEL CIRCUIT
Here, every component is connected by its own loop of wire, so the same voltage is applied to all, and each component could have a different current through it. This means one could be removed without affecting the others.

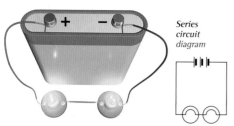

Series circuit diagram

▲ SERIES CIRCUIT
Here, the components are connected by a single loop of wire, so the same current flows through all. None is connected to both sides of the battery, so each could have a different voltage across it. If a component is removed, the current stops. Fairy lights are a series circuit. They stop working if one bulb is loose.

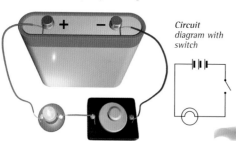

Circuit diagram with switch

▲ CONTROLLING THE CURRENT
Electrical circuits allow us to control electricity so that it does something useful. The simplest electrical control is a switch. An ordinary light switch is used to break a circuit to stop current flowing. Without it, lights would have to stay on all day. Computers are made from millions of electronically controlled switches.

ELECTRICITY IN THE HOME ▼
Modern homes depend on electrical circuits. They carry power to the electric motors in toasters, refrigerators, DVD players, and many other machines. Electricity also supplies heat and light. Forcing current through something with **RESISTANCE** turns electrical energy into heat. The result may be a red glow that makes toast, or the brilliant white of a light bulb.

circuits

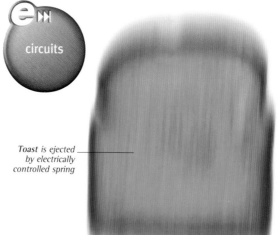

Toast is ejected by electrically controlled spring

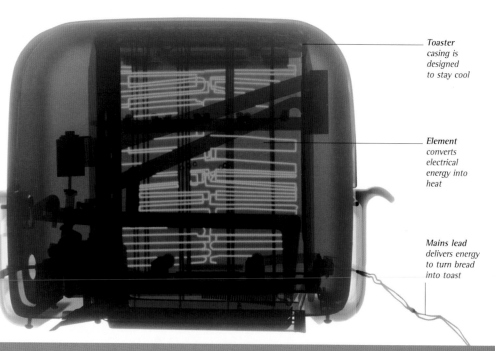

Toaster casing is designed to stay cool

Element converts electrical energy into heat

Mains lead delivers energy to turn bread into toast

PROTECTING CIRCUITS

Fuses protect wires from too much current, which could make them hot enough to start a fire. A fuse is a thin wire in a fireproof casing. Too much current makes it melt, breaking the circuit safely. An earth wire protects people from electric shock if the metal casing of an electrical machine accidentally gets connected to the electricity supply. Instead of current flowing to earth through a person when they touch the machine (and possibly killing them), it flows harmlessly through the earth wire.

The most simple electricity supply uses only two wires, so it needs only two-pin plugs and sockets.

Most electricity supplies use three-pin plugs and sockets. They have an earth wire to make sure the metal parts do not cause electric shocks.

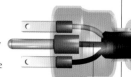

Earth wire is connected to a metal pin, then to a metal rod in the ground

Earth wire is secured by a screw terminal

Fuse is designed to blow (and break the circuit) if the appliance takes too much current

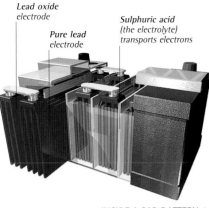

Lead oxide electrode
Pure lead electrode
Sulphuric acid (the electrolyte) transports electrons

INSIDE A CAR BATTERY ▲
A car battery can be used and recharged by the car's alternator for years. It can also deliver the huge current needed to start the car. Its electrodes of lead and lead oxide are immersed in dilute sulphuric acid.

BATTERIES

A battery turns chemical energy into electrical energy. It consists of one or more cells. Each cell contains two electrodes (pieces of metal or another substance), and a chemical called the electrolyte that transports electrons between them. The electrodes are made of different materials, so one gets more electrons than the other. The excess electrons can flow around a circuit connected to a battery as an electric current. Different kinds of battery are used for different purposes. Some, such as torch batteries, can be used only once. Others, including nickel-cadmium (NiCad) batteries and car batteries, can be recharged and used again.

ALESSANDRO VOLTA
Italian, 1745–1827

In 1800, the scientist Volta made the first battery, a pile of silver and zinc discs separated by salt-soaked card. His friend Luigi Galvani had noticed that a frog's leg twitched when in contact with two different metals. Galvani thought it was the frog that produced this electrical effect. Volta showed it was the metals.

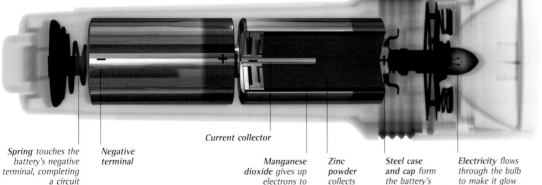

Spring *touches the battery's negative terminal, completing a circuit*

Negative terminal

Current collector

Manganese dioxide *gives up electrons to the zinc*

Zinc powder *collects electrons*

Steel case *and cap form the battery's positive terminal*

Electricity *flows through the bulb to make it glow*

◄ TORCH BATTERIES
Modern batteries have a steel case around a layer of manganese dioxide and a core of zinc powder. Both are coated in a strong alkali (the opposite of an acid) electrolyte. The manganese dioxide gives up electrons to the zinc. The electrons travel to the battery's negative end through a collector. The current stops when the chemicals are used up.

RESISTANCE

Electrons moving along a wire bump into lots of atoms, which slow the electrons down and make them lose energy. This effect is called resistance. It limits the current that can flow when a particular voltage is applied. The energy lost by the electrons makes the wire hotter – hot enough, perhaps, to light a room.

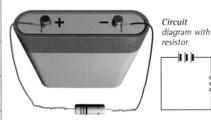

LIGHT WORK ►
Inside a light bulb is a filament – a length of very thin wire coiled up twice so it looks shorter and thicker than it really is. The filament is made from tungsten, a metal that withstands high temperatures. A 60W bulb filament has a resistance of 882Ω, allowing a current of 0.26A to flow when connected to a 230V supply.

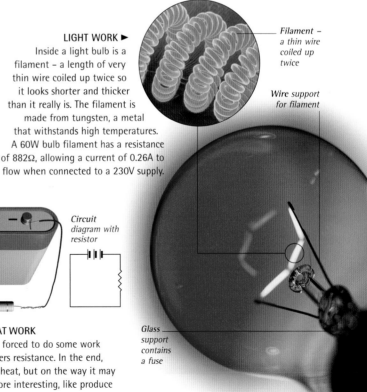

Filament – a thin wire coiled up twice

Wire support for filament

Glass support contains a fuse

Lamp holder connects the bulb to an electricity supply

CIRCUIT TERMS AND SYMBOLS

Voltage The electrical pressure that drives current through a circuit. It is measured between two points, one of which is often the surface of the Earth. Unit: volt. Symbol: V.
A single alkaline battery gives a voltage of 1.4 V.

Current The flow of electrical charge through a circuit. It is measured as the charge per second passing one point. Unit: ampère. Symbol: A.
Starting a car can draw a current of 200 A.

Resistance The property of a circuit that opposes the flow of current. It is measured as voltage divided by current. Unit: ohm. Symbol: Ω.
An ordinary torch bulb has a resistance of about 8Ω.

Power The rate at which energy is consumed or released by a circuit. It is measured as voltage times current. Unit: watt. Symbol: W.
An electric train uses about 3,000,000 W or 3 MW (megawatts).

Ohm's law (First stated by German physicist Georg Ohm.) The current through a circuit is given by the voltage across it divided by its resistance.
A triangle helps to show the three relationships:
$V = I \times R$, $I = V \div R$, and $R = V \div I$,
where V = voltage, I = current, and R = resistance.

V
I   R

Circuit diagram with resistor

▲ RESISTANCE AT WORK
A current is only forced to do some work when it encounters resistance. In the end, it just generates heat, but on the way it may do something more interesting, like produce music. The 33Ω resistor here would draw a current similar to that of a personal radio.

FIND OUT MORE ►► Computers 148–149 • Conductors 130 • Electronics 138–139

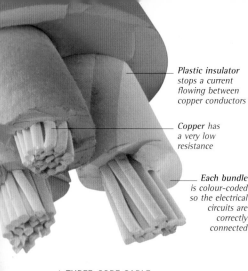

Plastic insulator stops a current flowing between copper conductors

Copper has a very low resistance

Each bundle is colour-coded so the electrical circuits are correctly connected

▲ THREE-CORE CABLE

This cable, shown close-up, is made up of three bundles of thin copper wire, each in a plastic cover. Copper is a conductor and plastic is an insulator. The two materials work together to guide power into electrical appliances.

CONDUCTORS

A conductor is a material that allows electric charge to move through it as an electric current. Usually the charge is carried by electrons, and the conductor is a metal. Metals make good conductors because the outer electrons of their atoms are loosely attached, and the electrons can drift through the metal when a voltage is applied. Some materials have all their electrons firmly fixed in place, so they do not conduct electricity well. A material like this is called an **INSULATOR**.

Arm of pylon carries cable at a safe height above the ground

Insulator made of glazed pottery stops the current leaking away

Ridges prevent a conducting film of water forming during rain

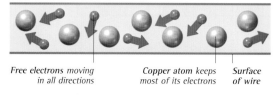

NO CURRENT FLOWING

Free electrons moving in all directions   *Copper atom keeps most of its electrons*   *Surface of wire*

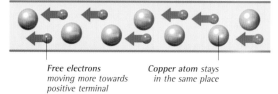

CURRENT FLOWING

Free electrons moving more towards positive terminal   *Copper atom stays in the same place*

◀ CONDUCTOR AT WORK

Usually, the free electrons in a conductor whizz around in all directions. When a voltage is applied, however, they move more towards the positive terminal (on the left here) than in any other direction.

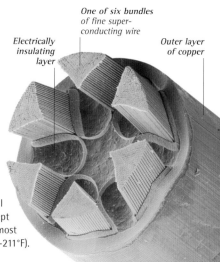

One of six bundles of fine super-conducting wire

Electrically insulating layer

Outer layer of copper

SUPERCONDUCTORS ▶

Some materials, called superconductors, have no resistance at all to the flow of current. Electrons move through them in a more organized way than in ordinary conductors. They are good for jobs like building huge electromagnets for medical scanners, but there is a problem. They only work if they are kept very, very cold. The highest temperature that even the most advanced superconductor can take is –135°C (–211°F).

INSULATORS

Insulators conduct electricity poorly or not at all. Their electrons are bound tightly and will only move if an extremely high voltage is applied. Insulators are essential in electrical engineering, stopping current flowing where it should not. Most common materials, except metals, are insulators, but not all are suitable for electrical engineering. The earliest practical insulators were air, pottery, glass, and rubber. All are still used, but most insulators today are plastics.

◀ HIGH-VOLTAGE INSULATOR

Some insulators have to work under extreme conditions. These electricity supply insulators have to withstand a voltage of 440,000 V (440 kV, or kilovolts) and stop current flowing from the power cables to earth even in the middle of a rainstorm. They also have to take the weight of the cables. Plastics are not good enough for a job like this, but a much more ancient material – pottery – takes the strain with ease.

conductors

FIND OUT MORE ▸▸ Ceramics 55 • Circuits 128–129 • Electromagnetism 134–135 • Metals 34–35 • Plastics 52–53

ELECTRICITY SUPPLY

Electricity comes to homes and workplaces through a huge network of power stations and cables. When the supply fails, we wonder how we ever managed without it. But electricity is not a source of energy, only a way of moving it around. Most of the energy comes from oil, gas, coal, or nuclear fuels. These sources will not last for ever. In future, more of our electricity will come from renewable sources, such as sunlight and wind.

Screens display network activity   *Information about generators at each power station*   *Computers keep track of events and warn of overloads*   *Maps show supply company's network and neighbouring power companies*

electricity supply

▲ NETWORK CONTROL CENTRE
Demand for electricity varies greatly from minute to minute. As electricity cannot easily be stored, supply networks must be ready to switch power to where it is needed at short notice. Control centres like this ensure that generators are started and running by the time a predictable surge occurs – for example, at the end of a television programme.

SOLAR POWER ►
At this solar power station, large mirrors are used to focus sunlight on to a tank of water. The water boils, and the steam that is given off drives an electric generator. Just one square metre (10 sq ft) of sunlight delivers enough power to run a one-bar electric fire. If we could build more solar power stations and capture enough of the Sun's energy as solar power, the world's future electricity supply would be more secure.

HOW ELECTRICITY DELIVERS ENERGY

POWER STATION
At a power station, heat from fuel or a nuclear reactor boils water to make steam. This goes through turbines – machines in which steam rushes past fanlike blades and makes them spin. The turbines turn huge generators, each able to produce enough power for 20 electric trains. The steam is cooled in big towers and turns back into water, which can be reused.

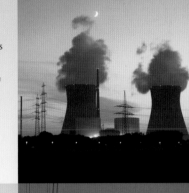

PYLONS
Electricity leaves the power station through metal cables on tall pylons. Power is sent out at a much higher voltage than that used in homes. This is because the higher the voltage, the lower the current needed for the same power. Lower currents allow thinner cables, cutting costs, but the high voltage means that huge insulators are needed for safety.

SUBSTATION
Electricity arriving at a city is not ready to use because its voltage is much too high. Transformers at substations reduce the voltage. At a big substation like this, the voltage is kept quite high because the electricity still has to travel around the city and to nearby country areas. Smaller, local substations will finally reduce the voltage down to the level we use in our homes.

CITIES
Big cities have complex electrical networks with miles of cable and many substations to deliver power to thousands of buildings. Some cities have overhead cables (such as Tokyo, Japan, where an earthquake could damage underground cables). In most cities, however, power travels in heavy cables that carry large electric currents under the streets.

IN THE HOME
When we switch on a cooker or heater in the home, the heat that became electricity at the power station is released again. It can even boil water, just as it did before. But there is a price to pay for the convenience of electricity. Only about a third of the heat from the fuel used to make electricity actually gets to our homes. The rest is wasted or lost on its journey.

FIND OUT MORE ▶▶ Earth's Resources **248–249** • Energy **76–77** • Energy Sources **86–87** • Generators **137**

MAGNETISM

Magnetism is what gives magnets their ability to attract objects made of iron or steel. A magnet creates around itself a region of space with special properties. This region is known as a **MAGNETIC FIELD**. When two magnets come near each other, their fields create forces that attract or repel. The Earth is itself a huge magnet, and the force its field exerts on other magnets makes them point in a north–south direction. This effect is used in the magnetic compass.

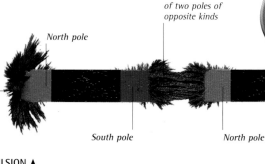

▲ LODESTONE
Magnetism was first discovered in a natural rock called magnetite, or lodestone. Its strange property of attracting iron objects was known nearly 3,000 years ago. Later, Chinese explorers discovered that a piece of lodestone, if able to move freely, would always point north. This led to the development of the compass.

▼ MAGNETIC MATERIALS
The most common magnetic material is steel, an alloy (mix) of iron, other metals, and carbon. Pure iron becomes magnetized in a magnetic field but does not stay magnetic. Steel can make a permanent magnet. Once it is magnetized, it stays magnetized.

Iron filings show how two poles of the same kind repel

North pole

Iron filings show the attraction of two poles of opposite kinds

North pole

magnetism

South pole

North pole

Horseshoe magnet made of steel

Iron filings attracted to both magnetic poles

ATTRACTION AND REPULSION ▲
The two ends of a magnet are always different from each other. The end that points north, if allowed to move freely, is called the north pole. The other end is the south pole. These magnetic poles behave rather like electric charges. Poles of opposite kinds attract each other, while poles of the same kind repel.

MAGNETIZED DOMAINS   UNMAGNETIZED DOMAINS

▲ MAGNETIC DOMAINS
Magnetic materials are made of thousands of tiny magnets called magnetic domains. Before the material is magnetized, all the little magnets point in different directions, so their effects cancel each other out. But a magnetic field can line them up so that they all point in the same direction. This turns the material into a magnet.

Horseshoe magnet bent from a strip of steel

Magnetic domains line up to create magnetic flux

◄ MAGNETIC FLUX
A magnet often has an iron keeper to help it stay magnetized. A magnet creates a magnetic flux, which is rather like a current flowing through an electrical circuit – although in a magnet, nothing actually moves. A high magnetic flux keeps the magnet's domains lined up. Materials that let a lot of flux through are said to have high permeability. Iron has high permeability, so it makes a good keeper.

Iron keeper becomes magnetized and helps the magnet stay magnetized

MAGNETIC FIELD

Every magnet is surrounded by an invisible, three-dimensional magnetic field. A field is a region in which something varies from point to point. In Earth's atmosphere, for example, wind speed and direction vary from place to place. In a magnetic field, the strength and direction of the magnetic effect varies in a similar way. The field is at its strongest near the magnet.

◄ MAGNETIC COMPASS
This modern compass has a pivoted magnetic needle. The needle points not to the Earth's geographical North Pole but to its magnetic north pole. This is in northern Canada, about 1,600 km (1,000 miles) from the North Pole. The magnetic north pole is currently moving northwards at about 40 km (25 miles) a year.

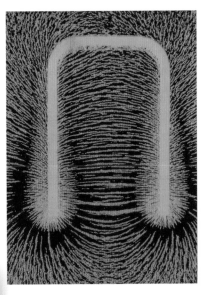

▲ FIELD AROUND A MAGNET
The idea of a field is based on the work of British scientist Michael Faraday (1791–1867) in the early 19th century. He sprinkled particles of iron around magnets to reveal what he called "lines of force" stretching from one pole to another. These helped him to explain many magnetic effects. We now see lines of force as indicating the direction of the field, with their spacing indicating its strength.

Magnetic field visualized as lines of force; red lines from the north magnetic pole

Blue lines from the south magnetic pole

EARTH'S FIELD ►
The Earth acts like a huge magnet, and its magnetic field (the magnetosphere) extends far into space. Although its centre is made of iron, it is too hot to be a permanent magnet. This is because high temperatures destroy magnetism. The field probably comes from molten, charged material circulating inside the Earth.

◄ MEASURING MAGNETISM
Scientists measure magnetic fields with an instrument called a magnetometer. The instrument can also be used to measure the magnetism in ancient rocks. As the rocks formed, they were magnetized by the Earth's field. Rocks of different ages may be magnetized in opposite directions, because the Earth's magnetic field has often reversed. By piecing together records from different places, scientists can work out how rocks have moved in the billions of years since the Earth was formed.

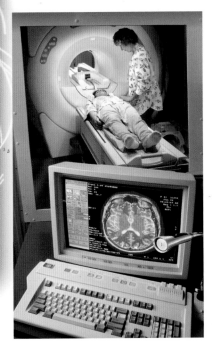

▲ MRI SCAN
Magnetic Resonance Imaging (MRI) lets doctors look deep inside people's bodies. Doctors put patients into a giant magnet and probe them with radio waves. The waves make molecules in the body vibrate. The rate of vibration depends on the type of molecule. Different parts of the body contain different molecules, so each part shows up clearly.

FIND OUT MORE ▶▶ Earth **176** • Earth's Structure **206–207** • Medical Technology **374–375** • Metals **34–35**

ELECTROMAGNETISM

Electromagnetism is a two-way link between electricity and magnetism. An electric current creates a magnetic field, and a magnetic field, when it changes, creates a voltage. The discovery of this link led to the invention of the **TRANSFORMER,** electric motor, and generator. It also, after more than 50 years of further work, explained what light is and led to the invention of radio.

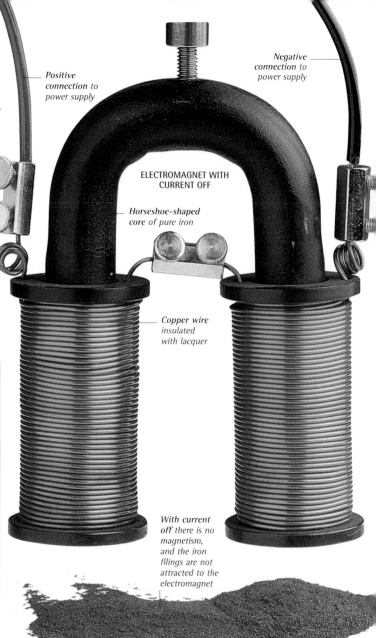

Positive connection to power supply

Negative connection to power supply

ELECTROMAGNET WITH CURRENT OFF

Horseshoe-shaped core of pure iron

Copper wire insulated with lacquer

With current off there is no magnetism, and the iron filings are not attracted to the electromagnet

HANS CHRISTIAN OERSTED
Danish, 1777–1851

It was this scientist who made the first, vital link between electricity and magnetism. Lecturing at the University of Copenhagen in 1820, he connected a battery to a wire that ran near a magnetic compass. The compass needle swung round, and Oersted realized that the current was producing magnetism. He published his revolutionary discovery in 1821.

▲ ELECTROMAGNET AT WORK
Electromagnets make it easy to handle scrap metal. When the current is switched on, it creates strong magnetism that picks up a load of steel. The crane swings round, the current is switched off, the magnetism disappears, and the steel drops where it is wanted.

MAGNETIC LEVITATION ▶
Travellers to Pudong Airport in China can ride at 430 kph (267 mph) on a train with no wheels. This Transrapid system, developed in Germany, uses electromagnets to suspend the train in thin air while a moving magnetic field from electromagnets in the track pushes the train along. Passengers have a smooth ride as the train floats above the magnetic guideway.

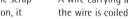

electromagnet

▲ ELECTROMAGNET
A wire carrying a current is surrounded by a magnetic field. If the wire is coiled, the fields from each turn of wire produce a stronger field. If the wire is wrapped around an iron core, the field gets stronger still. An electromagnet can be a single coil (called a solenoid) or bent double, with two coils (as above).

Positive *connection to power supply*

Negative *connection to power supply*

ELECTROMAGNET
WITH CURRENT ON

Wire links the two coils in a series

Current through windings creates a magnetic field

Force of magnetic field overcomes gravity and lifts the iron filings

With current on the iron filings are attracted to the electromagnet

ELECTROMAGNET IN ACTION ▲
When the current is switched on, the electromagnet becomes magnetic. But this does not happen instantly. A magnetic field is a store of energy, and it takes time to feed enough energy into it. This effect, known as inductance, can be used in electronics to control the rate at which things happen.

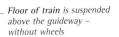

MICHAEL FARADAY
British, 1791–1867

When Oersted discovered that electricity produced magnetism, Faraday wondered if magnetism could produce electricity. In 1831 he showed that it can. He pushed a magnet into a coil of wire and found that a moving magnet created a current. American Joseph Henry discovered this around the same time.

Floor of train is suspended above the guideway – without wheels

Clearance above the guideway is 15 cm (6 in) to allow for snow

Electromagnets glide along 1 cm (⅜ in) below the guideway

TRANSFORMERS

A transformer uses electromagnetism to transfer power between two circuits. Power can take the form of high voltage and low current, or low voltage and high current. Transformers can convert one to the other, but only if the current is alternating, or continually reversing its direction. Low voltages for electronic circuits often come from mains voltage through a transformer.

ELECTRICITY SUPPLY TRANSFORMERS ▶
Transformers are essential for moving electricity around cheaply and safely. This is why modern systems use the alternating current that transformers require. At a generating station, huge transformers step up the voltage to transfer power along cables efficiently. The voltage is stepped down at local substations to a safer level for home use.

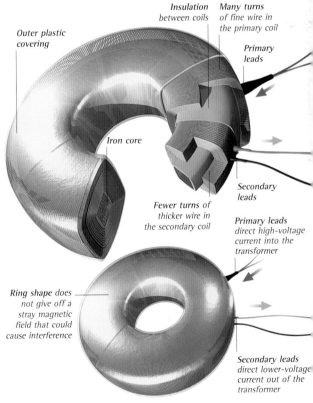

Outer plastic covering

Insulation between coils

Many turns of fine wire in the primary coil

Primary leads

Iron core

Fewer turns of thicker wire in the secondary coil

Secondary leads

Primary leads direct high-voltage current into the transformer

Ring shape does not give off a stray magnetic field that could cause interference

Secondary leads direct lower-voltage current out of the transformer

▲ MAINS TRANSFORMER
A transformer has two coils (windings) of wire round the same iron core, so that they share the same magnetic field. The changing field produced by alternating current in the primary winding causes a voltage across the secondary winding. If the secondary coil has fewer turns of wire than the primary, its voltage will be lower – as in this transformer from a hi-fi.

FIND OUT MORE ▶▶ Circuits **128–129** • Electricity Supply **131** • Magnetism **132–133**

Electric motor functions as a brake to reduce train speed

ELECTRIC MOTORS

Electric motors make things move. They convert electrical power into mechanical power using electromagnetic attraction and repulsion. There are many kinds of electric motor. Small motors can run on batteries to power toys. Larger motors use mains electricity to work kitchen gadgets. Factories use even bigger motors to power heavy machines. Trains and trams also use electric motors to push them along without smoke or noise.

▲ ELECTRIC TRAINS

Japanese Shinkansen ("bullet") trains, which can travel at 300 kph (187 mph), use electric motors. Electric motors are ideal for trains. As well as being clean and quiet, they can be placed all along the train instead of at just one end, as in a diesel locomotive. This helps the train get up to speed more quickly. The motors can also help slow it down, by acting as generators and turning motion back into electricity.

NIKOLA TESLA
American, 1856–1943

Tesla devised the motor most often used in factories. The rotor needs no electrical connection, so the motor is more reliable. To power the motor, Tesla invented a generator that produces three currents. These combine to create a rotating magnetic field that pushes the rotor around.

FLEMING'S LEFT-HAND RULE

An electric current and a magnetic field interact to produce motion in a direction at right-angles to both of them. British physicist Sir John Fleming (1849–1945) devised this simple way to show which way a wire in a motor will move.

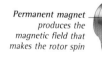

ThuMb shows direction in which wire will Move

First finger shows direction of magnetic Field (north to south)

SeCond finger shows direction of electric Current (positive to negative)

◄ DC MOTOR

This toy car is pushed along by the forces that arise when two magnets are placed close together. One is a permanent magnet. The other is a rotating electromagnet powered by a battery. The battery current flows only one way, so this is called a direct-current (DC) motor. A rotating switch, the commutator, keeps the rotor spinning.

Permanent magnet produces the magnetic field that makes the rotor spin

Commutator keeps the motor spinning by reversing the battery connections every half turn

Coils of wire act as an electromagnet

Permanent magnet alternately attracts and repels the rotor to make it spin continuously

electric motors

HOW A DC MOTOR WORKS ►

Current through coils of wire on a spindle makes the coils into a magnet. Its poles are attracted to the opposite poles of the surrounding magnet, turning the spindle. As the poles line up, the commutator reverses the battery connections. The poles now repel, making the spindle do another half turn. This reverses the connections again to keep the motor spinning.

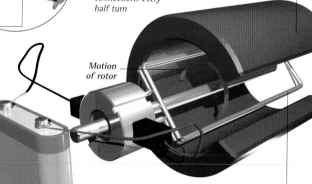

Motion of rotor

N

S

Battery supplies an electric current

Rotor turns into a magnet when the electric current is switched on

FIND OUT MORE ►► Electromagnetism 134–135 • Engines 92 • Generators 137 • Magnetism 132–133

GENERATORS

Generators convert energy from such sources as oil, gas, and wind into electrical energy. Like motors, they use the link between electricity and magnetism. A motor uses electric current to produce a magnetic field that creates motion; but a generator uses the changing magnetic field produced by motion to create an electrical voltage. Generators convert energy with little waste, but much energy is wasted when fuel is burned to work them.

HOW AN ALTERNATOR WORKS 1 ▶
One kind of generator is called an alternator. It produces alternating current – electric current that continually reverses its direction of flow. The alternator has coils of wire mounted on a spindle that turns inside a magnet (usually an electromagnet). The part that turns is called the rotor. As the rotor turns, its wires cut through the field of the magnet. This generates a voltage that drives current through the bulb.

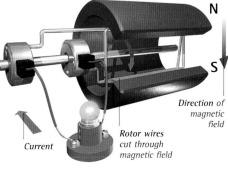

N

S

Direction of magnetic field

Rotor wires cut through magnetic field

Current

DIRECTION OF CURRENT FOR FIRST HALF-TURN

HOW AN ALTERNATOR WORKS 2 ▶
By the time the rotor has gone through half a turn, the direction in which the wires are moving through the field has reversed. This means that the voltage across the wires is reversed, and so is the current through the light bulb. This is how the alternator produces alternating current. Most generators are alternators, because alternating current can be used with transformers to change the voltage of an electricity supply.

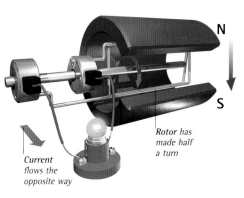

N

S

Rotor has made half a turn

Current flows the opposite way

DIRECTION OF CURRENT FOR SECOND HALF-TURN

WIND TURBINE ▶
A wind turbine is a modern, scientifically designed version of a windmill. Its gently turning blades, which rotate to face the wind, are connected to a gearbox. The gearbox turns a generator at the much higher speed needed for the efficient generation of electricity.

Gearbox speeds up rotation from the blades

Shaft connects the gearbox to the generator

Generator converts wind energy into electrical energy

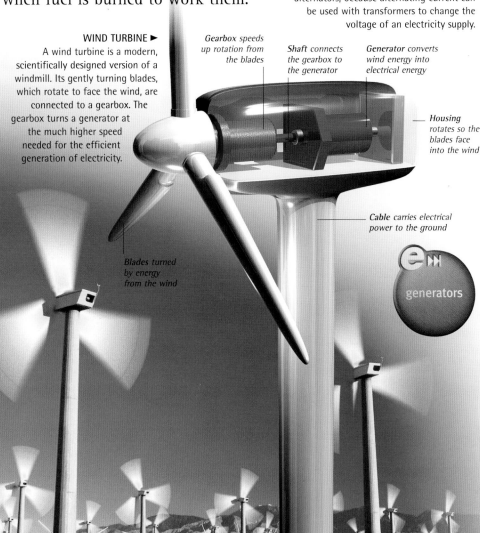

Housing rotates so the blades face into the wind

Cable carries electrical power to the ground

generators

Blades turned by energy from the wind

FLEMING'S RIGHT-HAND RULE

Fleming's rule works for generators, as well as motors. But motors and generators convert energy in opposite directions, so the current direction is reversed.

ThuMb shows direction of the wire's Motion

First finger shows direction of magnetic Field (north to south)

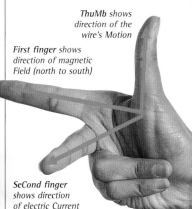

SeCond finger shows direction of electric Current (positive to negative)

◀ WIND FARM
It takes hundreds of wind turbines to match the power of one generator in a power station. This is why they are usually grouped together in large numbers on wind farms. Wind farms take up a lot of space, so in future they may be built out at sea. As winds are powered by the Sun, they will still be blowing and providing energy when fossil fuels, such as coal and oil, have run out. Using wind power also cuts pollution from burning these fuels, and avoids the dangers of nuclear power.

▶▶ Electromagnetism 134–135 • Energy Sources 86–87

ELECTRONICS

Electronics goes beyond simple electricity. Using the **TRANSISTOR** and other **COMPONENTS,** such as resistors and capacitors, electronics allows us to control large electric currents with small electric currents. This opens up a whole new world. Electronics can amplify sound, make radio waves, or handle computer data. **OPTOELECTRONICS** can use light to work a remote control or send messages across the globe.

TRANSISTOR TEAM
In 1947, at the Bell Telephone Laboratories in the USA, John Bardeen (1908–1991, left), Walter Brattain (1902–1987, right), and William Shockley (1910–1989, centre) invented a small, solid device that could amplify electrical signals. They called it a transistor. Until then, the only practical amplifiers were based on fragile glass tubes with a vacuum inside. The team won the Nobel Prize for Physics in 1956.

Microchip *controls the circuit's operation*

Capacitor *carries signals between different parts of the circuit*

Small transistor *handles low-power signals*

Large transistor *controls power to a motor*

electronics

◄ CIRCUIT BOARD
Electronic circuits are built by fixing components into a plastic board that has copper tracks on one side to link them together. The components are secured and connected by melting a metal called solder around their pins. Modern boards may have several layers of tracks. Some boards contain only part of a circuit. These plug into a mother board, which links several daughter boards to form a complete circuit.

Edge connector *plugs into mother board*

Resistor *mounted on pins because it gets hot*

COMPONENTS

Components are parts from which electronic circuits are built. Each component responds to electricity in a particular way. For example, capacitors block steady currents, while resistors let them through. By connecting up the right components, engineers can build anything from a door chime to a computer.

COMPONENT SYMBOLS

Electronics engineers use a visual language that gives every component its own symbol. Symbols are linked to show how a circuit is made.

-⋀⋀⋀-	Resistor
-⌒⌒⌒-	Inductor
	Transformer
-‖-	Capacitor
-⊣⊢-	Electrolytic capacitor
-▶⊢	Diode
	Light-emitting diode
	Bipolar transistor
	Field-effect transistor

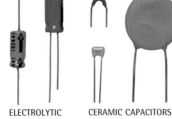

ELECTROLYTIC CAPACITORS   **CERAMIC CAPACITORS**

◄ CAPACITOR
Capacitors store electric charge. They contain two sets of insulated metal plates and can carry signals between two points without letting direct current through. Electrolytic capacitors hold more charge than other types but they need a steady voltage across them to work.

◄ RESISTOR
Resistors control currents and voltages. The current through a resistor is given by the voltage across it divided by its resistance. This means that a resistor can convert a voltage into the corresponding current. On the other hand, if a current is passed through a resistor, it can produce the corresponding voltage.

RESISTOR COLOUR CODE CHART

Resistors are made in standard resistance values. These cannot easily be printed on a resistor as numbers, because the resistor's body is too small. Coloured stripes are used instead. Common resistance values range from 10Ω (10 ohms) to 1MΩ (a million ohms)

Colour	Value	
	0	The first two stripes on a resistor stand for numbers. The third says how many zeros to add to these. The resistor shown on the far left, marked red (2), red (2), and brown (1 zero) has a value of 220Ω. The fourth stripe shows the resistor's accuracy. A gold stripe shows the resistor's actual value could be 5% more or 5% less than 220Ω.
	1	
	2	
	3	
	4	
	5	
	6	
	7	
	8	
	9	

Fourth stripe

Gold	±5%
Silver	±10%

No fourth stripe: ±20%

TRANSISTORS

Transistors make modern electronics possible. They allow tiny electric currents to control much bigger currents. This is called amplification. It makes the link between a small signal that says what we want to do and the electrical power that actually does it. The transistors inside a radio, for example, can amplify tiny signals from the aerial to produce loud sounds.

Thyristor used in lamp dimmers

Transistor for low-power signals

Power transistor used to amplify sound

High-powered transistor can control motors

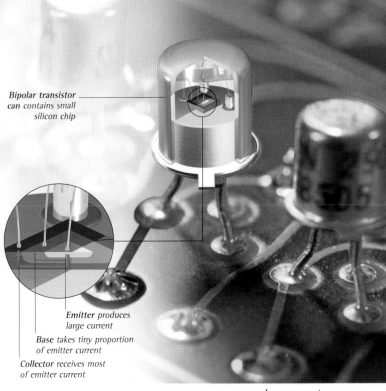

Bipolar transistor can contains small silicon chip

Emitter produces large current

Base takes tiny proportion of emitter current

Collector receives most of emitter current

INSIDE A TRANSISTOR ▶
Although most transistors are now found on microchips, many still come in individual packages like this bipolar transistor. Carefully sealed inside the can is a tiny silicon chip. This has three different regions – the emitter, base, and collector. Each has its own separate connection to the circuit board through a fine gold wire leading to a leg or pin.

◀ TRANSISTORS FOR DIFFERENT JOBS
Transistors come in different sizes. The transistors in computer chips are microscopic. Others may be 2.5 cm (1 in) across – big enough to control a motor or deliver high-energy sound. Those shown here (actual size) include a relative of the transistor called the thyristor. Between them, they can handle everything from light bulbs to radio waves.

HOW A TRANSISTOR WORKS ▶
Current between a transistor's emitter and collector is normally blocked by the base. But when a small current carrying a signal flows from the emitter to the base, lots of electrons can get through the base to form a much bigger current from the emitter to the collector. The current is a copy of the original signal, so the transistor amplifies it.

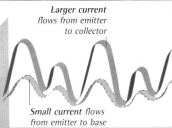

Larger current flows from emitter to collector

Small current flows from emitter to base

OPTOELECTRONICS

Optoelectronics links electronics with light. Its simplest device is the light-dependent resistor (LDR), used in lights that turn on by themselves at night. Light-emitting diodes (LEDs) are used for bike lamps and other signalling jobs, such as controlling a television. DVD players depend on the laser diode, an optoelectronic device that emits the very pure light needed to read the disc.

Light given out as current passes through LED

Positive connection

Negative connection

Plastic casing shaped to act as lens

▲ LIGHT-EMITTING DIODES
A light-emitting diode (LED) is a tiny chip of material in a plastic casing. It emits light when current flows through it. The light can be almost any colour, depending on what the diode is made of. Most LEDs contain the rare element gallium. When this is combined with nitrogen and another rare element, indium, it can give blue light. With arsenic and phosphorus it can give red light.

Metal tracks connect circuit components

Button, when pressed, sends signal to microchip

REMOTE CONTROL ▶
When you change TV channels, an LED in the remote control sends out invisible pulses of infra-red light. These are picked up by a light-sensitive transistor, or phototransistor, at the front of the TV. The pulses are generated by a microchip inside the control. They form a code that tells the television what to do.

Microchip

Batteries power the circuits and the LED

FIND OUT MORE ▶▶ Circuits **128–129** • Elements **22–23** • Lasers **112** • Lenses **115** • Microelectronics **142**

DIGITAL ELECTRONICS

The simplest kind of electronics, known as analogue electronics, works with continuous signals – a smoothly rising and falling sound wave goes through an analogue circuit as a smoothly rising and falling voltage. Digital electronics works differently. Using **SAMPLING**, it converts signals into strings of numbers that can be processed mathematically by electronic circuits.

Battery supplies power to the chip and display

Printed wiring carries control signals to the display

Printed circuit board holds and connects the components

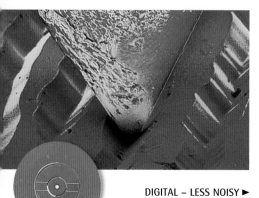

Analogue electrical signal is a smooth, continuous wave

◄ ANALOGUE – SIMPLE BUT RISKY
The wiggly grooves of a record (shown here in close-up with the pick-up stylus, or needle) mimic the shape of the original sound wave. The pick-up produces an electrical wave the same shape as the wiggles. Unfortunately, it copies faithfully everything it finds on the disc – so scratches come out as noisy clicks.

DIGITAL – LESS NOISY ►
The surface of a CD looks nothing like the original sound waves. Before sound is put on a CD, digital electronic circuits convert it into complicated on–off patterns, which are pressed into the plastic as a series of pits. The patterns allow the CD player to play just the music and leave out the scratches.

Digital signal consists of individual steps or pulses

ELECTRONIC CALCULATOR ►
Pocket calculators would not be possible without digital electronics. They handle numbers as electrical signals that are either on or off. This is because they do their maths with transistors – electronic switches that, like other switches, can only turn on or off. Numbers in this form can easily be processed by the calculator's **LOGIC CIRCUITS** to produce the right result.

CALCULATOR

SAMPLING

Before a signal can be handled by digital electronics, it has to be converted into digital form. In sampling, circuits called analogue-to-digital converters make thousands of measurements of the signal each second. The measurements are then converted into binary form, which is a way of writing numbers with just two digits – ideal for on–off electronic switches.

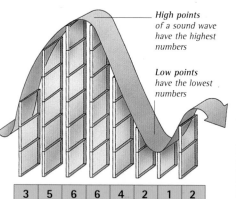

High points of a sound wave have the highest numbers

Low points have the lowest numbers

| 3 | 5 | 6 | 6 | 4 | 2 | 1 | 2 |

▲ FROM WAVES TO NUMBERS
Digital sound starts with a electrical circuit that samples an analogue sound signal about 44,000 times each second. This rate of sampling is needed to capture the highest frequencies (speeds of vibration) in the original sound wave. The circuit stores the value of each sample for just 20 millionths of a second – the time it takes to convert it into binary form.

3	5	6	6	4	2	1	2
011	101	110	110	100	010	001	010

▲ NUMBER CODE
The numbers from the sampled sound wave are handled in binary code, which uses only two digits – 1 and 0. In electronics, the corresponding code is "on" and "off". Samples in this form can be sent as pulses. On a CD, samples use a complex error-correcting code to make the CD more resistant to scratches.

SIGNAL PROCESSING ►
Sampling is not limited to sound. It is used to convert the picture in a camera phone into digital form. The picture is sliced into thousands of tiny square samples, called pixels. A small computer inside the phone works on the samples to produce a simplified picture, which can be sent to someone else more quickly than the original picture. Their phone changes the samples back into a picture again.

LOGIC CIRCUITS

Computers function by breaking big problems down into thousands of smaller ones. They then solve these little problems one by one until the job is done. All the actual work is done by logic gates – circuits that obey the rules of logic. Each logic gate obeys a single, simple rule, such as saying that C is true only if A and B are true. With enough gates, computers can solve any problem that is strictly logical. Most of the millions of transistors at the heart of a computer are in logic gates.

Microchip (in black protective casing) carries out calculations

Keys operate contacts on the circuit board

Contacts send numbers to the microchip

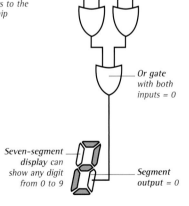

LOGIC CHIP ▲
The large, complex chips in digital circuits are often supported by smaller logic chips. Each of these devices contains only a few logic gates. The gates are made from transistors and resistors formed on the surface of silicon. Shown here is part of a logic chip containing three AND gates. The wires around the edge connect it to the rest of the circuit.

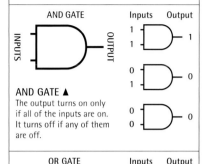

AND GATE ▲
The output turns on only if all of the inputs are on. It turns off if any of them are off.

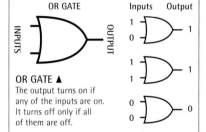

OR GATE ▲
The output turns on if any of the inputs are on. It turns off only if all of them are off.

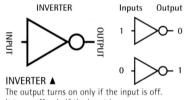

INVERTER ▲
The output turns on only if the input is off. It turns off only if the input is on.

0010 = 2   *0011 = 3*

Inverter output = 0   *Inverter output = 0*

Or gate with both inputs = 0   *Or gate with one input = 1*

Seven-segment display can show any digit from 0 to 9

Segment output = 0   *Segment output = 1*

GOTTFRIED LEIBNIZ
German, 1646–1716

A binary system was known in ancient China, but mathematician Gottfried Leibniz wrote about it in 1703. He was, however, mainly interested in its philosophical meaning because, before computers, it was not of much practical use. Leibniz was also one of the inventors of calculus, a branch of mathematics that is very important in science.

digital electronics

▲ CONVERTING CODES
Logic circuits are good at converting one code into another. The illustration shows part of a circuit that converts the digits 0–9, expressed in binary form, into numerals that people can read. Three gates and an inverter control the lower right segment of a seven-segment number display. The choice of gates and the way they are connected ensures that this segment is switched on for every digit except 2.

▲ ON DISPLAY
Many video recorders and DVD players show numbers and other information using devices like this, called a vacuum fluorescent display. An individual segment glows green when a voltage is applied to one of the wires at the top. The voltage comes from a logic circuit similar to the one shown on the left.

FIND OUT MORE ▶▶ Cameras 118–119 • Electronics 138–139 • Radio 143 • Sound Reproduction 108–109 • Television 144–145

MAKING A MICROCHIP

PURE SILICON
Silicon, which is extracted from sand, is melted in a furnace. A tiny seed crystal of silicon is added to the red-hot molten silicon. A big crystal grows around it and is slowly pulled out, forming a long, sausage-shaped crystal of pure silicon. The silicon sausage is then cut into very thin slices, called wafers. Silicon is used because its electrical properties can be changed by adding impurities.

MASKING AND ETCHING
Each wafer is heated to create a layer of silicon dioxide, which is then given a light-sensitive coating and exposed to ultraviolet light through a mask, like the one shown here. The light hardens the coating in some places. In others, the oxide can be etched away, leaving a pattern of naked silicon ready for doping.

DOPING
The silicon wafers are heated in a furnace full of a gas containing another element, such as arsenic. This process, called doping, adds impurities to the silicon, altering its electrical properties. Different combinations of heat and chemicals form transistors and other components on the silicon. Each wafer goes through many stages of masking, etching, and doping.

QUALITY CONTROL
Computer-controlled testing equipment puts each wafer through a set of tests to make sure that every chip is working properly. Even though operators wear protective suits during the manufacture of the chips, some chips are spoiled by just a speck of dust. Failures are marked so that they can be recycled once the wafer has been cut into chips.

PACKAGING
Metal pins are connected to the chip by welding fine gold wires to the pins and chip. The chip is then encased in a protective plastic or ceramic package, leaving the pins sticking out. The pins are then soldered into a thin plastic board with copper tracks "printed" on it. This connects several chips to form a circuit. Some chips, such as memory chips, are placed in sockets.

MICROELECTRONICS

Microelectronics shrinks circuits to microscopic size. It is the power behind technology from computers to mobile phones. It came from one crucial invention – a way of making transistors and other components on the surface of silicon. A microchip (also called a silicon chip or an integrated circuit) is a complete circuit, just a few millimetres square. Microchips are cheap and reliable, and have made electronic equipment affordable, efficient, and smaller.

◄ X-RAY OF A CD PLAYER
Without microelectronics, the complex calculations that decode the music on a CD player would have to be done by a stack of separate circuits. The CD player would therefore be the size of a refrigerator and very expensive. Instead, the calculations are done by a single chip, so you can buy the player cheaply and slip it in your pocket. Other chips control the player's operation and information display.

Signal processor turns CD code into music

Logic chip helps to control the CD player

micro electronics

ROBERT NOYCE
American, 1927–1990

Engineer Robert Noyce devised the microchip that was the direct ancestor of those used today. He made use of a process invented by his co-worker Jean Hoerni, which created transistors on a flat silicon surface. Noyce realized this process was ideal for making microchips, and worked out how to link the transistors together with a film of metal.

Radio aerial sends out data

Microchip stores identity code

◄ SECRET IDENTITY
This packet of razor blades carries a tiny microchip. The chip has a radio aerial and responds to radio waves by sending out information stored inside it. It can speed up in-store checkouts by removing the need to scan barcodes. Chips of this kind are called radio-frequency identification (RFID) chips. They are already used to identify lost pets.

FIND OUT MORE ►► Ceramics 55 • Chemical Reactions 30–31 • Electronics 138–139

RADIO

Radio depends on electricity and magnetism working together to make waves that travel through space at the speed of light. When you tune in your radio, you hear sounds that have taken a ride, in electrical form, on these radio waves. **DIGITAL RADIO** is a stream of digital data carrying dozens of programmes and travelling on hundreds of different radio waves.

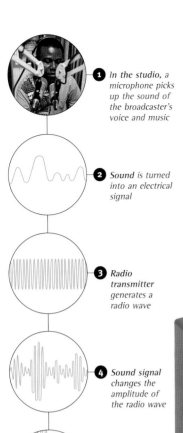

1 *In the studio, a microphone picks up the sound of the broadcaster's voice and music*

2 *Sound is turned into an electrical signal*

3 *Radio transmitter generates a radio wave*

4 *Sound signal changes the amplitude of the radio wave*

5 *Modulated radio wave is amplified (made stronger)*

6 *Transmitter aerial launches the radio wave into space*

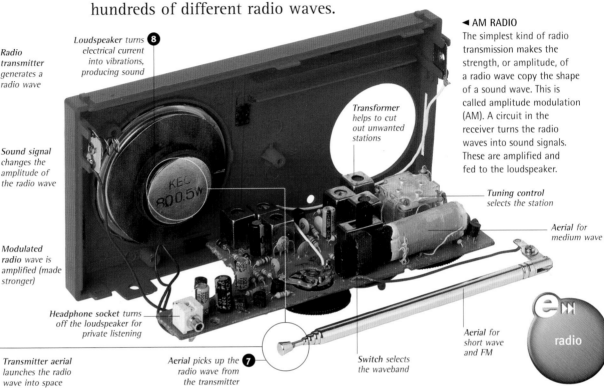

Loudspeaker turns electrical current into vibrations, producing sound **8**

Transformer helps to cut out unwanted stations

Headphone socket turns off the loudspeaker for private listening

Aerial picks up the radio wave from the transmitter **7**

Switch selects the waveband

Tuning control selects the station

Aerial for medium wave

Aerial for short wave and FM

◀ **AM RADIO**
The simplest kind of radio transmission makes the strength, or amplitude, of a radio wave copy the shape of a sound wave. This is called amplitude modulation (AM). A circuit in the receiver turns the radio waves into sound signals. These are amplified and fed to the loudspeaker.

AM SPECTRUM ▶
A single radio transmission contains many different radio waves mixed together. A spectrum sorts them out according to their frequency (speed of vibration). In this AM broadcast, the frequency you tune to is the peak in the middle. The smaller peaks are other frequencies that carry the programme.

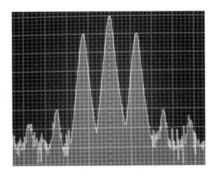

◀ **FM SPECTRUM**
Frequency modulation (FM) varies the frequency of a radio wave, instead of its amplitude, to transmit sound. FM is less open to interference than AM, but reflections of its waves can cause distortion. FM has a wider spectrum than AM and is transmitted at higher radio frequencies.

DIGITAL RADIO

Digital radio converts each programme into digital codes, made of 1s and 0s, representing sound. It then puts together a block of codes from each programme in turn, to form a "multiplex". This huge stream of data is divided between hundreds of radio channels. The receiver picks up all the channels at once, extracts the blocks belonging to the required programme, sticks them together again, and turns them back into sound.

DIGITAL ADVANTAGES ▶
A digital radio tunes to only one set of frequencies, so the radio does not need to be retuned as you move around. Digital audio broadcasting (DAB) also resists interference because it uses so many different frequencies at once. Interference usually affects only a few frequencies, so there is little effect on the programme overall.

FIND OUT MORE ▶▶ Electromagnetism **134–135** • Energy Waves **98–99** • Sound **100–101**

TELEVISION

Television converts the image from a camera lens into a stream of data that can be sent down a cable or broadcast by radio waves. It uses technology that has been developing for over a century. Many homes now get television signals from a satellite orbiting the Earth. The most important recent development, **DIGITAL TELEVISION**, allows people to watch a wider range of programmes and to interact with their TVs.

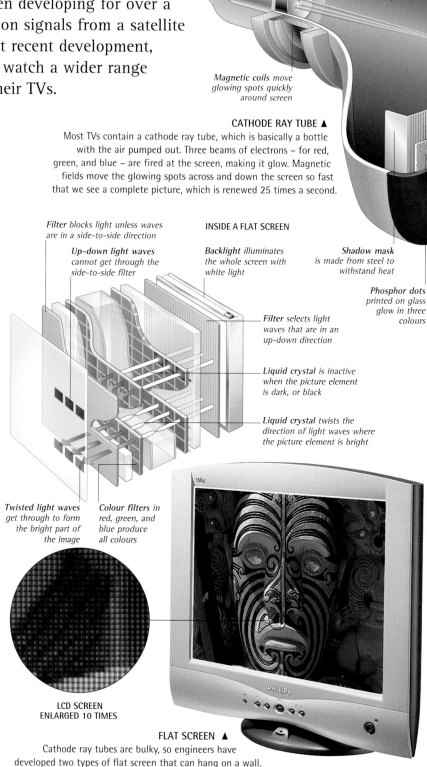

Electron beams from three separate guns

Vacuum inside tube allows electrons to move freely

Magnetic coils move glowing spots quickly around screen

CATHODE RAY TUBE ▲

Most TVs contain a cathode ray tube, which is basically a bottle with the air pumped out. Three beams of electrons – for red, green, and blue – are fired at the screen, making it glow. Magnetic fields move the glowing spots across and down the screen so fast that we see a complete picture, which is renewed 25 times a second.

TRANSMITTING TELEVISION SIGNALS

TELEVISION STUDIO
In the studio, a lens shines an image on to light-sensitive microchips inside a camera. The brightness of each point of the image is read from the chips to form a signal that goes to the control room. It is combined with signals from other cameras to form the complete programme. This is usually recorded, ready for broadcasting at a later date.

INTO SPACE
Giant dishes at Earth stations are used to export television programmes from the country where they were made so that people in other countries can see them. Programmes are beamed up to a satellite, which sends them to a station in the receiving country. Earth stations also send programmes to satellites that broadcast directly to homes.

SATELLITE
Satellites like this can send TV programmes across oceans or into homes. Each satellite is like a television station on a tower 35,800 km (22,200 miles) high. Its position above the Earth never changes, making it easy to beam programmes up to it, and to receive them when the satellite sends them back to a different point on Earth. Live news is often sent by satellite.

TERRESTRIAL TV
Most people still get television signals from towers based on Earth. This is called terrestrial television. The transmitting aerial is placed high up to get its signal to as many people as possible. Terrestrial TV cannot deliver as many channels as satellite television, even when digital technology is used, because it works at lower radio frequencies.

Filter blocks light unless waves are in a side-to-side direction

Up-down light waves cannot get through the side-to-side filter

INSIDE A FLAT SCREEN

Backlight illuminates the whole screen with white light

Shadow mask is made from steel to withstand heat

Phosphor dots printed on glass glow in three colours

Filter selects light waves that are in an up-down direction

Liquid crystal is inactive when the picture element is dark, or black

Liquid crystal twists the direction of light waves where the picture element is bright

Twisted light waves get through to form the bright part of the image

Colour filters in red, green, and blue produce all colours

LCD SCREEN ENLARGED 10 TIMES

FLAT SCREEN ▲
Cathode ray tubes are bulky, so engineers have developed two types of flat screen that can hang on a wall. Plasma screens contain thousands of tiny lamps in which electricity makes gas produce a red, green, or blue glow. Liquid crystal displays (LCDs) use thousands of tiny red, green, and blue filters in front of a white light the size of the screen.

FIND OUT MORE ▶▶ Artificial Satellites **189** • Cinema **120–121** • Colour **122–123** • Radio **143**

▲ STRIPED SCREEN
In many cathode ray tubes, the shadow mask has vertical slots and the screen has its colours arranged in vertical stripes, as shown in this picture of a TV screen enlarged five times. Tubes like this give brighter pictures.

Electron guns arranged in a triangle

◄ CREATING COLOUR
A colour cathode ray tube combines red, green, and blue images from three separate electron guns. These light up tiny phosphor dots printed on the glass screen in groups of three. From a distance, the red, green, and blue dots merge so that the eye sees the image in full colour.

DIGITAL TELEVISION

Ordinary television transmits a new image 25 times a second, even if nothing in the picture is changing. Digital television sends out unchanging parts of the image just once. Receivers repeat these parts until they need to change them. As useless information is not transmitted, there is room for more TV channels.

television

▲ INTERACTIVE TV
Digital television set-top boxes and integrated TV sets contain computers that decode programmes. These can be used to provide other services, such as interactive TV. Viewers press remote control buttons to send commands through their phone line. They can then receive a different view of a football match, prices on a shopping channel, or the World Wide Web.

VIDEO

Video cameras are small television cameras that can capture images on a portable recording device, such as an in-built tape or the camera's digital memory. Television pictures were not often recorded until 1956, when the first practical videotape machine was invented. Now compact cameras are used to capture the latest news from all around the world, family holidays, and special occasions.

HOME CAMCORDER ►
Home cameras have become smaller and smaller as microelectronics has produced better chips and compact image sensors. Early camcorders used cathode ray tubes both to form the image and to display it. Now, most use CCD (charge-coupled device) sensors and colour liquid crystal displays (LCDs).

video

Camcorder is small enough to fit in the palm of a hand

LCD shows what is being recorded

Camera on tripod needs no operator

Broadcasting box collects and processes signals

Solar cells make electricity from sunlight for power

◄ PROFESSIONAL VIDEO
Professional cameras use wider tape to capture more detail. Their high-quality lenses and tough bodies allow broadcasters to record or transmit excellent images from almost anywhere. The latest cameras record images in digital form, which uses half the amount of tape because unnecessary information is not recorded.

CCD SENSOR ►
The heart of a modern video camera is its CCD (charge-coupled device) sensor. This microchip turns an image from the camera lens into a electrical signal. Thousands of tiny, light-sensitive elements on the sensor's surface charge up with electricity when they are exposed to the image. Each element then transfers its electrical charge to its neighbour until the all the charges that form the image have been read out in sequence.

FIND OUT MORE ►► Cameras 118–119 • Microelectronics 142

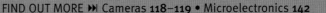

MAKING A TELEPHONE CALL

DIALLING THE NUMBER
Pressing the keys on a telephone sends signals through wires to a local telephone exchange. A numbering plan stored in a computer at the exchange tells the exchange when a complete number has been dialled. If the phone you are calling belongs to a different exchange, your exchange sends signals to other exchanges to set up a route for your call.

THE LOCAL LOOP
Most calls from fixed phones travel to the local exchange through copper wires. Each phone has its own line card – a circuit that is permanently connected to the phone. This responds with a dialling tone when you pick up the phone. It also converts your call into electrical pulses, so that it can be handled by computers that route the call.

OPTICAL FIBRE CABLE
Nearly all calls between big cities now travel as laser light through thin glass fibres, called optical fibres. The laser switches rapidly on and off to send out high-speed digital codes. Clever coding squeezes as many different calls as possible into each optical fibre, but allows them to be sorted out again when they arrive at the next telephone exchange.

MICROWAVE LINK
Some calls, particularly those to isolated areas, make part of their journey by riding on a beam of microwaves. These very short waves are focused by a dish-shaped reflector on a tower and sent from point to point in a straight line. Microwave links are quick and cheap to set up, as there is no need to dig tunnels or erect poles to carry fibres or wires.

HOME AT LAST
Eventually the call reaches the local exchange that handles the telephone you have dialled. There, it is directed to that phone's line card and the signal is changed back to analogue form. A pulsing current sent down the line rings the phone. When the phone is picked up, a switch in the receiver completes a circuit that cuts off the ringing current and connects the call.

TELECOMMUNICATIONS

Telecommunications began more than 160 years ago, with telegraphs and telephones working through wires. We still use wires – known as landlines, or the fixed network – but now a web of **OPTICAL FIBRES**, radio, and satellite links connects every place in the world. You can control this machine yourself, simply by picking up a telephone.

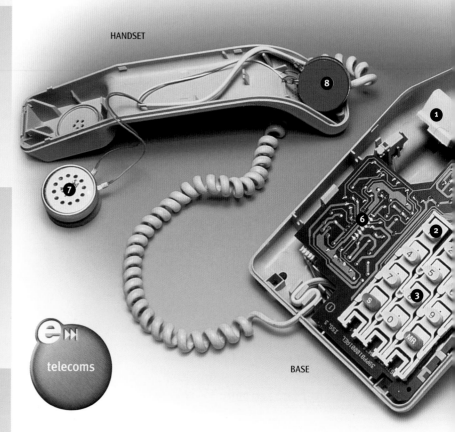

HANDSET

telecoms

BASE

OPTICAL FIBRES

Light can be used to send signals – for example, with a torch. However, light sent through air is stopped by objects in its path. An optical fibre traps light inside a thin strand of glass. The light is reflected back from the surface of the glass and cannot escape. An optical fibre can direct pulses of laser light for many miles. Some fibres amplify the light to send signals around the world.

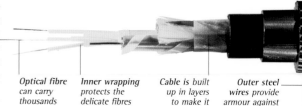

UNDERSEA CABLE ▶
Delicate optical fibres are heavily protected when laid on the sea bed. Each cable contains several fibres. Some may not be needed at first, but these "dark fibres" will be brought into use when calls on the cable route increase.

Optical fibre can carry thousands of calls

Inner wrapping protects the delicate fibres

Cable is built up in layers to make it flexible

Outer steel wires provide armour against shark bites

MOBILE COMMUNICATIONS

Mobile communications allows direct radio contact with people on the move, connecting them immediately, even in an emergency. Radio was originally invented more than a century ago as a way of communicating with ships. Transistors and microchips now make it possible to get powerful radio equipment into cars and small boats. Unlike a **MOBILE PHONE**, mobile communication does not rely on the fixed telephone network.

SOME MAJOR OPTICAL FIBRE LINKS		
NAME	*DISTANCE*	*CAPACITY**
FLAG FEA (Japan–UK)	14,000 km (8,700 miles)	163,840
Japan–US Cable Network	10,500 km (6,500 miles)	655,360
FLAG FA-1 (UK–USA)	7,000 km (4,350 miles)	1,310,720
Atlantic Crossing (UK–USA)	27,000 km (4,350 miles)	10,737,418

* equivalent simultaneous phone calls

HOW A TELEPHONE WORKS

1 *Switch* operates when the handset is picked up, telling the local exchange you want to make a call

2 *Pressing a key* sends out two tones, identifying which row and column that key is in

3 *Correct sequence of keys* must be pressed or else the exchange will not recognize the tones and route the call

4 *Copper wires* inside plastic cable carry speech to and from the exchange in the form of electrical waves

5 *Electronic circuits* adjust and amplify (make louder) speech signals so they are easier to hear

6 *Other circuits* use a pulsing current from the exchange to work a loudspeaker, making the phone ring

7 *Speaker* in the earpiece vibrates and recreates the sound of the person's voice on the other end

8 *Disc* in the mouthpiece vibrates, copying the vibrations of your voice as an electrical signal that goes to an amplifier in the phone

◄ EMERGENCY SERVICES
Fire, police, and ambulance services all have their own radio networks. Some can handle data, such as maps, as well as speech. Messages are sent out from a central transmitter to several vehicles, all of which use one channel to reply – so communications have to be short, and are not private.

MOBILE PHONES

Mobile phones use radio and landlines to transmit calls. A call is picked up by a nearby base station, which passes the call through landlines to another base station or to a fixed telephone. Base stations are low-powered, so they do not interfere with each other, allowing millions of people to talk using only a few frequencies.

Camera lens for taking photographs

MINIATURE MIRACLE ►
A pocket-sized mobile phone contains more than one computer as well as a microwave radio transmitter and receiver. When you switch it on, the phone finds the nearest base station and logs on so that the system knows where it is. If you start to move out of range, the phone finds another base station and, if necessary, retunes itself.

Liquid crystal display shows pictures that are sent and received

Keypad sends signals to the phone's computer

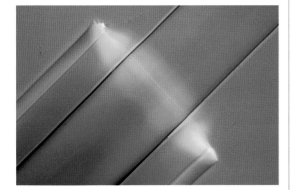

▲ STRUCTURE OF A FIBRE
Optical fibre glass is so pure than you could see through a mile of it. It is even more transparent to the invisible laser light that it carries. The inner core is covered with a layer of less heavy glass, and the light is reflected (and so trapped) where the two kinds of glass meet. A plastic coating on the outside makes the fibre tougher and easier to handle.

cellcoms

◄ AERIAL IN DISGUISE
As the number of people using a mobile network increases, it has to be divided into smaller regions that each contain their own base station. This means more aerials, some of which are disguised as trees.

FIND OUT MORE ►► Electronics 138–139 • Radio 143 • Telecommunications 146

COMPUTERS

A computer is an electronic machine that obeys instructions telling it how to present information in a more useful form. Its **HARDWARE** is the actual machine, including parts such as the screen. The hardware stores instructions as a computer program, or **SOFTWARE**. Hardware and software work together to change basic data into something people can use. A long list of numbers, for example, can be presented as a colourful picture.

HARDWARE

The body of the computer and the devices that plug into it, such as the keyboard, are called its hardware. The body contains the parts that store and process information. These include the hard disk, which stores programs and files permanently. Faster, electronic memory holds the data being processed. A chip called the processor does most of the work, helped by others that do special jobs, such as displaying images.

Screen has more than two million separate coloured spots

Web cam can send out pictures over the Internet

Monitor uses new technology to make it thin and light

computers

PERSONAL COMPUTER ▶
Today's personal computer may have a big colour screen, loudspeakers, and possibly a camera. It is thousands of times more powerful than computers built around 30 years ago, which were so bulky they could fill a whole room. This improvement is due to the microprocessor (invented in 1971), which replaced hundreds of separate computer parts with a single microchip.

Speakers driven by sound circuits inside the computer

Processor and hard disk hidden inside the computer's base

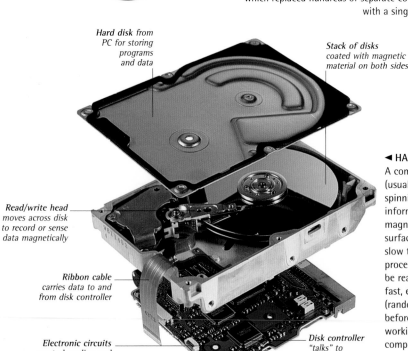

Hard disk from PC for storing programs and data

Stack of disks coated with magnetic material on both sides

Read/write head moves across disk to record or sense data magnetically

Ribbon cable carries data to and from disk controller

Electronic circuits control reading and writing data to disk

Disk controller "talks" to computer

◀ HARD DISK
A computer's hard disk (usually several disks spinning together) stores information permanently as magnetic spots on the disks' surface. The hard disk is too slow to keep up with the processor, so all data has to be read from the disk into fast, electronic RAM (random-access memory) before use. RAM chips stop working as soon as the computer is switched off, so new data needed again must be saved on the hard disk.

BITS AND BYTES

Bit	Smallest unit of information
Byte	Eight bits
Kilobyte	1,024 bytes
Megabyte	1,024 kilobytes
Gigabyte	1,024 megabytes

Computers store and process information in the form of bits. A bit can stand for one of just two different things, such as "yes" and "no". For example, a hard disk stores information as magnetic spots with the magnetism pointing up or down. When bits are grouped together, they allow more choices. Every extra bit doubles the possibilities, so a byte can stand for 256 different things. A modern PC can handle billions of bits per second and store up to 120 gigabytes (over 1,000 trillion bits) on its hard disk.

SOFTWARE

A computer needs software, which consists of sets of instructions called programs, to tell it what to do. Different programs allow people to write letters, play games, or connect to the Internet. Software is written in special languages by computer programmers. The languages are then translated into instructions that can be understood by the computer's microprocessor.

▲ COMBINING IMAGES
To put the bee on the flower, the program holds both in memory, together with data describing the outline of the bee. It then finds all points of the flower that lie within this outline and replaces them with matching data from the image of the bee.

▲ COMPUTER LANGUAGE
This screen shows a small part of a program that can change images. It is written in a computer language called C, which must be translated into a code before the computer can use it. Computer languages have strict rules and it is easy to make mistakes. Programs therefore go through many cycles of correction and testing before use.

▲ CHANGING COLOURS
To change yellow to blue, a graphics program looks through all the codes that represent the image. Whenever it finds the code for yellow, it changes it to the code for blue.

CD tray slides out so CDs can be inserted and played

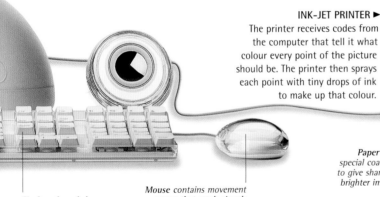

INK-JET PRINTER ►
The printer receives codes from the computer that tell it what colour every point of the picture should be. The printer then sprays each point with tiny drops of ink to make up that colour.

Paper has special coating to give sharper, brighter image

Screen displays image of what is being printed

Keyboard sends key codes to the computer

Mouse contains movement sensor that sends signals to the processor

◄ WEARABLE PERSONAL COMPUTER
Not all computers are used by people sitting at desks. Some can be worn like a pair of glasses, such as this minicomputer. An image generated by the computer (carried in a pocket or on a belt) is projected straight into the wearer's eye. Computers like this leave the wearer's hands free to do another job. For example, technicians servicing an aircraft can have the plane's service manual displayed to them as they work.

VIRTUAL WORLD ►
This flight simulator game requires complicated hardware and software. The virtual world is stored as a list of numbers specifying all the points in the world and how they are linked. The computer works out how this would look and generates numbers specifying the colour of every point on the screen at a rate of 25 times a second.

FIND OUT MORE ►► Colour **122–123** • Digital Electronics **140–141** • Internet **152–153** • Microelectronics **142**

COMPUTER NETWORKS

A computer network links several computers. Together, they can do much more than a single computer. Office networks allow people to work as a team. At home, a network allows two or more computers to share a printer. Local-area networks (LANs) like these can be linked into wide-area networks (WANs) that may cover a country or span the globe.

LINE NETWORK

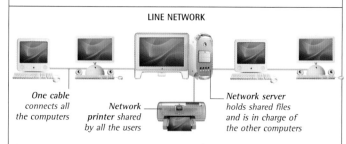

One cable connects all the computers

Network printer shared by all the users

Network server holds shared files and is in charge of the other computers

A local-area network is usually connected with cables similar to those used for telephones. The most popular network system is Ethernet, which allows communication at up to 12 megabytes per second. An Ethernet network can be a straight line network (also called a "bus" network), or a star network. One computer, called the network server, controls communications within the network.

STAR NETWORK

RING NETWORK

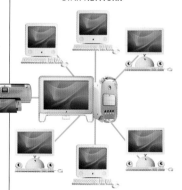

In a star network, each computer is connected to the server by its own cable. It is more reliable than a line network because a broken wire affects only one or two computers.

A ring network has its computers in a loop. Data travels right around the ring back to the device that sent it. The device it was sent to changes part of the data to show it has arrived safely.

◀ ROUTERS

Networks often contain devices that make them work better. In large networks, the flow of information can be controlled by routers and bridges. Routers send data to where it is needed. Bridges link two smaller networks and can prevent parts of one network seeing data from parts of the other. The simplest device, a hub, connects several computers to a shared resource, such as a printer.

COMPUTER NETWORK ▶
This diagram shows how the US National Science Foundation's huge computer network is connected. The white lines form the network's "backbone", which covers the whole of the United States by linking many smaller, regional networks. Using these connections, a scientist anywhere in the USA can make use of the Foundations' big, expensive supercomputers hundreds of miles away.

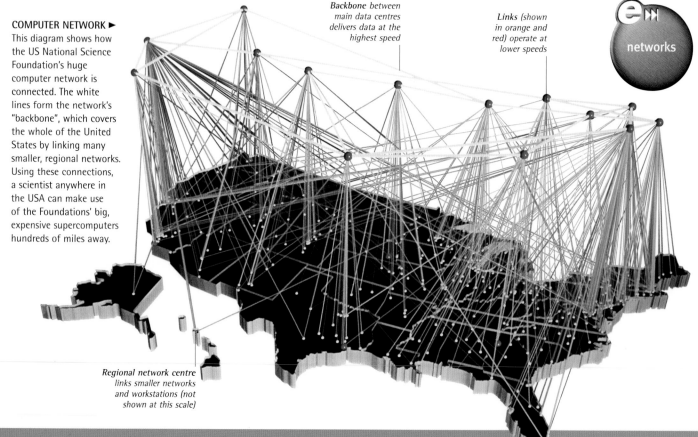

Backbone between main data centres delivers data at the highest speed

Links (shown in orange and red) operate at lower speeds

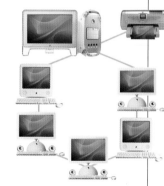
networks

Regional network centre links smaller networks and workstations (not shown at this scale)

SUPERCOMPUTERS

A supercomputer is a computer that works thousands of times faster than the best desktop personal computer. It does this by combining many processors together so that they can all work on different parts of a problem at the same time. People use supercomputers when they want millions of detailed results very quickly. These monster machines are now at work on all kinds of things, from predicting storms to designing next year's cars.

▼ FORECASTING A HURRICANE

Supercomputers can predict the track of a spiralling hurricane in time to warn people of its approach. The giant machines are fed with thousands of atmospheric measurements to produce pictures showing how the weather will develop over the next few hours. Changes in a small part of the atmosphere can have a big effect on the weather. Doing the maths fast enough to keep ahead of events can be done only by a supercomputer.

super computers

HARDWARE ▲

The first serious supercomputer was produced by US engineer Seymour Cray in 1976. This shows a later model from 1982. The round shape came from the need to keep all wires as short as possible, while the padded seats around the base concealed the cooling system needed to stop the machine catching fire.

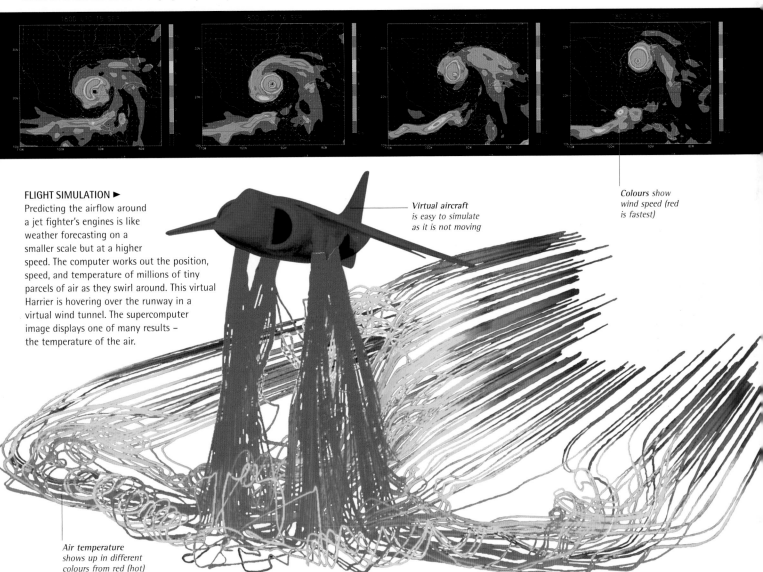

Colours show wind speed (red is fastest)

FLIGHT SIMULATION ►

Predicting the airflow around a jet fighter's engines is like weather forecasting on a smaller scale but at a higher speed. The computer works out the position, speed, and temperature of millions of tiny parcels of air as they swirl around. This virtual Harrier is hovering over the runway in a virtual wind tunnel. The supercomputer image displays one of many results – the temperature of the air.

Virtual aircraft is easy to simulate as it is not moving

Air temperature shows up in different colours from red (hot) to blue (cool)

FIND OUT MORE ▸▸ Computers 148–149 • Computer Networks 150 • Weather 238–239

INTERNET

The Internet is a computer network covering the whole world. We can use it to search through three billion pages of the **WORLD WIDE WEB**, or to keep in touch with people by **EMAIL**. Unlike other networks, the Internet is not under anyone's control. It is held together by a set of standards, or rules, that set out how computers connected to it should exchange information.

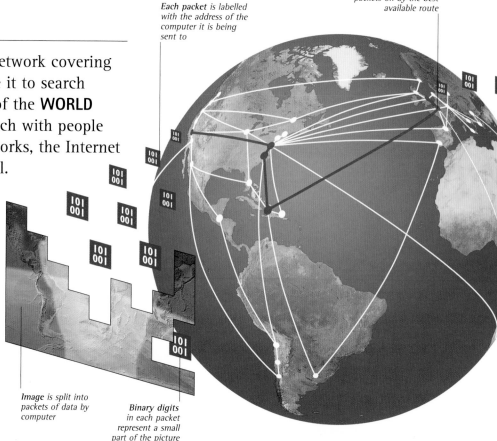

Each packet is labelled with the address of the computer it is being sent to

Routers read the address and pass packets on by the best available route

Image is split into packets of data by computer

Binary digits in each packet represent a small part of the picture

PACKET SWITCHING ►
Data is sent across the Internet as small packets. Each packet travels by the best route available, avoiding busy or broken links. Computers linked to the Internet handle data using agreed protocols (procedures). The most important are TCP (transfer control protocol) and IP (Internet protocol).

◄ INTERNET ACCESS
People on the move or without their own computer can connect to the Internet at an Internet café. They pay a small fee to use one of the café's machines. Public wireless points make it even easier to get connected. A wireless-enabled computer can access the Internet through a radio link in the café or another public place. You can also connect to the Internet through a mobile phone.

Palmtop computer can slip in a pocket

Screen is large enough to read simple web pages

Shopping page shows goods and prices

CONNECTING TO THE INTERNET ►
Your computer does not connect directly to the Internet. Usually, it connects through telephone wires to an Internet service provider (ISP). Your computer is linked with one of theirs, which has a unique address on the network. Anything you want to see goes to this address first, then to your computer.

5 *Website computer (web server) sends the web file back to your ISP by a similar route*

6 *ISP passes the file on to your computer, and your browser converts it into an on-screen page*

4 *Series of routers and communication links send the number around the world*

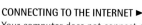

1 *Home computer connects to ISP through telephone line*

2 *Address of the website you want goes to a computer at the ISP*

3 *Domain name server translates the address of the computer that holds the website into a number*

INTERNET SHOPPING ▲
With a hand-held computer like this, a shopper can order goods by Internet from anywhere in the world. People still like to browse around actual shops, but if they need to compare prices or get something unusual then a computer can offer more than the biggest store.

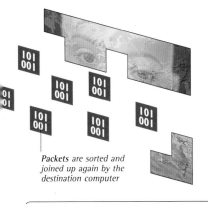

Packets are sorted and joined up again by the destination computer

HISTORY OF THE INTERNET

1963	Arpanet created to link US research computers
1970	Packet switching first used
1978	TCP/IP protocols established for communication and data exchange
1983	TCP/IP made compulsory, effectively creating the Internet
1990	World Wide Web protocol created

EMAIL

Email (short for electronic mail) is an electronic postal service. It was invented in 1971, and works on any computer network but is now mostly used on the Internet. A message starts from and ends up at a mail client – the program used to write and read emails on a computer. In between, it is handled by one or more mail transfer agents. These are computers that pass the email on until it gets to the electronic mailbox specified by its email address. An email is not private because it may pass through many computers before it arrives, giving other people a chance to read it.

EMAIL ADDRESS

The part of an email address after the @ is the domain name, which tells mail transfer agents where to send the email. The part to the right of the dot (here, "com", for a company) is the top-level domain. To the left of the dot is the company name. The complete domain name identifies a particular mail server. The name to the left of the @ identifies the user of a particular mailbox on that server.

Separator symbol indicates the start of the domain name

sunshine@dk.com

Username chosen by user, selecting their mailbox on the server

Domain name used by the operator of the mail server

WORLD WIDE WEB

The World Wide Web is a collection of billions of files held in a huge number of computers, all linked to the Internet. The files may contain words, pictures, sounds, or almost anything else. They are linked to each other by hypertext – a way of making a word or picture in one file call up another file anywhere in the world.

Internet

TIM BERNERS-LEE
British, 1955-

MARC ANDREESSEN
American, 1971-

The World Wide Web owes its existence to British scientist Tim Berners-Lee. He worked at a research lab called CERN in Switzerland. Frustrated by the difficulty of working with information scattered all over the Internet, he developed hypertext software to link it up. The result was the World Wide Web, which came into public use in 1991. Marc Andreessen created the first easy-to-use web browser in 1993.

Web address box shows site, file, and method of handling (such as HTML)

Page is created by browser from the HTML file

Title bar shows the title of the page

Browser logo is animated while the page is loading

WWW ADDRESS

A web address or URL (short for uniform resource locator) tells the browser where to find a file and how to treat it. A slash (/) marks the start of the file's name. The "http" says the information is to be handled as hypertext.

http://www.dk.com/web-server.htm

Protocol name says how the file is to be treated

File path (after the first slash) indicates the required file

Domain name indicates the name of the website and server

File extension shows the type of file (here, HTML)

DK dk.com

Discover more at *dk.com* - click a flag to visit a chosen country

United Kingdom · United States · Australia & New Zealand · India · Germany · South Africa · Canada

BROWSING AND SEARCHING ▲
Websites such as this are written in a computer language called hypertext mark-up language (HTML). A computer program called a web browser translates HTML into a neat layout of text and pictures on your screen. To see a web page, you type its address into the address box. If you need to find pages about a particular subject, you can use a search engine. Search engines keep a constantly updated index of every word in billions of documents. They produce a list of pages that might be suitable, and you click on any pages you want to see.

▲ HTML CODE
This screen shows the layout on the left in hypertext mark-up language (HTML). One part of the code indicates that a word or phrase is a link to another document. Clicking on the link takes you straight to that page.

FIND OUT MORE ▶▶ Computer Networks **150** • Computers **148–149** • Telecommunications **146–147**

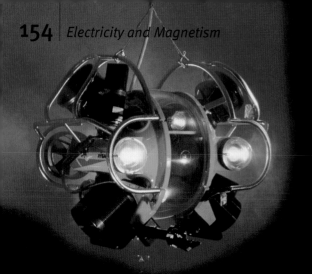

ROBOTS

Robots are machines that behave a bit like people, and can perform difficult or repetitive tasks. **HUMANOID ROBOTS** even look like people, and can move about and do different jobs without human help. Many robots cannot quite do this. Some need people to guide them, or do just one specific job. Some cannot move. But even these robots will help to improve the movement, senses, and intelligence of robots yet to come.

robots

▲ UNDERWATER EYE
Robots are good at exploring the oceans. They do not need air, and can survive deep water pressure that would crush a human diver. Some are little submarines that can gather data unaided. Others, like this Hyball ROV (remotely operated vehicle), are attached to a ship and controlled by a human. They are ideal for inspecting oil rigs.

HOME HELP ▶
Domestic robots work in ordinary homes. Some do only one repetitive job, such as mowing grass or vacuuming floors. Some, such as this ER2, can respond to words and alert police or relatives if something goes wrong in the house. This makes them useful for elderly people.

Arm can pivot up and down and extend telescopically in and out

Air lines feed compressed air to motors that move the joints

Cable supplies heavy current needed for welding

Welding head melts metal to join parts

Robot built out of parts from a construction kit

Control panel for selecting different game programs

Ball emits infrared signals so robots can locate it

Rear wheel turns to change direction

Striker flicks ball forwards

▲ ROBOT FOOTBALLER
By 2050, robots could be taking on football's World Cup holders – and winning. This is the aim of Robocup, an international project to develop robots with the many skills needed to play football. There is still a long way to go. This little robot, built from a construction kit, can get possession of the ball but only kick it into a goal defended by a single opponent.

INDUSTRIAL ARM ▶
About a million robots work in factories worldwide. Most are computer-controlled mechanical arms fixed to the floor. Industrial robots do jobs like welding car bodies and packing goods into boxes. They cannot see, so everything they need has to be in exactly the right place. Unlike human workers, they never get tired and rarely make mistakes.

HUMANOID ROBOTS

Robots that look like people are called humanoids.
They are harder to make than fixed arms or machines
on wheels, because they have to balance and walk on
two legs. They also need advanced senses, intelligence,
and power systems that will keep going all day.
However, they can use tools and fit in spaces
designed for humans, so engineers are working
hard to develop them.

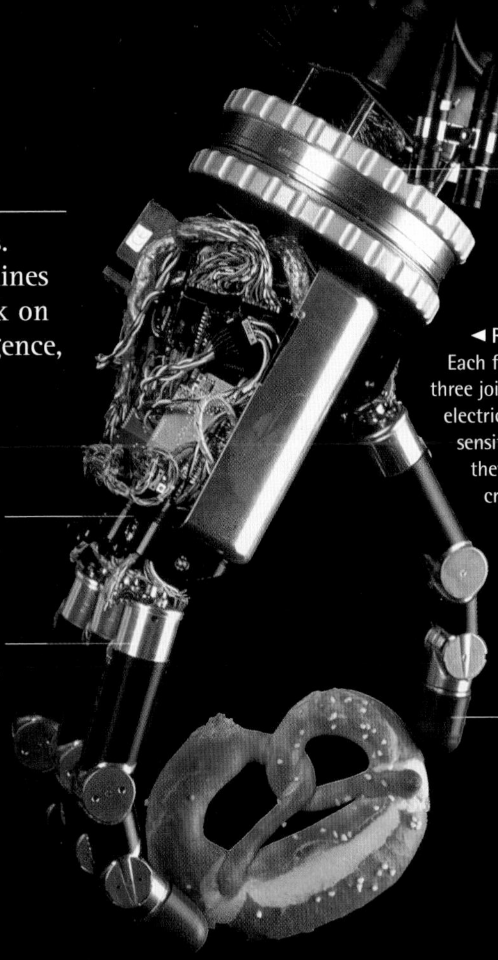

Computer sends
signals to the
hand, adjusting
the strength of
the grip

◄ ROBOTIC HAND
Each finger of this robot hand has
three joints and is moved by its own
electric motor. The fingers also have
sensitive tips that can tell how hard
they are gripping. This stops them
crushing delicate objects or
dropping heavy ones. Artificial
hands are used for research
into the way real hands work,
and to help people who have
lost their hands.

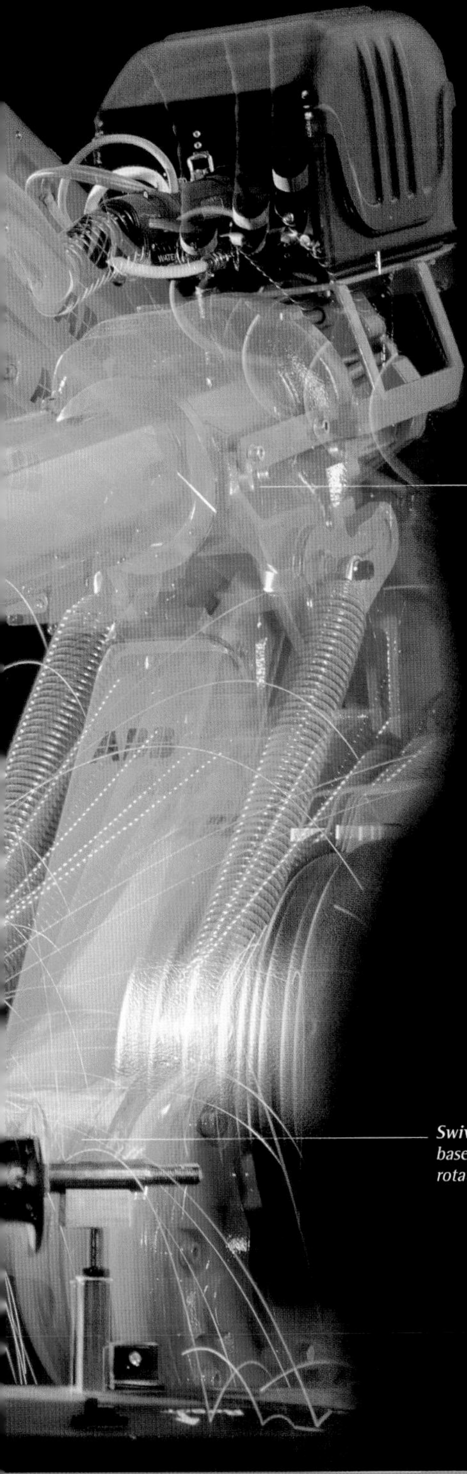

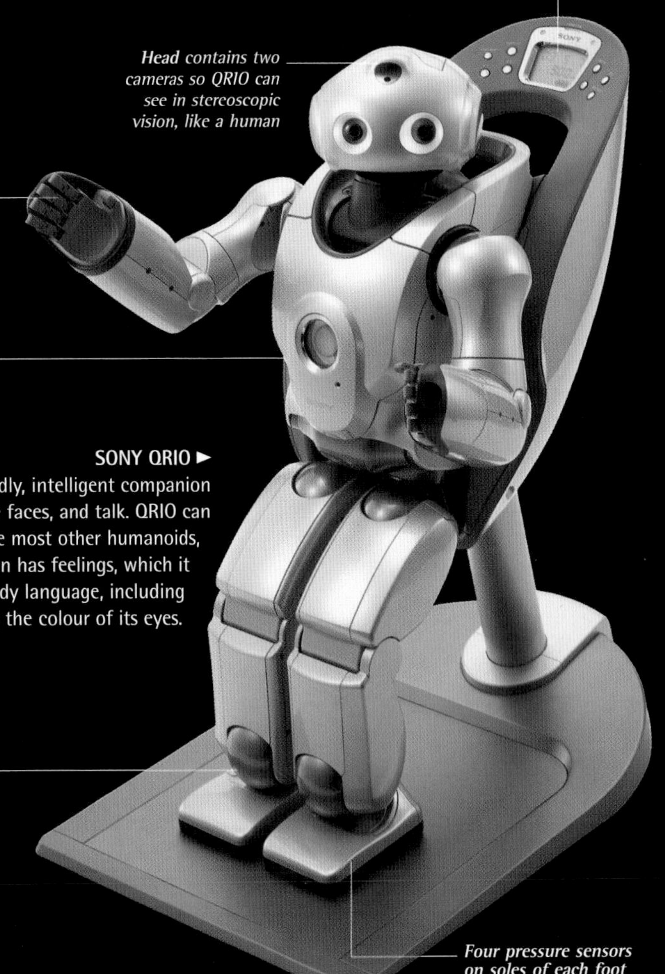

Each finger is
controlled by its
own electric motor

Hand is jointed,
just like a
human hand

Sensors in finger
tips send signals
back to stop
further pressure

Battery recharger
charges QRIO's in-built
battery cells, so it can
run for over an hour

Elbow joint allows
robot arm to move
up and down

Head contains two
cameras so QRIO can
see in stereoscopic
vision, like a human

Limbs and joints are
given movement
commands by QRIO's
central computer

Lightweight body built
from high-strength
magnesium alloy

SONY QRIO ►
Sony's experimental QRIO is a friendly, intelligent companion
and helper. It can dance, recognize faces, and talk. QRIO can
walk on uneven surfaces and, unlike most other humanoids,
get up again if it falls over. It even has feelings, which it
expresses through words and body language, including
changing the colour of its eyes.

Swivel joints at
base allow robot to
rotate in a circle

Jointed body parts
and built-in
computer allow
QRIO to walk
smoothly and stably

Four pressure sensors
on soles of each foot
help QRIO to walk

FIND OUT MORE ▶▶ Artificial Intelligence 156 • Computers 148–149 • Machines 88–89

ARTIFICIAL INTELLIGENCE

Artificial intelligence gives machines the ability to solve a problem, such as recognizing a face, even when there is not enough information to solve it using logic alone. We find it easy to tell people apart, but machines have to work hard to do it. More difficult problems, such as driving a car, are still beyond their reach. Intelligence clearly demands more than just logic. Research aims to give machines feelings, too.

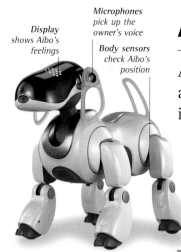

Display shows Aibo's feelings

Microphones pick up the owner's voice

Body sensors check Aibo's position

▲ ROBOT DOG
Sony's Aibo robotic dog was introduced in 1999. It uses advanced computer software to give it abilities that seem natural. Aibo's basic instincts are to sleep, explore, exercise, and play. It can also express joy, sadness, anger, surprise, and fear with lights, sounds, and gestures. Aibo recognizes its owner and comes when it is called.

CYNTHIA BREAZEAL
American, 1969-

Kismet's creator started with a degree in electrical and computer engineering from the University of California. She worked on Kismet in the Artificial Intelligence Lab at the Massachusetts Institute of Technology, and now directs its Media Lab Robotic Life group. Her aim is to create AI robots that work alongside people.

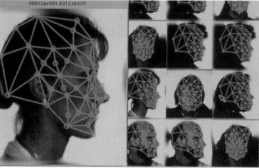

▲ FACE RECOGNITION PROGRAM
Face recognition programs on computers work by measuring prominent features of the face, such as the pupils of the eyes and the tip of the nose. The distances and angles between these are different for every face. By looking at enough features, the program can spot a known face even when the image is poor or the person is disguised.

Two pairs of cameras give humanlike stereo vision

Sensor gives humanlike sense of balance

Head can turn from side to side and tilt

Neck joint allows the head to nod

Touch sensors give COG a "skin"

Springy arm joints protect both the robot and the people around it

Crank handle turned by COG using a natural swing of the arm

◄ KISMET SHOWING HAPPINESS
Kismet was one of the first robots that responded to people in a natural way. It was designed by US engineer Cynthia Breazeal in 1999. The robot can move its ears, eyebrows, eyelids, and jaw, and can bend its lips up or down to smile or frown. It also responds to speech with babbling sounds.

FIND OUT MORE ▶▶ Computers 148–149 • Robots 154–155

NANOTECHNOLOGY

Nanotechnology gives us the ability to make incredibly small objects. Some of its methods came from microchip technology. Shapes are printed on to the surface of silicon, which is then etched away to make microscopic wheels or even micromotors. Other methods work with individual atoms to make even smaller objects. Although it is not used much yet, nanotechnology promises a future in which machines too small to see are part of our everyday world.

◄ COG
COG is a robot without legs that learns how to move by handling objects. Its intelligence comes from several computer programs that work together like parts of the brain. Rodney Brooks, director of the Artificial Intelligence Laboratory of the Massachusetts Institute of Technology in the USA, started the COG project in 1994 to see how artificial intelligence is affected by experience in the real world.

▲ MICROGEARS
These gears were made by etching silicon in the same way as a microchip. Sixty of them would fit on the head of a pin. As the gears were etched from the surface downwards, they have black triangular holes where the teeth below were shaped.

MINIATURE ENGINE ►
In a few years' time, silicon microengines like this one could replace laptop computer batteries. Liquid fuel burns inside the tiny combustion chamber to spin a central rotor, which turns a generator. A tank of fuel for the engine would weigh no more than a standard laptop battery, but could power the computer for 10 times as long.

Combustion chamber in which a mixture of fuel and air burns

Rotor spins in jet of hot gas from burning fuel

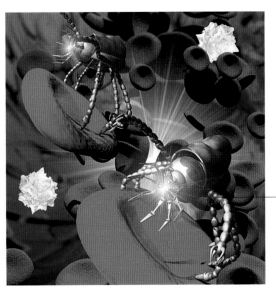

Imaginary microbot on a red blood cell

▲ NANO MEDICAL
Scientists are already working on structures thousands of times smaller than micromotors. To make them, they use atoms like builders use bricks. One day they might be able to build robots as small as the cells that make blood red. Here, a pair of imaginary microbots check out a patient's blood.

CARBON NANOTECHNOLOGY ►
Carbon atoms can form molecules shaped like tubes and also ball-shaped molecules, known as buckyballs. This tube has some buckyballs rolling along inside it. Carbon nanotubes can be either electrical conductors or insulators, and are 10 times stronger than steel. The biggest nanotubes are only a millimetre or so long, but they are ideal for building microscopic electrical machines.

Buckyballs fit inside a nanotube

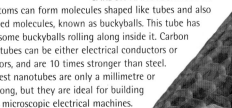

Carbon nanotube could be used as an electrical or mechanical part

FIND OUT MORE ►► Atoms 24–25 • Generators 137 • Medical Research 376 • Molecules 28–29

SPACE

UNIVERSE

Everything that exists – stars, planets, galaxies, and all that lies between – makes up the Universe. Scientists believe the Universe is 4 per cent ordinary matter, 23 per cent dark matter, and 73 per cent dark energy. Almost nothing is known about dark energy, but this is the name given to something that appears to exert a force making the Universe expand. Forces, such as gravity and the laws of physics and chemistry, determine what matter is like and how it behaves.

▲ MATTER IN THE UNIVERSE
Looking up at the heavens, we see matter in the form of stars, planets, and in the glowing clouds of gas and dust known as nebulas. These are visible forms of matter. However, astronomers believe that up to 90 per cent of the Universe's matter is invisible matter, known as dark matter.

▼ OUR PLACE IN THE UNIVERSE
The Earth we live on seems big and very important to us. But in the Universe as a whole, it is a tiny, and very insignificant, speck of rock. To put things into perspective, Earth is just a small planet in the Solar System, part of a family of bodies that circle round the Sun. The Sun is just one of billions of stars in a great star island that makes up our Galaxy. And this Galaxy is just one of billions that make up a Universe bigger than most of us can imagine.

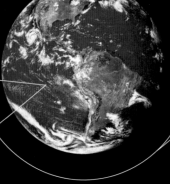

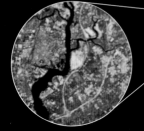

▲ CROWDED CITY
In the last 100 years cities have grown to house nearly half of the 6.6 billion humans that live on Earth. The largest cities cover areas tens of kilometres wide.

▲ OUR PLANET
Earth looks beautiful from space. It appears mainly blue because of its vast oceans – water covers more than 70 per cent of our planet. Wisps of white clouds fleck the atmosphere. The continents appear brown and green. Earth is 12,756 km (7,926 miles) across at the equator.

▲ THE SOLAR SYSTEM
Eight planets, and other smaller bodies, circle the Sun to make up the Solar System. Distances across the Universe are so vast that they are measured in light years – the distance light travels in one year, or 9.46 million million km (5.8 million million miles). The Solar System measures about 3 light years across.

*The central bulge
is made up of a dense
mass of stars*

*The light
from distant galaxies
takes billions of years
to reach Earth*

▲ OUR GALAXY
The Sun is one of at least 200 billion stars in the Galaxy, our local star island in space. It sits in one of the Galaxy's spiral arms, about 25,000 light years from the centre. The Galaxy measures about 100,000 light years across but is only around 2,000 light years thick.

THE EXPANDING UNIVERSE ▶
Our Galaxy is one of tens of billions of galaxies in the Universe. Galaxies are found in groups, or clusters, which in turn gather together to form superclusters. These interconnecting superclusters and the spaces, or voids, between them make up the Universe. Astronomers believe that almost all the galaxies are rushing away from us – and each other – at high speed, and they move faster the further apart they are. This tells us that the Universe is expanding. Astronomers believe that an explosion, known as the Big Bang, started off this expansion 13.7 billion years ago.

Universe

◀ WHERE SPACE BEGINS
The Earth is wrapped inside a layer of air we call the atmosphere. At a height of about 300 km (200 miles) above Earth, few traces of air remain. Astronauts in orbiting spacecraft have spectacular views of this region and are able to see the blue of the Earth's atmosphere gradually merge into the inky blackness of empty space.

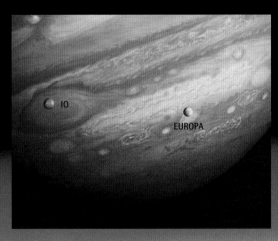

▲ UNIVERSAL FORCES
The moons Io and Europa are seen here travelling across the face of Jupiter. They are kept in orbit by Jupiter's powerful gravitational pull. Gravity is a dominant force in the Universe, and holds together systems such as galaxies. It is one of the four fundamental forces, and is the only one that can act over vast distances. Electromagnetism is the force that acts between all substances with an electric charge. The strong force and the weak force occur only in the nuclei of atoms.

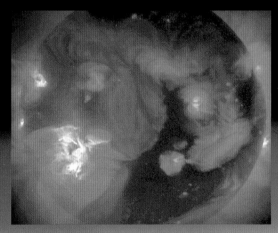

▲ ENERGY AND RADIATION
This X-ray image of the Sun reveals that its atmosphere, shown in red, is in fact hotter than its surface, which appears black. Stars like our Sun pour out energy into the Universe as different types of rays, or radiation. The full range of rays is called the electromagnetic spectrum. Within the spectrum there are light rays we can see and infrared rays we can feel as heat, but gamma rays, X-rays, ultraviolet rays, and radio waves can only be detected with specialized instruments.

FIND OUT MORE ▶▶ Atmosphere **234–235** • Big Bang **162–163** • Energy Waves **98–99** • Forces **64–65** • Gravity **72**

BIG BANG

Astronomers believe that the Universe came into being about 13.7 billion years ago, in an explosion known as the Big Bang. In an instant, space and the building blocks of matter were created, and time began. From that moment, the Universe began to expand, and continues to expand today. Over billions of years, matter formed into large, complex structures that continue to evolve.

BIG BANG	13 BILLION YEARS AGO: FIRST ATOMS FORM	12 BILLION YEARS AGO: GALAXIES BEGIN TO FORM	11 BILLION YEARS AGO: FORMATION OF MILKY WAY			
TIME BEGINS	13 BILLION YEARS AGO	12	11	10	9	8

▲ THE MOMENT OF EXPLOSION

The Big Bang created an incredibly hot Universe a fraction of the size of an atom. It immediately started to cool and expand, for a brief moment growing at a tremendous rate, in a process called inflation. In less than a millisecond, the first matter was created but, for thousands of years, the Universe was dominated by radiation.

▲ THE FIRST THREE MINUTES

The first forms of matter created were the tiniest and most basic particles of matter, such as quarks. Scientists today try to recreate what happened by smashing particles in a particle accelerator and studying the tracks. As the Universe cooled, these particles combined to form protons and neutrons, which later joined to form the nuclei of atoms.

ATOMS ▲

After 300,000 years, nuclei began to capture electrons and form the first atoms. This cosmic microwave map reveals what the Universe was like after 380,000 years. The red and yellow areas are slightly warmer than the blue and green ones and are a sign that matter was clumping.

REDSHIFT

By analysing the spectrum of light from a star or galaxy, astronomers can tell how fast it is moving, and whether it is moving towards or away from Earth. If an object is moving away from Earth, its light shifts to longer, redder wavelengths, an effect known as redshift. We know the Universe is expanding as almost all galaxies show redshift, and are rushing apart.

GALAXIES NEAR AND FAR ▶

The two galaxies in this picture seem to be close together in space, but in fact they are very far apart. The large galaxy is 80 million light years away, the small one 1 billion light years distant. Analysing the redshifts of their light reveals that the more distant galaxy is travelling much faster than the other.

Big Bang

Astronomers believe that the Big Bang took place 13.7 billion years ago, and that galaxies began to form 1-2 billion years later. Our Solar System was not created until about 4.6 billion years ago, with primitive single-celled life appearing on Earth about 1 billion years later. It was not until around 600 million years ago that an explosion of life occurred, in the Cambrian Period of Earth's history. The first dinosaurs evolved 230 million years ago, and man's earliest ancestors just 4 million years ago.

4.6 BILLION YEARS AGO: SOLAR SYSTEM FORMED	3.6 BILLION YEARS AGO: LIFE BEGINS ON EARTH	230 MILLION YEARS AGO: FIRST DINOSAURS

6	5	4	3	2	1	PRESENT TIME

▲ THE UNIVERSE TAKES SHAPE
As matter was drawn together by gravity, the first stars and galaxies were born. This Hubble Space Telescope picture shows galaxies 2.2 billion light years away, and many more remote galaxies beyond. The gravity of the cluster, including its invisible **DARK MATTER**, act like a lens to magnify the images of the more distant galaxies.

▲ THE EVOLVING UNIVERSE
The composition of the Universe continues to change. These two galaxies are colliding, and flinging out streams of stars. The Universe is also still expanding. Because of an effect called **REDSHIFT**, astronomers know that almost all galaxies are accelerating away from each other. It is not the galaxies themselves that expand — they are held together by gravity — but the vast distances between the galaxies.

DARK MATTER

The Universe is made up of matter and energy. Stars and galaxies are forms of matter that we can see. But there are also forms of matter that we cannot see, called dark matter. We know that dark matter exists because of the effects of its gravity. Astronomers believe that dark matter might account for up to 90 per cent of the matter in the Universe.

◄ INVISIBLE ATTRACTION
This picture of a galaxy cluster is made up of an image from the Hubble Space Telescope, showing galaxies in red, and a map showing regions of dark matter in blue. When the two images are combined, they reveal that dark matter is found where the galaxies clump. The gravity of the dark matter helps hold the cluster together.

FIND OUT MORE ▶▶ Atoms **24–25** • Gravity **72** • Light **110–111** • Matter **10–11** • Universe **160–161**

GALAXIES

Stars are not scattered evenly throughout the Universe. Instead, they are grouped together in great star islands, called galaxies. All the stars we see in the sky belong to our home galaxy, the **MILKY WAY**. Some galaxies are tiny and contain only a few million stars, but many contain hundreds of billions of stars. Galaxies are classed into three broad groups, according to their shape: elliptical (oval), spiral (if they have spiral arms), and irregular.

EDWIN HUBBLE
American, 1889–1953

While working at Mount Wilson Observatory in California, astronomer Edwin Hubble was the first to discover, in 1923, that there are other galaxies beyond our own. Today we still use Hubble's original method of classifying galaxies into spirals, ellipticals, and irregulars.

◄ SPIRAL GALAXY ESO 510-G13
A spiral galaxy is roughly disc-shaped and has a bulge in the middle. The disc is formed by arms that curve out from the central bulge. The stars in the central bulge are relatively old. Most star formation takes place on the spiral arms, which are full of gas and dust. In this sideways view of a slightly warped spiral galaxy, dark dust lanes are visible in the disc.

galaxies

Spiral arm contains gas and dust clouds and mainly hot, young stars

The central bulge is packed with old red and yellow stars, which glow the brightest

◄ SPIRAL ARMS
In this face-on view of a spiral galaxy, the arms are quite distinct and well separated. This galaxy is called M100 and is one of the finest spirals we know. It is classed as an Sc galaxy — S standing for spiral and c referring to its wide-open arms. Sa and Sb galaxies have more closed-up arms. Some spirals, called barred spirals (SB), have arms that spiral out from a straight bar of stars.

GALAXY CLUSTER ▶
The Virgo cluster contains over 2,000 galaxies. Only a small part of it is seen here. Galaxies form in clusters which are held together by the gravity of the galaxies and invisible dark matter. Within clusters, galaxies move around. Small galaxies orbit larger ones and sometimes merge. In some places, clusters of galaxies are concentrated in a supercluster.

▲ IRREGULAR GALAXY
The Large Magellanic Cloud is one of our nearest galactic neighbours in space. It is an example of an irregular galaxy, which means it has little definite structure. It is some 160,000 light years away, and is less than a third as wide as our own Galaxy.

▲ ELLIPTICAL GALAXY
M87, found in the Virgo cluster of galaxies, is an example of an elliptical galaxy. These galaxies lack the curved arms of spirals and can be round or oval in shape. Some of the largest galaxies are elliptical. M87 may be as big as 500,000 light years across.

▲ ACTIVE GALAXY
This face-on spiral galaxy is called a Seyfert, and is a type of galaxy that has a very bright centre. It is classed as an active galaxy because it gives out exceptional energy. Other active galaxies include radio galaxies, quasars, and blazars.

MILKY WAY

The galaxy that is our home is called the Milky Way Galaxy, or just the Galaxy. It measures about 100,000 light years across. Our local star, the Sun, is one of at least 200 billion stars in the Galaxy, and lies in one of the Galaxy's spiral arms. We also call the faint band of light that arches across the night sky the Milky Way. This band is a just a section of our Galaxy.

Spiral arm   Satellite galaxy

MILKY WAY ▶
On a clear dark night, the faint band of the Milky Way can be seen in the sky. The Galaxy appears as a band because it is a flat disc and, from our position in a spiral arm, we look through the disc side-on. With binoculars or a telescope, we can see the Milky Way's mass of stars, seemingly packed close together. Dark lanes among the stars show where dust clouds are blocking the light from other distant stars.

▲ ANDROMEDA, OUR NEIGHBOUR
Our Galaxy is part of a small cluster of galaxies we call the Local Group. We know of about 40 galaxies in the Local Group, and the largest is the Andromeda Galaxy. It is a huge spiral galaxy, half as wide again as the Milky Way, and contains around 400 billion stars. Although it lies 2.5 million light years away, it is still visible to the naked eye. Andromeda has two satellite galaxies, both small elliptical galaxies, that orbit it as it travels through space.

FIND OUT MORE ▶▶ Stars **166–167** • Sun **170–171**

STARS

Like our Sun, stars are massive globes of intensely hot gas. They produce huge amounts of energy, which is given off as heat and light. The bright stars form patterns, which we call the **CONSTELLATIONS**. All the stars lie so far from Earth and from each other that the distances are measured in light years. The light from our nearest star, Proxima Centauri, takes over four years to reach us. This means it lies over four light years away.

ASTEROPE

TAYGETA

MAIA

CELAENO

PLEIONE

ELECTRA

ATLAS

ALCYONE

MEROPE

▼ STAR SIZES
Stars vary widely in size. Our Sun is quite a small star, known as a yellow dwarf. Red giant stars are typically 30 or more times bigger in diameter than the Sun and supergiants are hundreds of times bigger. In contrast, the Sun is 100 times bigger than the tiny dense stars called white dwarfs.

*Red giant
30 times
bigger than
Sun*

*Blue-white star
7 times bigger
than Sun*

*Supergiant
hundreds of times
bigger than Sun*

*Sun,
a yellow dwarf*

▲ STARLIGHT IN THE PLEIADES
This cluster of bright stars is known as the Pleiades. Like all stars, they give off energy from their surface as light and heat; their energy is produced in a central core. There, nuclear reactions take place in which the nuclei (centres) of hydrogen atoms fuse (join) together to make helium nuclei. This nuclear fusion process produces enormous amounts of energy.

DISTANCE TO THE STARS

If you hold up a finger and look at it with first one eye, then the other, your finger appears to move in relation to objects in the background. This happens because your line of sight from each eye is slightly different. The effect is called parallax. In the same way, as Earth orbits the Sun, our line of sight to the stars changes. Nearer stars shift relative to those further away. Astronomers measure the shift of a star's position at different times of year and can then calculate how far away it is.

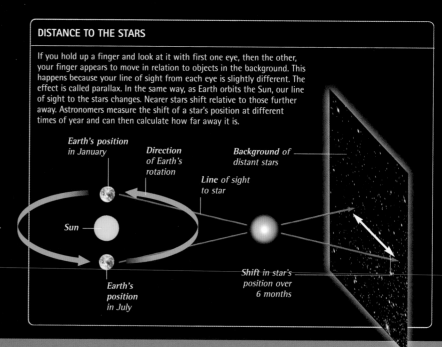

*Earth's position
in January*

*Direction
of Earth's
rotation*

*Background of
distant stars*

*Line of sight
to star*

Sun

*Shift in star's
position over
6 months*

*Earth's
position
in July*

CONSTELLATIONS

Many stars form patterns that we can recognize. We call these patterns the constellations, and many of them are named after real or mythological animals. Astronomers recognize 88 constellations, and divide the sky up into areas around each constellation. Although the groups of stars appear to be close together, they can be hundreds of light years apart. From Earth, they just appear to be grouped because they all lie in the same direction in space.

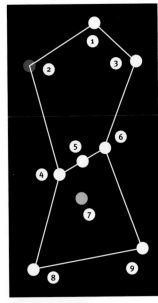

PATTERNS OF STARS ▶
This is the main star pattern in the constellation Orion. The numbers are keyed to the names of the stars that are marked in the photograph of the constellation below. On the star maps used by observers, the brightest stars in a constellation are often linked to form a recognizable shape, although this does not always resemble the figure its name suggests.

Heka (Meissa) **1**

Betelgeuse, **2**
a supergiant star that looks red in the sky

3 *Bellatrix*

▼ MIGHTY HUNTER
The constellation of Orion takes its name from a great hunter in Greek mythology. To early astronomers the pattern of the brightest stars looked like a hunter holding a club. The three stars in the middle make up Orion's Belt. The Orion Nebula forms part of the sword that hangs from his belt.

6 *Mintaka*
5 *Alnilam*
4 *Alnitak*

7 *Orion Nebula, a glowing cloud of gas*

9 *Rigel, a very hot blue-white giant star*

Orion's Belt

Orion Nebula

CONSTELLATION OF ORION

Saiph **8**

▲ BINARY STAR
Although our Sun travels through space alone, most stars have one or more companions. Sirius, the brightest star in the sky, is an example of a binary star. Sirius A is the star that gives out the most light, but in this X-ray picture, its companion, Sirius B, looks brightest because it gives out the most X-rays. Sirius B is a white dwarf star – a small, very hot star.

▲ ORION IN THE SKY
The constellation Orion is one of the most easily recognizable in the heavens. It straddles the celestial equator, so it can be seen well by observers in both the Northern and Southern Hemispheres. Northern observers see it best in winter skies and southern observers in summer skies. Betelgeuse and Rigel are its brightest stars.

FIND OUT MORE ▶▶ Atoms 24–25 • Nebulas 68 • Nuclear Energy 85

NEBULAS

The spaces between the stars are not completely empty, but are filled with clouds of gas and dust called nebulas. We can see nebulas when they glow or when they reflect light from nearby stars. Sometimes we cannot see them, but we know they are there because they block out the light from stars behind them. Dark nebulas include giant large molecular clouds, where new stars are formed.

e ▸▸ nebulas

CLOUDS AMONG THE STARS ▶

This picture shows vast clouds of gas and dust — nebulas — among the stars in the constellation Sagittarius. There are three main types of nebula, which can all be seen here. Emission nebulas appear red or pink. This is because they are mostly hydrogen gas, which glows red when it is excited (given extra energy) by nearby stars. Reflection nebulas appear blue, because they reflect light from nearby stars. Dark nebulas are regions where dust is blotting out distant stars.

LIFE CYCLE OF STARS

Stars are born in dark molecular clouds. Within these clouds, matter clumps together as it collapses under gravity. Within these clumps, even denser masses are formed, called cores. In the centre of a core, the matter becomes increasingly compressed and heats up. It begins to give off heat and light as a protostar. When the temperature of the protostar reaches 10 million°C (18 million°F) or so, nuclear fusion reactions begin, and the star begins to shine. It will shine steadily for millions or billions of years, but eventually it will start to die. Whether a star becomes a red giant or a supergiant depends on its mass.

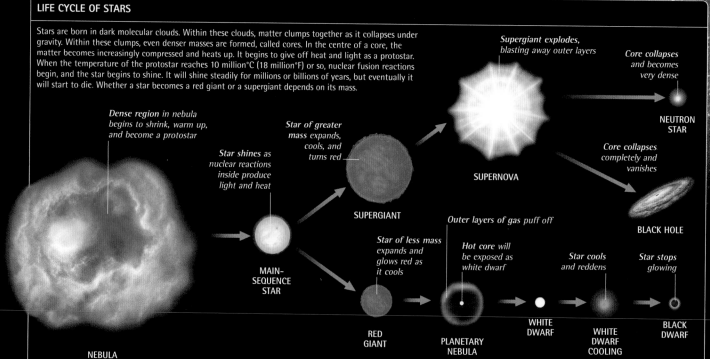

Dense region in nebula begins to shrink, warm up, and become a protostar

Star shines as nuclear reactions inside produce light and heat

Star of greater mass expands, cools, and turns red

Supergiant explodes, blasting away outer layers

Core collapses and becomes very dense

SUPERGIANT

SUPERNOVA

NEUTRON STAR

Core collapses completely and vanishes

BLACK HOLE

Star of less mass expands and glows red as it cools

Outer layers of gas puff off

Hot core will be exposed as white dwarf

Star cools and reddens

Star stops glowing

MAIN-SEQUENCE STAR

RED GIANT

PLANETARY NEBULA

WHITE DWARF

WHITE DWARF COOLING

BLACK DWARF

NEBULA

FIND OUT MORE ▸▸ Black Holes 69 • Nuclear Energy 85 • Stars 66–67 • Supernovas 69

SUPERNOVAS

A massive star dies in an explosion called a supernova. Only the collapsed core remains. If the core is very dense it becomes a neutron star which rotates rapidly, sending out beams of energy. If these beams reach Earth they are picked up as pulses, and the body is called a pulsar. A supernova also occurs if a white dwarf star in a binary pair blows up when material from the other star falls on it.

JOCELYN BELL BURNELL
British, 1943–

Astronomer Bell Burnell discovered the first pulsar when working as a research student at the Cambridge radio observatory. On 6 August, 1967, she picked up an unusual radio signal: pulses repeating every 1.337 seconds. Astronomers later identified the pulses as coming from a rapidly rotating neutron star.

SUPERNOVA 1987A ▶
On 23 February, 1987 a brilliant new star seemed to blaze in the Large Magellanic Cloud. It was easily visible to the naked eye. In fact, it was not a new star, but an existing star (called Sanduleak −69°202) that had exploded as a supernova.

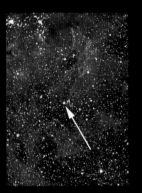

SANDULEAK −69°202
BEFORE EXPLOSION

SUPERNOVA 1987A
AFTER EXPLOSION

supernovas

THE CRAB SNR ▲
The glowing cloud called the Crab SNR was caused by a supernova first recorded by Chinese astronomers in AD 1054. When the supernova exploded it blasted a great cloud of gas into space. Astronomers call such a cloud a supernova remnant (SNR). Inside the Crab SNR is a pulsar, flashing on and off 30 times a second.

FIND OUT MORE ▶▶ Galaxies 164–165 • Nebulas 68

BLACK HOLES

black holes

When the core left after a supernova explosion is made of more than three times the Sun's mass, it collapses in on itself to become a black hole. Such regions have so much gravity that not even light can escape their incredibly powerful pull. Astronomers can't see black holes, but they can detect them. This is because matter spiralling into a black hole emits X-rays, which can be seen by X-ray telescopes.

SPAGHETTIFICATION

If you were unfortunate enough to wander near a steep black hole, its enormous gravity would pull you in, in the same way that it sucks in light and matter. As you got nearer, the strength of gravity would increase so rapidly that it would tug at your feet more than your head, and you would stretch out long and thin, like a piece of spaghetti.

Body looks redder as gravity affects light waves

SUPERMASSIVE BLACK HOLES ▶
Supermassive black holes seem to be responsible for the exceptional energy output of active galaxies such as quasars. They have a mass millions of times greater than the Sun's. Matter drawn in from surrounding gas clouds or stars forms an accretion disc, which emits light and other radiation. A central jet emits energy too.

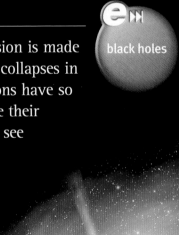

FIND OUT MORE ▶▶ Galaxies 164–165 • Gravity 72 • Nebulas 68

THE SUN

Dominating our corner of space is a star we call the Sun. It travels through space with a family of planets, moons, and other bodies, which form the Solar System. The Sun is huge — over 100 times wider than Earth. It has the most mass of any object in the Solar System, 750 times more than all the other bodies put together. Sometimes the Moon passes in front of the Sun during the day and a **SOLAR ECLIPSE** occurs.

▲ LOOPS IN THE ATMOSPHERE
Layers of gas surround the Sun, forming an atmosphere. The atmosphere's outer layer is called the corona (crown). In ultraviolet light, it is revealed to be full of loops of hot gas. These coronal loops may rise as high and as wide as 500,000 km (300,000 miles). The inner layer of the Sun's atmosphere is a pinkish colour and is called the chromosphere (colour-sphere).

WARNING! Never look directly at the Sun, especially through a telescope or binoculars. Its glare may blind you.

SPOTS ON THE SUN ▶
From time to time, dark patches called sunspots appear on the Sun's surface. These regions are around 1,500°C (2,700°F) cooler than the rest of the surface. They vary in size from a few thousand kilometres up to 100,000 km (62,000 miles). Sunspots may last for a few hours or several weeks. The number of sunspots rises and falls over a period of about 11 years — this is known as the sunspot cycle.

Penumbra is the lighter region
Umbra is the darkest, central region

STORMY SURFACE ▲
The surface of the Sun is a seething mass of gases, like a stormy sea. One of the reasons for this is its powerful magnetic field, which can be thousands of times stronger than Earth's. Close up, the surface appears covered in speckles, called granulations, with dark sunspots and light areas caused by explosions called solar flares.

Sun

A loop prominence, a fountain of fiery gas, is created by forces in the Sun's magnetic field

Photosphere, the surface of the Sun, which emits light

Granulations are rising pockets of hot gas 1,000 km (600 miles) across

INSIDE THE SUN

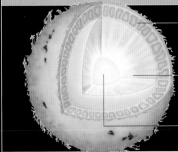

Convective zone, where rising gas carries energy to the surface

Radiative zone, where energy travels out from core

Core produces energy

ESSENTIAL DATA

Diameter at equator	1,400,000 km (865,000 miles)
Distance from Earth	149.6 million km (93 million miles)
Mass (Earth=1)	330,000
Average density	1.41 x water's density
Rotation period	25.4 days (at equator)
Surface temperature	5,500°C (9,930°F)
Core temperature	15,000,000°C (27,000,000°F)
Age	4.6 billion years

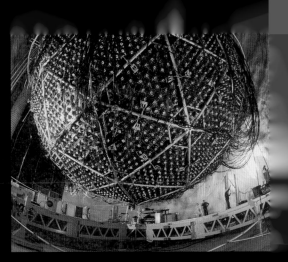

▲ NEUTRINOS FROM THE SUN

Scientists at the Sudbury Neutrino Observatory in Ontario, Canada, use a tank deep underground to detect particles called neutrinos. Neutrinos are produced at the centre of the Sun and other stars, and by studying them astronomers can learn more about the cores of stars. Neutrinos pass through matter, such as the Earth, and are detected underground because there is less interference from other particles.

SPACE WEATHER ▶

The Sun constantly emits streams of electrically charged particles known as the solar wind. The solar wind is mainly responsible for the weather conditions in space around Earth. As this picture shows, the Sun sometimes ejects a huge blast of particles, called a coronal mass ejection. This makes the solar wind stronger and can cause magnetic storms on Earth that affect compasses and disrupt radio signals.

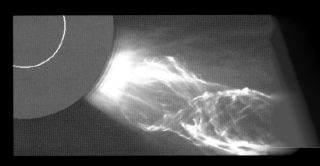

SOLAR ECLIPSE

Sometimes, as the Moon circles around the Earth, it passes directly in front of the Sun and blocks out its light. This is known as a solar eclipse. If the Moon only partly covers the Sun, we see a partial eclipse. If it covers the Sun completely, we see a total eclipse. A total eclipse is rare – usually the Moon passes slightly above or below the line between the Sun and the Earth. When a total eclipse occurs, day turns to night and the air becomes cold. A total eclipse can last for up to 7½ minutes but is usually shorter.

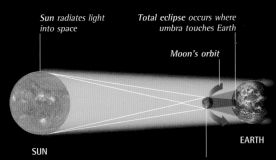

Sun radiates light into space

Total eclipse occurs where umbra touches Earth

Moon's orbit

SUN

EARTH

Moon moves between Sun and Earth and casts shadow on Earth

SEEING THE CORONA ▶

From Earth, observers cannot usually see the corona, or outer atmosphere of the Sun, because the photosphere (surface) is so bright. During a total eclipse, however, the moon blocks out the surface and we can see the Sun's atmosphere. It appears as a milky halo around the Moon and extends millions of kilometres out into space. Its temperature can reach 3 million °C (5.4 million °F).

IN THE MOON'S SHADOW ◢

During a solar eclipse the Moon casts a shadow on the Earth. Observers see a total eclipse if they are in the umbra, the central and darkest part of the shadow. Observers in the part-shadow, or penumbra, see a partial eclipse. As the Moon orbits the Earth the umbra traces a path across the Earth's surface known as the path of totality (total darkness) up to 270 km (170 miles) wide

FIND OUT MORE ▶▶ Big Bang 162–163 • Earth's structure 206–207 • Moon 177 • Stars 166–167

SOLAR SYSTEM

Our tiny corner of the Universe is dominated by a star we call the Sun. Trapped in the gravity of the Sun is a huge family of bodies — **PLANETS, MOONS,** dwarf planets, asteroids, Kuiper Belt Objects, and comets — which hurtle with it through space. This family is our Solar System. The effects of the Sun — its heat, gravity, light, and particles — extend to about halfway to the next nearest star, Proxima Centauri.

THE BIRTH OF THE SOLAR SYSTEM

The Solar System is around 5 billion years old. It formed out of a huge cloud of gas and dust called the solar nebula. Under gravity, the cloud collapsed and the material formed the Sun and a disc of matter in which the planets were born.

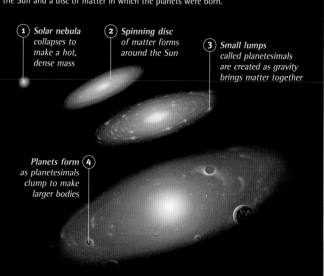

① *Solar nebula collapses to make a hot, dense mass*

② *Spinning disc of matter forms around the Sun*

③ *Small lumps called planetesimals are created as gravity brings matter together*

④ *Planets form as planetesimals clump to make larger bodies*

▼ ORBITS IN THE SOLAR SYSTEM

The planets travel in space around the Sun in paths known as orbits. The orbits are not circular, but elliptical (oval) in shape. This diagram shows the orbits of the planets roughly to scale. All the planets orbit the Sun in much the same plane (level). The planets also all travel in the same direction — anticlockwise here. Comets, however, loop round the Sun from any angle and have elongated orbits.

Solar System

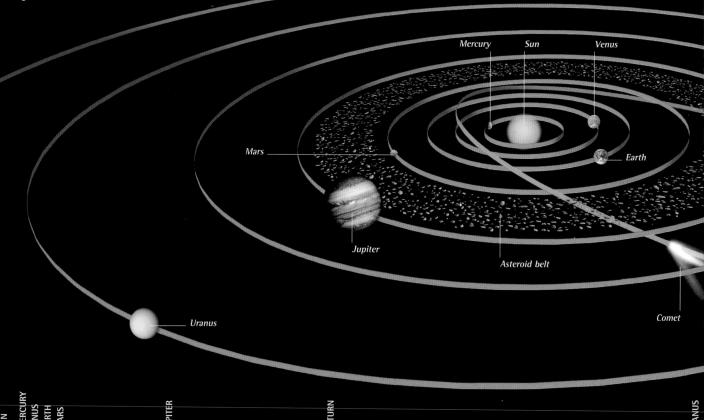

Mercury

Sun

Venus

Mars

Earth

Jupiter

Asteroid belt

Comet

Uranus

SUN	MERCURY VENUS EARTH MARS	JUPITER	SATURN	URANUS
0	MILLION KM/MILES	500 KM (300 MILES)	1,000 KM (600 MILES)	1,500 KM (900 MILES)	2,000 KM (1,240 MILES)	2,500 KM (1,550 MILES)

NEPTUNE   URANUS

EARTH   MERCURY

MARS   VENUS

SATURN   JUPITER

SUN

PLANETS

Eight planets orbit the Sun at different distances. The four inner planets are balls of rock and metal. The four outer ones are giant balls of gas and liquid. The time it takes a planet to orbit the Sun is its orbital period (its year). Planets also rotate (spin round) as they travel. The time it takes a planet to rotate once is its rotation period (its day). Three dwarf planets also orbit the Sun: Ceres within the Asteroid belt, and Pluto and Eris beyond Neptune.

▲ COMPARING THE PLANETS
The planets vary widely in size. Earth is one of the smallest, just 12,756 km (7,926 miles) in diameter. More than 1,300 Earths could fit inside the largest planet, Jupiter. However, the Sun makes up 99.9 per cent of the mass of the Solar System. The planets are not upright in relation to their orbits around the Sun. The axis (the line around which it turns) of each planet is tilted at a different angle.

MOONS

A moon is a body that orbits a planet. Altogether we know of more than 160 moons in the Solar System. Earth has one, the Moon, while Jupiter has at least 63. Most of these moons are small asteroids captured by Jupiter's gravity, but its largest moon Ganymede, with a diameter of 5,268 km (3,266 miles), is bigger than Mercury.

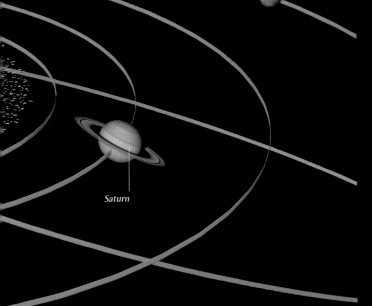

Neptune

Saturn

▲ HOW OUR MOON FORMED
No one is certain how the Moon formed, but many astronomers believe that it was born when a body the size of Mars collided with the young Earth over 4 billion years ago. In the collision, material from the two bodies was heated up, became molten, and was thrown out into space. In time, the material clumped together to form the Moon.

▼ DISTANCES IN THE SOLAR SYSTEM
The scale below shows the distances between the planets. The four inner planets are grouped closely together compared with the four outer ones. Smaller bodies including the dwarf planets Pluto and Eris lie beyond Neptune, some as far as 16 million million kilometres (10 million million miles) away.

NEPTUNE

3,500 KM (2,200 MILES)   4,000 KM (2,500 MILES)   4,500 KM (2,800 MILES)

FIND OUT MORE ▶▶ Asteroids **184** • Comets **185** • Earth **176** • Jupiter **179** • Moon **177** • Sun **170–171**

MERCURY

Mercury is the planet closest to the Sun, and experiences scorching temperatures by day. But it has virtually no atmosphere to trap the heat, so is freezing at night. It is a rocky planet, just over a third the diameter of Earth. Its surface is covered with craters, which make it look similar to parts of the Moon. These craters were formed when the planet was bombarded with meteorites long ago.

MERCURY'S SURFACE ▶
The craters that cover almost all of the surface of Mercury are generally shallower than those on the Moon. They vary in size from a few metres to hundreds of kilometres across. In between, there are relatively smooth lava-covered plains crossed by cliffs and ridges.

▲ CALORIS BASIN
The biggest feature known on Mercury is the Caloris Basin, which measures about 1,300 km (800 miles) across. It was created when a space rock 100 km (60 miles) wide crashed into the planet. This Mariner 10 spacecraft image shows half the basin coming in from the top, ringed by the mountains formed by the impact.

Red areas, around Mercury's equator, are hottest

Purple areas are the coolest, out of direct sunlight

Beethoven, the largest crater after the Caloris Basin measures 640 km (400 miles) across

Polar regions include areas that are always shaded from the Sun

Discovery Rupes is one of several prominent ridge systems

EXTREME TEMPERATURES ▶
In this heat map of Mercury, red shows the hottest areas, where the planet faces the Sun. The purple areas are the coldest. Temperatures vary from 450°C (843°F) in the Sun to –180°C (–292°F) in the dark. Mercury's slow rotation means that some parts have 176 days of sunlight, then 176 days of darkness.

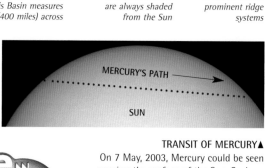

MERCURY'S PATH →

SUN

TRANSIT OF MERCURY▲
On 7 May, 2003, Mercury could be seen crossing the surface of the Sun. Such an event, called a transit, occurs once or twice a decade. This image was put together using pictures taken by the SOHO spacecraft at regular intervals over the five-hour transit.

Mercury

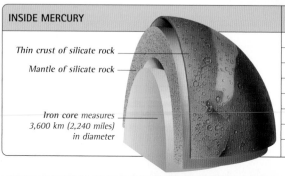

INSIDE MERCURY	ESSENTIAL DATA		SIZE COMPARISON
Thin crust of silicate rock	Diameter at equator	4,880 km (3,032 miles)	
Mantle of silicate rock	Average distance from Sun	57.9 million km (36 million miles)	
	Orbital period	88 days	
	Rotation period	58.7 days	
	Mass (Earth=1)	0.06	
	Gravity (Earth=1)	0.38	
Iron core measures 3,600 km (2,240 miles) in diameter	Average surface temperature	167°C (332°F)	EARTH
	Number of moons	0	MERCURY

VENUS

Venus is only a little smaller than Earth, is made up of rock, and has an atmosphere, but it is otherwise quite unlike our planet. Its surface is covered with volcanoes, the atmosphere is crushing, and the temperature is hotter than an oven. Venus completes its orbit in 225 Earth days, but turns slowly, taking 243 Earth days to rotate once. This means that a day on Venus is longer than its year.

Venus

MAPPING VENUS ▶
This false-colour picture of Venus's surface is based on radar images sent back by the Magellan spacecraft. It shows features of the planet's northern hemisphere. The reddish areas show the highest ground, the blue ones low plains and valleys. Lava plains cover four-fifths of the surface.

◀ CLOUDY ATMOSPHERE
The atmosphere of Venus is much thicker than that of Earth and is made up mostly of carbon dioxide. Its pressure is nearly 100 times Earth's atmospheric pressure. The dense clouds in the atmosphere are made up of sulphuric acid droplets. The atmosphere traps heat like a greenhouse, keeping the surface temperature within a few degrees of a scorching 464°C (867°F).

▼ VOLCANIC LANDSCAPE
Maat Mons is one of the largest volcanoes on Venus. Around 200 km (125 miles) across, it rises 9 km (5 ½ miles) in height. Probably extinct, it has erupted repeatedly in the past, spewing out vast quantities of runny lava that formed the surrounding volcanic plains.

Ishtar Terra is the second largest highland area, or continent, on Venus

Sedna Planitia is one of the low-lying valleys that cover much of Venus

Maxwell Montes mountain range rises up to 12 km (7 ½ miles)

INSIDE VENUS

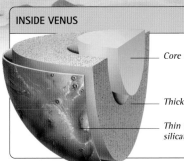

— *Core of iron and nickel*

— *Thick rocky mantle*

— *Thin crust of silicate rock*

ESSENTIAL DATA

Diameter at equator	12,104 km (7,521 miles)
Average distance from Sun	108.2 million km (67.2 million miles)
Orbital period	224.7 days
Rotation period	243 days
Mass (Earth=1)	0.82
Gravity (Earth=1)	0.9
Average surface temperature	464°C (867°F)
Number of moons	0

SIZE COMPARISON

EARTH

VENUS

FIND OUT MORE ▶▶ Earth **176** • Interplanetary Missions **198–199** • Volcanoes **212–213**

EARTH

The planet we live on is unique in the Solar System because it provides just the right conditions to support life. It is neither too hot nor too cold; it has plentiful supplies of liquid water; and it has oxygen in its atmosphere. The third planet from the Sun, Earth is made mainly of rock. At the centre is a large core of iron, partly molten. Movements in the core give Earth a strong magnetic field, which extends into space to form a great magnetic bubble called the magnetosphere.

INSIDE EARTH

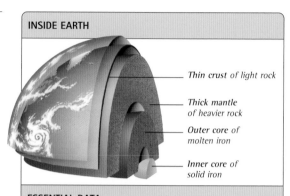

- **Thin crust** of light rock
- **Thick mantle** of heavier rock
- **Outer core** of molten iron
- **Inner core** of solid iron

ESSENTIAL DATA

Diameter at Equator	12,756 km (7,926 miles)
Average distance from Sun	149.6 million km (93 million miles)
Orbital period	365.25 days
Rotation period	23.93 hours
Mass (Earth=1)	1
Gravity (Earth=1)	1
Surface temperature	-70˚C to 55˚C (-94˚F to 131˚F)
Number of moons	1 (the Moon)

◄ NORTHERN LIGHTS
Shimmering curtains of coloured light brighten the skies in Alaska. This light display is called the aurora borealis, or Northern Lights. Similar displays occur in the far Southern Hemisphere. They happen when electrically charged particles in Earth's magnetosphere are disturbed by a magnetic surge from the Sun. The particles flow to the poles where they glow as they mix with air.

BLUE PLANET ►
From space, Earth appears mainly blue — the colour of the oceans. The surface of the Earth is a layer of solid rock, known as the crust. Where the crust is above sea level, it becomes land. A layer of gas — the atmosphere — cocoons our planet. It is made up mainly of nitrogen and oxygen — the gas essential for life. Clouds of water vapour swirl in the atmosphere.

JAMES VAN ALLEN
American, 1914–2006

An astrophysicist who worked on the first US satellite Explorer 1 in 1958. The satellite sent back new data about Earth's magnetic field, and Van Allen identified concentrations of electrically charged particles around Earth, now known as the Van Allen belts.

Atmospheric activity affects the climate on Earth

Oceans cover more than 70 per cent of the Earth's surface

Earth

The continent of South America is one of Earth's great landmasses

FIND OUT MORE ►► Atmosphere **234–235** • Earth's Structure **206–207** • Plate Tectonics **208–209**

MOON

Circling round Earth once a month, the Moon is our planet's only natural satellite. As the Moon orbits Earth, our view of it constantly changes, following a cycle known as the **PHASES OF THE MOON**. Like Earth, the Moon is made up of rock, but is only about a quarter as big across. Because it is so small, its gravity is low (about one-sixth of Earth's), and it has no atmosphere.

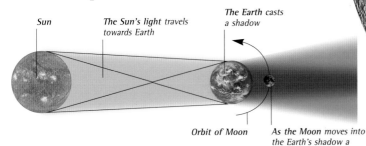

Sun
The Sun's light travels towards Earth
The Earth casts a shadow
Orbit of Moon
As the Moon moves into the Earth's shadow a lunar eclipse occurs

▲ THE MOON IN ECLIPSE
Two or three times a year the Moon enters the shadow Earth casts in space. This happens when the Sun, Earth, and Moon line up and is called a lunar eclipse. During a total eclipse, when the Sun, Earth, and Moon line up exactly, the Moon does not disappear but takes on a reddish hue as it is lit by light from the Sun that is bent by Earth's atmosphere.

PHASES OF THE MOON

Over a month we see the Moon appear to change shape. These different shapes, or phases, occur because, as the Moon circles Earth, we see more or less of the half of its surface that is lit by the Sun. The Moon takes 29.53 days to go through its phases.

THE LUNAR CYCLE ▶
With the Moon directly between Sun and Earth, the side facing us is dark. We call it a New Moon. As the Moon moves on, we see more and more of its face lit up, until we see it all at Full Moon. Afterwards, we see less and less until it disappears at the next New Moon.

Moon

The lunar surface photo labels: SEA OF SHOWERS, SEA OF SERENITY, OCEAN OF STORMS, COPERNICUS, SEA OF TRANQUILITY, APOLLO 11 LANDING SITE, SEA OF NECTAR, SEA OF VAPOUR, SEA OF CLOUDS, BRIGHT HIGHLANDS, TYCHO

▲ THE LUNAR SURFACE
From Earth, we only ever see one side of the Moon – the near side. The dark areas are great dusty plains, called maria (seas). The bright areas are highlands hundreds of kilometres across and covered with craters. The hidden far side of the Moon is more heavily cratered, but has no large seas.

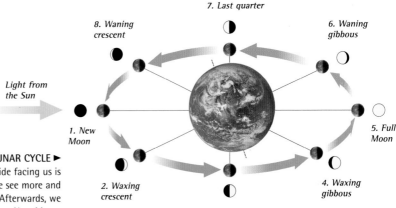

7. Last quarter
8. Waning crescent
6. Waning gibbous
Light from the Sun
1. New Moon
5. Full Moon
2. Waxing crescent
4. Waxing gibbous
3. First quarter

INSIDE THE MOON	ESSENTIAL DATA		SIZE COMPARISON
Inner core of solid iron	Diameter at equator	3,476 km (2,160 miles)	
Outer core of partly molten iron	Average distance from Earth	384,400 km (238,900 miles)	
	Orbital period	27.32 days	
	Rotation period	27.32 days	
Thick rocky mantle	Time to go through phases	29.53 days	
	Mass (Earth=1)	0.01	
Thin rocky crust	Gravity (Earth=1)	0.17	EARTH MOON
	Average surface temperature	-20˚C (-4˚F)	

FIND OUT MORE ▶▶ Earth **176** • Solar System **172–173** • Sun **170–171**

MARS

The fourth planet from the Sun, Mars is one of Earth's closest neighbours. Mars has a slight atmosphere (mainly carbon dioxide), and ice caps at the poles. Channels on the surface suggest that water may have flowed on the planet in the past. Although freezing now, astronomers believe that Mars was once warmer and some think it may have supported some form of life.

Mars

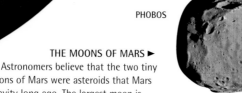

Valles Marineris is a canyon system, 8 km (5 miles) deep in places and 4,500 km (2,800 miles) long

◄ THE RED PLANET
Mars appears red because of the iron oxide (rust) in the rocks and dust on its surface. Features such as the Valles Marineris canyon can be seen through the thin atmosphere. There are also sandy deserts, large cratered regions, channels, and volcanoes.

PHOBOS

THE MOONS OF MARS ►
Astronomers believe that the two tiny moons of Mars were asteroids that Mars captured in its gravity long ago. The largest moon is Phobos, an irregular lump around 26 km (16 miles) in diameter. Deimos is even smaller — only about 16 km (10 miles) wide. Both moons are covered with craters and are thought to be made up of rock rich in carbon.

DEIMOS

▲ OLYMPUS MONS
Clouds ring the summit of Mars' gigantic volcano, Olympus Mons (Mount Olympus). Some 600 km (400 miles) across at the base, Olympus Mons rises to a height of 24 km (15 miles). This makes it by far the biggest volcano we know in the Solar System. It is located near the equator, close to three other large volcanoes on a bulge called the Tharsis Ridge.

▼ ON THE SURFACE
The surface of Mars is covered with fine reddish sandy material and strewn with small rocks. The Pathfinder probe sent back this image when it landed on the planet in 1997. It shows a region called Ares Vallis, which some astronomers believe was flooded with water long ago. The two hills on the horizon are called the Twin Peaks.

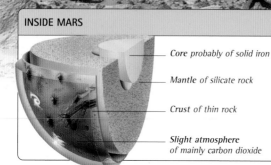

INSIDE MARS
Core probably of solid iron

Mantle of silicate rock

Crust of thin rock

Slight atmosphere of mainly carbon dioxide

ESSENTIAL DATA

Diameter at equator	6,794 km (4,222 miles)
Average distance from Sun	227.9 million km (141.6 million miles)
Orbital period	687 days
Rotation period	24.63 hours
Mass (Earth=1)	0.11
Gravity (Earth=1)	0.38
Average surface temperature	−63°C (−81°F)
Number of moons	2 (Phobos and Deimos)

SIZE COMPARISON

EARTH

MARS

FIND OUT MORE ►► Interplanetary Missions **198–199** • Solar System **172–173**

JUPITER

Jupiter is the biggest planet in our Solar System, eleven times bigger in diameter than Earth and two and a half times more massive than all the other planets put together. Jupiter has no solid surface. Beneath the gas clouds lies hot, liquid hydrogen, then a layer of hydrogen in a form similar to liquid metal, and finally a rocky core. Jupiter has a faint ring around its equator made of microscopic dust particles.

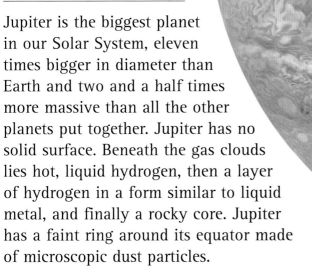

CALLISTO

GANYMEDE

EUROPA

IO

▲ JUPITER'S ATMOSPHERE

Alternate dark and pale bands streak Jupiter's atmosphere. The dark bands are called belts and the pale ones are called zones. The different colours reflect the presence in the atmosphere of different chemicals, such as sulphur, ammonia, and phosphorus compounds.

Jupiter

▲ THE GALILEAN MOONS

In 1609, the Italian astronomer Galileo first spied Jupiter's four largest moons: Io, Europa, Ganymede, and Callisto. The largest Galilean moon, Ganymede, with a diameter of 5,268 km (3,273 miles), is also the biggest moon in the Solar System.

THE GREAT RED SPOT ▶

For over 300 years astronomers have observed Jupiter's Great Red Spot. Space probes have shown it to be a violent super-hurricane, where winds swirl anticlockwise at high speeds. It measures around 40,000 km (25,000 miles) across.

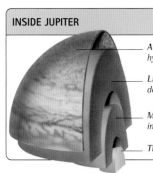

Jupiter's northern aurora is revealed in ultraviolet light by the Hubble Space Telescope

Southern aurora

◀ JUPITER'S LIGHTS

There are light displays, called auroras, at Jupiter's poles, like the Northern and Southern Lights we see on Earth. But on Jupiter the displays are much more dramatic. Lightening bolts 10,000 times more powerful than any seen on Earth also light up the planet.

INSIDE JUPITER	ESSENTIAL DATA		SIZE COMPARISON

INSIDE JUPITER

Atmosphere of mainly hydrogen and helium

Liquid hydrogen forms a deep, planet-wide ocean

Metallic hydrogen is hydrogen in the form of a liquid metal

Tiny core, probably of rock

ESSENTIAL DATA

Diameter at equator	142,984 km (88,849 miles)
Average distance from Sun	778.4 million km (483.7 million miles)
Orbital period	11.87 years
Rotation period	9.93 hours
Mass (Earth=1)	318
Gravity (Earth=1)	2.36
Cloud-top temperature	-110°C (-166°F)
Number of moons	At least 63

SIZE COMPARISON

EARTH

JUPITER

FIND OUT MORE ▶▶ Astronomy 186 • Earth 176 • Solar System 172–173

SATURN

A system of shining rings makes Saturn a very distinctive planet. We see the rings from different angles at different times as the planet circles the Sun every 29.5 years. Second in size only to Jupiter, Saturn is also made up mainly of gas. It is the lightest (least dense) planet, and would float if placed in water. Like Jupiter, Saturn has bands of clouds in a deep atmosphere, but they are much fainter than Jupiter's.

▼ GLORIOUS RINGS

From Earth, just three rings – A, B, and C – can be seen around Saturn, but there are several other rings both inside the C ring and further out beyond the A ring. The complete system spans a distance of more than 400,000 km (248,500 miles) out from the edge of the planet.

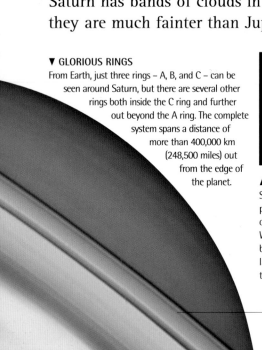

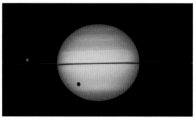

▲ THE RINGS EDGE-ON

Saturn's rings are broad, but thin. In some places they may be a kilometre (half a mile) or more thick, but in others just 10 m (33 ft). When we see the rings edge-on, they are barely visible. In this picture Titan, Saturn's largest moon, seen just above the rings on the left, casts its shadow on the planet.

A ring
has a small gap called the Encke Division

INSIDE SATURN

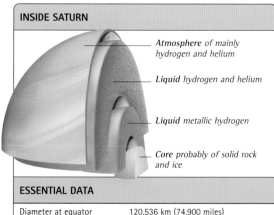

Atmosphere of mainly hydrogen and helium

Liquid hydrogen and helium

Liquid metallic hydrogen

Core probably of solid rock and ice

ESSENTIAL DATA

Diameter at equator	120,536 km (74,900 miles)
Average distance from Sun	1,427 million km (887 million miles)
Orbital period	29.46 years
Rotation period	10.66 hours
Mass (Earth=1)	95
Gravity (Earth=1)	1.07
Cloud-top temperature	-140°C (-220°F)
Number of moons	At least 60

SIZE COMPARISON

EARTH

SATURN

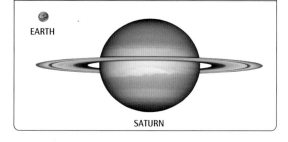

▲ RING MATERIAL

Saturn's rings are made up of chunks of ice and dust whizzing round the planet at high speed. They vary in size from particles as small as sand grains to huge boulders. No one is sure where this material came from. It could be the remains of ancient moons or maybe the debris of comets that came too close.

Cassini Division contains few ring particles

B ring is the brightest ring

C ring is nearly transparent

Saturn

URANUS

German-born English astronomer William Herschel discovered Uranus in 1781. It was the first planet discovered that is not easily visible with the naked eye. It lies twice as far from the Sun as Saturn. Uranus is unusual because it spins on an axis tilted at 98° and so appears to spin on its side. This may be because Uranus collided with another large object as it was forming.

Red light reveals hydrogen in the atmosphere

Uranus

WILLIAM HERSCHEL
German, 1738-1822

Herschel moved to England in 1757 where he worked as a musician, but also began to build superb reflecting telescopes. In 1781, he discovered an object he first thought was a comet, but was a new planet, Uranus. He later built the largest telescope in the world at that time, and discovered hundreds of nebulas.

THE RINGS OF URANUS ▶
The rings of Uranus were discovered in 1977. As the planet passed in front of a star the rings could be seen against the bright background. In 1986, Voyager 2 imaged the rings and 11 were identified. Here they are seen in infrared light by the Hubble Space Telescope.

◀ BLAND ATMOSPHERE
The atmosphere of Uranus is a greenish-blue colour. It is almost completely featureless in ordinary light. There are no signs of the cloud bands visible on Jupiter and Saturn. Computer enhancement of this Voyager 2 picture shows a smog-like haze (in red) at the south pole.

Rings circle around Uranus's equator

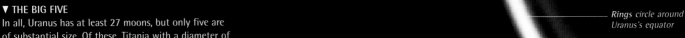

▼ THE BIG FIVE
In all, Uranus has at least 27 moons, but only five are of substantial size. Of these, Titania with a diameter of 1,578 km (981 miles) is the largest and Miranda with a diameter of 470 km (290 miles) the smallest. The smaller moons include asteroids captured by the planet's gravity.

OBERON

TITANIA

UMBRIEL

URANUS

MIRANDA

ARIEL

INSIDE URANUS

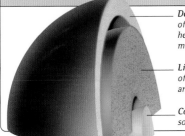

Deep atmosphere of hydrogen and helium, with some methane

Liquid mantle of water, ice, methane, and ammonia

Core probably of solid rock

ESSENTIAL DATA

Diameter at equator	51,118 km (31,764 miles)
Average distance from Sun	2,871 million km (1,784 million miles)
Orbital period	84 years
Rotation period	17.24 hours
Mass (Earth=1)	14.5
Gravity (Earth=1)	0.89
Cloud-top temperature	-197°C (-323°F)
Number of moons	At least 27

SIZE COMPARISON

EARTH

URANUS

FIND OUT MORE ▶▶ Interplanetary Missions **198–199** • Solar System **172–173**

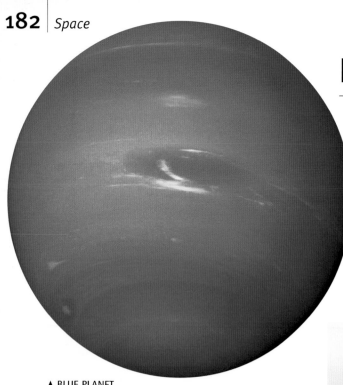

NEPTUNE

Almost the same size and colour as Uranus, Neptune orbits about 1.6 billion km (1.0 billion miles) further from the Sun than Uranus. German astronomer Johann Galle first spotted the planet in 1846 after mathematicians had calculated where it should be. Neptune has more extreme weather than Uranus, which is thought to be caused by heat from deep within its core. Winds blow up to 2,000 kph (1,250 mph), faster than on any other planet.

▲ BLUE PLANET

A Voyager 2 image of Neptune showing clouds ringing a large storm region known as the Great Dark Spot. Neptune's deep blue colour is caused by methane in the atmosphere. Methane also condenses high in the atmosphere to form wisps of cloud, rather like cirrus clouds in Earth's atmosphere.

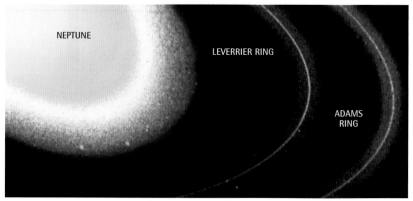

▲ NEPTUNE'S RINGS

Like Uranus, Neptune has a faint set of rings made of tiny dust particles circling its equator. There are four main rings — two broad but faint, and two bright but narrow. The two bright rings, Adams and Leverrier, are named after John Couch Adams and Urbain Leverrier, the mathematicians who calculated where the planet could be found.

Neptune

INSIDE NEPTUNE

Atmosphere of mainly hydrogen and helium, with some methane

Liquid mantle of water, ice, methane, and ammonia

Core probably of solid rock

ESSENTIAL DATA

Diameter at equator	49,532 km (30,785 miles)
Average distance from Sun	4,498 million km (2,795 million miles)
Orbital period	164.8 years
Rotation period	16.11 hours
Mass (Earth=1)	17.2
Gravity (Earth=1)	1.13
Cloud-top temperature	-200°C (-328°F)
Number of moons	13

SIZE COMPARISON

EARTH

NEPTUNE

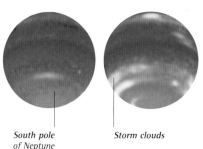

South pole of Neptune

Storm clouds

◄ CHANGING WEATHER

Compared with Uranus, Neptune has a very active atmosphere — in other words, a lot of weather. And the weather is constantly changing. These two images, taken by the Hubble Space Telescope, show the dramatic change in weather patterns over six years. In 2002, the weather was much more cloudy and stormy than it was in 1996.

TRITON'S GEYSERS ►

This Voyager 2 image shows Neptune's largest moon, Triton. It has a diameter of 2,710 km (1,685 miles) and has the coldest surface of any body in the Solar System, -235°C (-391°F). Made up of rock and ice, Triton is covered with frozen nitrogen and methane. A pinkish snow-like substance lies over the south polar region and dark streaks show where strange geysers spew out dust. Astronomers think that Triton and the planet Pluto are very similar.

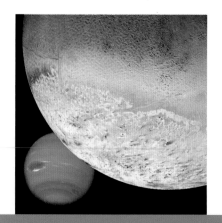

FIND OUT MORE ►► Pluto **83** • Solar System **172–173** • Uranus **180–181**

DWARF PLANETS

A dwarf planet is almost round and orbits the Sun as part of a belt of objects. There are three known dwarf planets in the Solar System: Eris, Pluto, and Ceres. Eris and Pluto are in the Kuiper Belt, and Ceres is in the Asteroid belt. The "dwarf planet" class of objects was introduced in 2006 after the discovery of Eris. It is larger than Pluto which was then considered a planet.

CLYDE TOMBAUGH
American, 1906–1997

Tombaugh made a systematic photographic survey of the night sky from the Lowell Observatory in Arizona, USA. Pictures of the same area of sky taken nights apart revealed moving objects. He eventually discovered Pluto on 18 February 1930, after 7,000 hours of investigation.

PLUTO ▶
Pluto was classed as a planet from its discovery until 2006. It is a frozen world of rock and ice just 2,304 km (1,432 miles) across, with a surface temperature of about -230°C (-382°F). This image is an artist's impression, but the New Horizons spacecraft, due to arrive in 2015, will hopefully capture quality images of it.

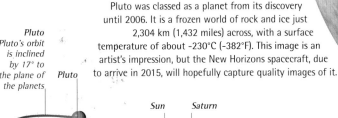

Pluto
Pluto's orbit is inclined by 17° to the plane of the planets

Pluto *Sun* *Saturn* *Uranus* *Neptune*

A thin nitrogen-rich atmosphere surrounds Pluto when it is closest to the Sun and the heat evaporates its surface ice

▲ PLUTO'S ORBIT
Pluto's unusual orbit takes it much further above and below the orbits of the eight planets. It is also much more elliptical. For 20 years of its 248.6 year orbit Pluto is closer to the Sun than Neptune.

▲ CERES
Ceres is not only categorised as a dwarf planet, but is also the largest asteroid. It orbits the Sun every 4.6 years in the Asteroid belt. It was discovered in 1801 and was the first known asteroid. It is about 960 km (596 miles) across and made mainly of rock, with significant amounts of water ice.

▲ ERIS
Eris is thought to be made of rock and ice, with an ice surface. Its exact size is uncertain but Eris is probably a few hundred kilometres bigger than Pluto. When discovered in 2005, Eris was almost 10 billion miles away and the most distant object ever seen orbiting the Sun.

▲ PLUTO AND ITS MOONS
Pluto (centre) has three moons: Charon (just above Pluto), and the smaller moons Nix (near top) and Hydra (top). Charon was discovered first in 1978. Hydra and Nix were both discovered in May 2005. Charon is about 1,180 km (730 miles) across, which makes it about half Pluto's size. Both Pluto and Charon spin every 6.38 days and Charon also orbits Pluto in 6.38 days. This means the two objects keep the same face to each other at all times.

EARTH

CERES PLUTO ERIS

dwarf planets

FIND OUT MORE ▸▸ Neptune **182** • Solar System **172–173**

ASTEROIDS

Billions of rocky lumps, called asteroids or minor planets, circle the Sun between the orbits of Mars and Jupiter. They occupy a broad band around 180 million km (112 million miles) wide, known as the asteroid belt. Most of the **METEORITES** that bombard the Earth from space appear to be asteroid fragments. There is also a region of small bodies in the outer Solar System, beyond Neptune, called the **KUIPER BELT.**

◄ ASTEROID IDA
Asteroid number 243, called Ida, is an elongated lump of rock 56 km (35 miles) long. In 1993 it was photographed by the Galileo spacecraft. Amazingly, this small rocky body has a moon 1.6 km (1 mile) across circling round it. Ida seems to be made up entirely of rock, but asteroids can also be made up mainly of metal, or a mixture of rock and metal.

Craters and fractures scar Ida's surface

asteroids

◄ NEAR EARTH ASTEROIDS
Most asteroids circle the Sun within the asteroid belt, but others have orbits that take them out beyond the orbit of Saturn or in towards Earth. Asteroids that pass close to Earth are known as Near Earth Asteroids, but there is no imminent danger of collision. At its closest, asteroid Eros (left) passes about 22 million km (14 million miles) from Earth.

METEORITES

Every day small space rocks rain down on Earth and reach the ground as meteorites. Most meteorites are made up of rock and are called stones. Others are mainly iron and are called irons. Stony-irons are a mixture of rock and iron. Most meteorites are tiny, but occasionally large ones tens or even hundreds of metres across hit the Earth.

KUIPER BELT

Just beyond Neptune's orbit to about 12 billion km (7.4 billion miles) from the Sun is a flat belt of small icy objects known as the Kuiper Belt. Past this is the Oort Cloud, a sphere of comets surrounding the planets of the Solar System. The material in both these areas is thought to be debris left over from the formation of the Solar System 4.6 billion years ago.

◄ SNOW-COVERED CRATER
Around 50,000 years ago an iron meteorite about 46 m (150 ft) across gouged out this huge crater in the Arizona Desert, in the USA. Called the Arizona Meteor Crater or Barringer Crater, it measures roughly 1,265 m (4,150 ft) across and 175 m (575 ft) deep. If a meteorite of similar size struck a city today, it would cause immense devastation.

SIKHOTE-ALIN METEORITE ►
This fragment is one of hundreds found in Siberia after a meteorite fall in 1947. It is part of an iron meteorite weighing 300 tonnes, which broke up in Earth's atmosphere before landing.

▲ KUIPER BELT OBJECTS
More than one thousand Kuiper Belt Objects (KBOs), hundreds of kilometres across, have been found since 1992, but there are estimated to be tens of thousands in total. KBOs seem to be icy bodies very similar to comets, and short-period comets are believed to originate from the Kuiper Belt. Astronomers now believe that the dwarf planets Pluto and Eris are in fact large KBOs. Neptune's moon Triton may also once have been a KBO, but was then captured by Neptune's gravity.

FIND OUT MORE ►► Atmosphere **234–235** • Solar System **172–173**

EDMOND HALLEY
English, 1656–1742

Halley was an astronomer and mathematician who became the second Astronomer Royal. He is best known for discovering that some comets are regular visitors to Earth's skies. He correctly predicted that the comet he had seen in 1682 would return again in 1758. It did, and was named Halley's comet after him.

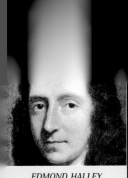

Sometimes icy lumps left over from the birth of the Solar System visit our skies. We see them as comets. Although tiny, comets release vast clouds of gas and dust as they approach the Sun and heat up. The clouds form a bright head and long tails, often millions of kilometres long. When Earth passes through the dust from past comets we see **METEOR** showers.

COMET HALE-BOPP ►
One of the brightest comets of the 20th century, Hale-Bopp blazed in the night sky for weeks during the spring of 1997. Its bright coma (head) hid a nucleus about 30–40 km (20–30 miles) across. The effects of sunlight and of the solar wind strung out the gas and dust released by the comet into long tails.

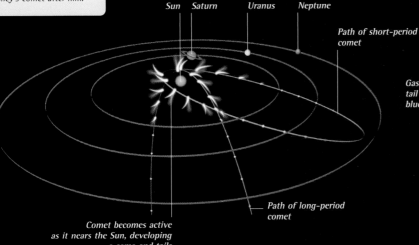

Sun
Saturn
Uranus
Neptune
Path of short-period comet
Path of long-period comet
Comet becomes active as it nears the Sun, developing a coma and tails

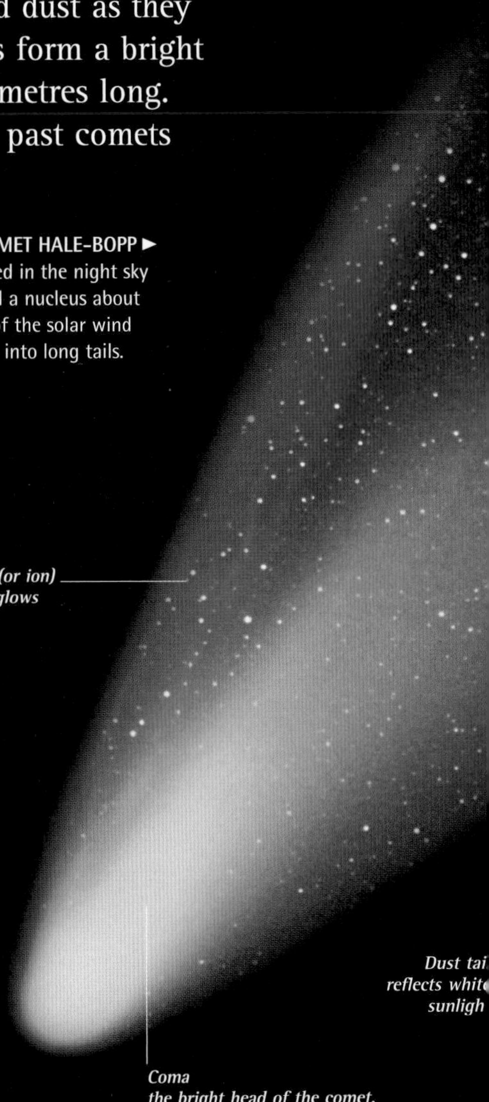

Gas (or ion) tail glows blue
Dust tai reflects whit sunligh
Coma the bright head of the comet, which hides the tiny nucleus

COMET ORBITS
Comets may head in towards the Sun from any direction. They have highly elliptical (oval) orbits. Comets may take just a few years or thousands to circle the Sun. Some seem to come from reservoirs of icy bodies in the Kuiper Belt, or from further out in a region called the Oort Cloud. Unseen for most of the time, comets become visible only when they approach the Sun.

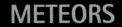

e ⏭ comets

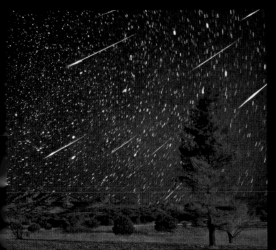

METEORS

On a clear night you may see little streaks of light in the sky, which are often called falling or shooting stars. But these streaks are properly called meteors. They are caused by little rocky specks plunging through the atmosphere towards Earth. Friction with the air makes them so hot that they burn up into dust. As much as 200 tonnes of meteor dust falls to Earth every day.

◄ LEONID METEOR SHOWER
When Earth crosses the orbit of a comet dozens of meteors per hour may be seen. This is called a meteor shower. Showers are named after the part of the sky they appear to come from. For example, the Leonid shower in mid-November seems to come from the constellation Leo. It takes place when Earth passes through dust from comet Tempel-Tuttle. Every 30 years or so it puts on an exceptional display of hundreds and thousands of meteors per hour, as seen here.

FIND OUT MORE ⏭ Atmosphere 234–235 • Solar System 172–173

ASTRONOMY

The scientific study of the stars and other objects in space is called astronomy. Astronomers observe the Universe using telescopes, which focus light from distant objects and make them clearer. Different types of telescope also reveal rays of light that are invisible to the human eye. On the ground, radio telescopes capture radio waves. Telescopes in space study rays that cannot pass through Earth's atmosphere.

▲ ANCIENT STARGAZERS

People began stargazing and recording their observations over 5,000 years ago. This 17th-century Indian painting shows a man with the tools of astrology: the study of the sky for signs that might influence life on Earth. He has an astrolabe for sighting the stars, zodiacal tables giving information about the constellations of the zodiac, and an hour-glass to tell the time.

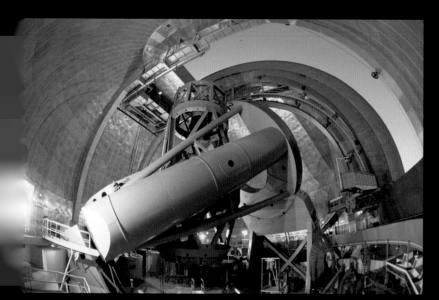

◄ GIANT TELESCOPES

The Hale Telescope at Palomar Observatory in California, USA, was completed in 1948, and was the first giant telescope. It has a light-gathering mirror 508 cm (200 in) across. There are two types of telescope that study light. Reflecting telescopes capture light with a mirror, and refracting telescopes use a lens. All modern professional telescopes are reflectors.

astronomy

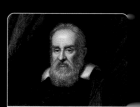

GALILEO
Italian, 1564-1642

Galileo was a physicist, mathematician, and an astronomer. In 1609 he built a telescope and became one of the first to use one to look at the heavens. He saw mountains on the Moon, spots on the Sun, observed the phases of Venus, and watched Jupiter's four large moons. In later life, his support for the view of Copernicus that Earth and the planets circle round the Sun brought him into conflict with the Church.

◄ RADIO ASTRONOMY

A radio astronomer works at the control console of the 100-m (330-ft) radio telescope at Effelsberg, Germany. Many important discoveries in astronomy have been made with radio telescopes, such as active galaxies and supernova remnants, pulsars, gas between the stars, and even echoes from the Big Bang.

FIND OUT MORE ▶▶ Lenses 115 • Reflection 113 • Refraction 114 • Space Observatories 196–197 • Telescopes 117

OBSERVATORIES

Optical observatories study the visible radiation (light) from objects in space, and most are located high up on mountaintops where the air is thinner, drier, and less polluted than at lower altitudes. But there are also observatories with telescopes that can detect invisible forms of radiation, such as gamma rays, infrared rays, and radio waves. Observatories in space are also used to detect these rays, as well as X-rays and ultraviolet rays.

MAUNA KEA OBSERVATORY ▲
Mauna Kea Observatory in Hawaii sits on the summit of an extinct volcano. Inside the open dome is one of the giant Keck telescopes. There are two Keck telescopes, each with a mirror 10 m (33 ft) across. A single mirror of this size would bend under its own weight. Instead, the mirrors are made up of 36 hexagonal (six-sided) sections which can be adjusted to get the best views of space.

EUROPEAN SOUTHERN OBSERVATORY ▶
The most powerful telescope on Earth is the Very Large Telescope (VLT) of the European Southern Observatory in Chile. The VLT is made up of four telescopes working together. Each has a mirror 8.2 m (27 ft) across – a billion times more powerful than the naked eye. The VLT is returning some of the best images of the Universe seen so far.

observatories

▼ VERY LARGE ARRAY
In New Mexico, USA, there is a group of 27 radio telescopes called the Very Large Array. The telescopes work together, acting as one big dish around 27 km (17 miles) across. Radio signals are collected by the dish and reflected onto a central antenna. The signals are then fed to a receiver and are processed to produce pictures, called radio images.

Each telescope runs on tracks so it can focus on different parts of the sky

Dish reflector 25 m (82 ft) across collects radio waves

Radio antenna receives signals focused on it by the reflector

FIND OUT MORE ▶▶ Energy Waves **98–99** • Space Observatories **196–197** • Telescopes **117**

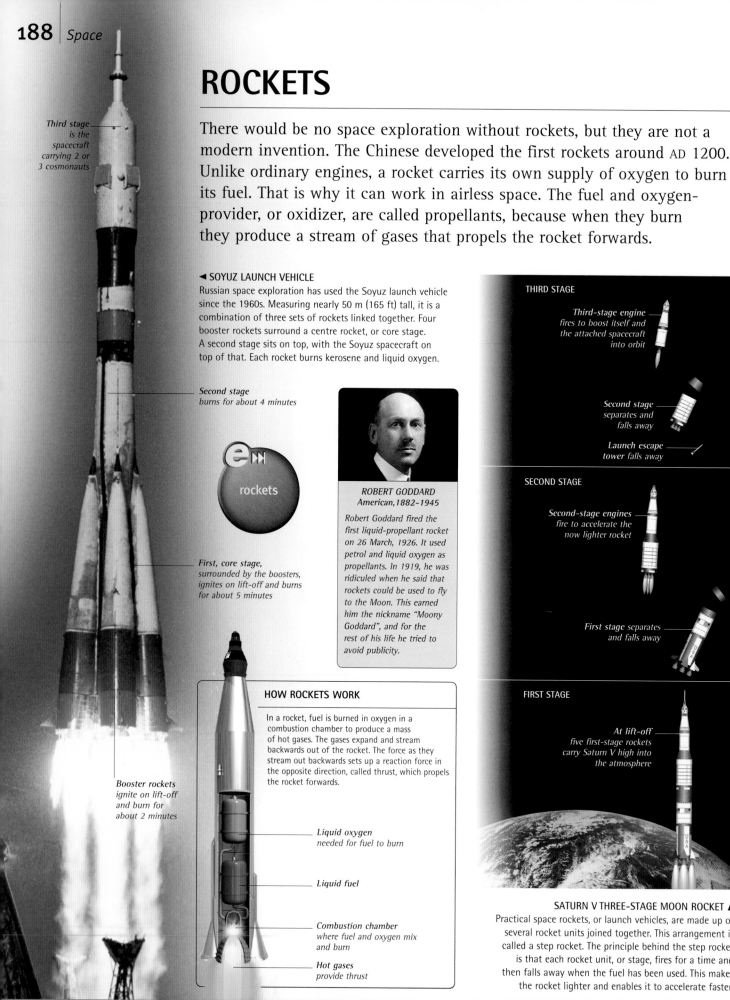

ROCKETS

There would be no space exploration without rockets, but they are not a modern invention. The Chinese developed the first rockets around AD 1200. Unlike ordinary engines, a rocket carries its own supply of oxygen to burn its fuel. That is why it can work in airless space. The fuel and oxygen-provider, or oxidizer, are called propellants, because when they burn they produce a stream of gases that propels the rocket forwards.

Third stage is the spacecraft carrying 2 or 3 cosmonauts

Second stage burns for about 4 minutes

First, core stage, surrounded by the boosters, ignites on lift-off and burns for about 5 minutes

Booster rockets ignite on lift-off and burn for about 2 minutes

◀ SOYUZ LAUNCH VEHICLE

Russian space exploration has used the Soyuz launch vehicle since the 1960s. Measuring nearly 50 m (165 ft) tall, it is a combination of three sets of rockets linked together. Four booster rockets surround a centre rocket, or core stage. A second stage sits on top, with the Soyuz spacecraft on top of that. Each rocket burns kerosene and liquid oxygen.

rockets

ROBERT GODDARD
American, 1882-1945

Robert Goddard fired the first liquid-propellant rocket on 26 March, 1926. It used petrol and liquid oxygen as propellants. In 1919, he was ridiculed when he said that rockets could be used to fly to the Moon. This earned him the nickname "Moony Goddard", and for the rest of his life he tried to avoid publicity.

HOW ROCKETS WORK

In a rocket, fuel is burned in oxygen in a combustion chamber to produce a mass of hot gases. The gases expand and stream backwards out of the rocket. The force as they stream out backwards sets up a reaction force in the opposite direction, called thrust, which propels the rocket forwards.

Liquid oxygen needed for fuel to burn

Liquid fuel

Combustion chamber where fuel and oxygen mix and burn

Hot gases provide thrust

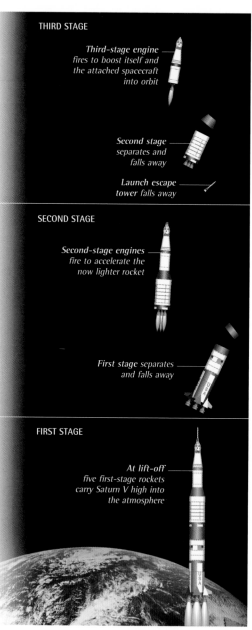

THIRD STAGE

Third-stage engine fires to boost itself and the attached spacecraft into orbit

Second stage separates and falls away

Launch escape tower falls away

SECOND STAGE

Second-stage engines fire to accelerate the now lighter rocket

First stage separates and falls away

FIRST STAGE

At lift-off five first-stage rockets carry Saturn V high into the atmosphere

SATURN V THREE-STAGE MOON ROCKET ▲

Practical space rockets, or launch vehicles, are made up of several rocket units joined together. This arrangement is called a step rocket. The principle behind the step rocket is that each rocket unit, or stage, fires for a time and then falls away when the fuel has been used. This makes the rocket lighter and enables it to accelerate faster.

FIND OUT MORE ▶▶ Forces 64–65 • Space Travel 190–191

SATELLITES

An object that circles another in space is called its satellite. Earth has one natural satellite, the Moon, but a swarm of artificial satellites. The USSR launched the first artificial satellite, Sputnik 1, in October 1957. Today, thousands of satellites circle Earth in different **ORBITS**, doing all kinds of jobs, such as relaying phone calls and TV broadcasts, and monitoring weather.

HOW SATELLITES WORK ▶
Satellites are built of the lightest materials possible to make them easier to launch. They can carry a wide variety of instruments, such as cameras, telescopes, radiation sensors, and radio equipment. Panels of solar cells provide electricity to power the instruments.

Solar panel
made up of thousands of solar cells converts the energy in sunlight into electricity

Satellite structure
built of ultralightweight materials

Dish antenna
used for transmitting and receiving radio communications to and from Earth

satellites

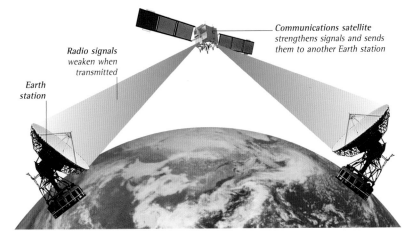

Communications satellite strengthens signals and sends them to another Earth station

Radio signals weaken when transmitted

Earth station

▲ SATELLITE COMMUNICATIONS
Earth stations use huge dish antennae to beam radio waves up to communication satellites. The satellites beam them back down to other Earth stations in the same or a different country. Many national and most international telephone calls, emails, and fax messages are now handled by networks of communications satellites, many of them in geostationary orbit.

▲ SATELLITE IMAGING
The effects of logging in the forests of British Columbia, Canada, are shown in this remote-sensing satellite image. The yellow patches show where trees have been cut down. Images such as these provide a useful way of monitoring changes in the environment and Earth's natural resources.

ORBITS

Satellites circle in space around Earth in a variety of paths, or orbits. They are kept in orbit by achieving a balance between their speed and the force of gravity: speed pushes them outward, while gravity pulls them inward. The speed needed to stay in orbit at a given height is called orbital velocity. For orbit at a few hundred kilometres, the orbital velocity is 28,000 kph (17,500 mph).

SATELLITES AND THEIR ORBITS	
Low-Earth orbit	Mobile communications, reconnaissance, astronomy
Polar orbit	Weather, navigation
Highly elliptical orbit	Communications at northern latitudes
Geostationary orbit	Communications, weather

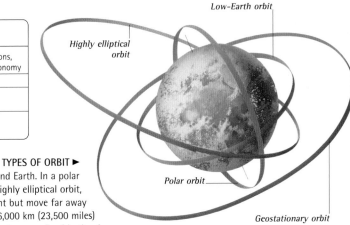

Low-Earth orbit

Highly elliptical orbit

Polar orbit

Geostationary orbit

TYPES OF ORBIT ▶
Satellites follow different orbits around Earth. In a polar orbit, they circle over the poles. In a highly elliptical orbit, they swoop close to Earth at one point but move far away at another. In a geostationary orbit, 36,000 km (23,500 miles) high, they circle every 24 hours and appear fixed in the sky. A low-Earth orbit is only a few hundred kilometres above Earth.

SPACE TRAVEL

Less than four years after the launch of the first satellite into space, Sputnik 1 in October 1957, human space travel began. Since then, American astronauts have walked on the Moon, and Russian cosmonauts have remained in space for more than a year at a time in space stations. Today, astronauts are launched by both rockets and the **SPACE SHUTTLE**. In the future, astronauts are likely to return to the Moon to set up bases, and even travel to Mars to explore the secrets of our neighbouring planet.

▲ FIRST SPACEMAN
Yuri Gagarin was first man to travel in space, on 12 April, 1961, orbiting Earth once. The first American to orbit Earth, John Glenn, flew into space on 20 February, 1962.

▼ EXPLORING THE MOON
The last Apollo Moon-landing mission, Apollo 17, took place in December 1972. Eugene Cernan stands by the lunar rover, used to carry the astronauts and their equipment. Cernan and Harrison H. Schmitt landed on the Moon's surface in a lunar module. In all, six Apollo spacecraft made successful landings on the Moon.

Lunar rover, powered by electric motors, had a top speed of 16 kph (10 mph).

▲ FIRST MEN ON THE MOON
On 20 July, 1969, Apollo 11 made the first Moon landing. Neil Armstrong (left) and Buzz Aldrin (right) walked on the Moon. Michael Collins (centre) manned the command module.

▲ MISSION CONTROL
After launch, all American manned space missions are controlled from Mission Control at Houston, Texas. Controllers oversee all aspects of mission operations, such as checking spacecraft engineering data, following in-flight experiments, and communicating with the crew.

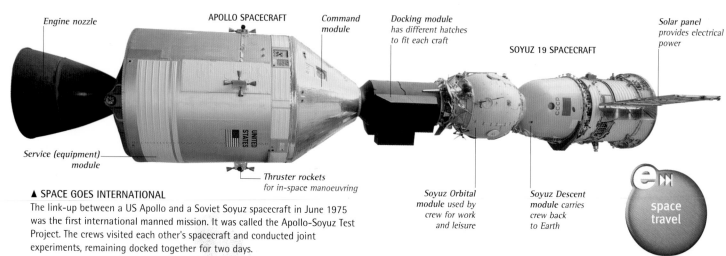

Engine nozzle

APOLLO SPACECRAFT

Command module

Docking module has different hatches to fit each craft

SOYUZ 19 SPACECRAFT

Solar panel provides electrical power

Service (equipment) module

UNITED STATES

Thruster rockets for in-space manoeuvring

Soyuz Orbital module used by crew for work and leisure

Soyuz Descent module carries crew back to Earth

▲ SPACE GOES INTERNATIONAL
The link-up between a US Apollo and a Soviet Soyuz spacecraft in June 1975 was the first international manned mission. It was called the Apollo-Soyuz Test Project. The crews visited each other's spacecraft and conducted joint experiments, remaining docked together for two days.

space travel

SPACE SHUTTLE

When it was first launched, on 12 April, 1981, the space shuttle began a new era in space flight. Until then, all launch vehicles had been expendable – they could be used only once. But the space shuttle is re-usable – most parts can be used again. The shuttle is made up of twin booster rockets, a winged orbiter which carries the crew, and an external fuel tank.

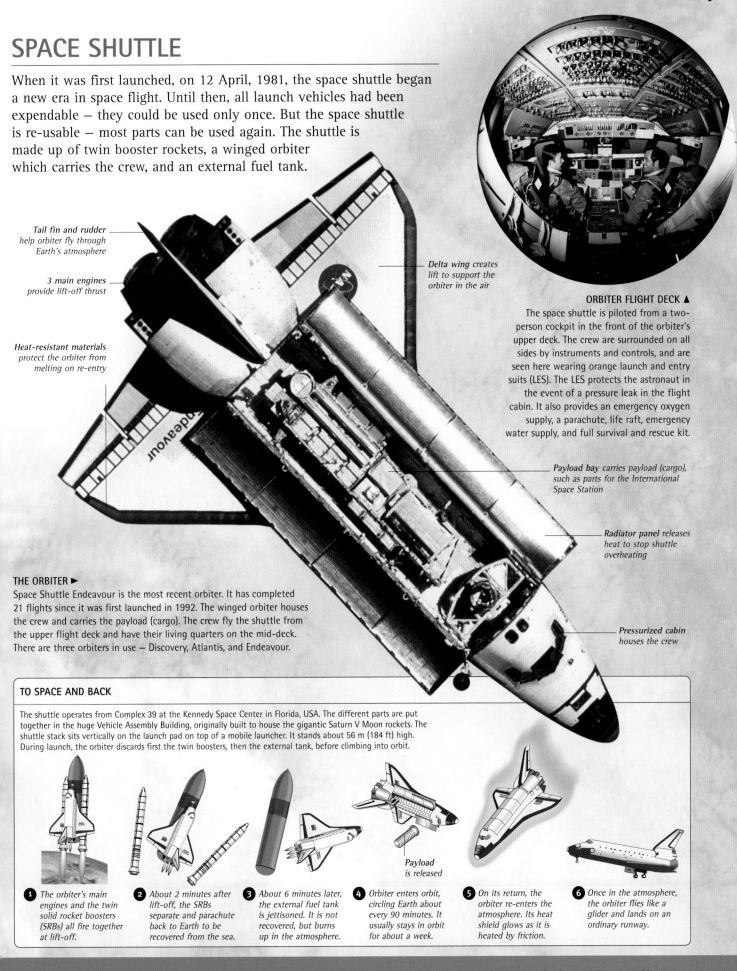

Tail fin and rudder
help orbiter fly through Earth's atmosphere

3 main engines
provide lift-off thrust

Heat-resistant materials
protect the orbiter from melting on re-entry

Delta wing *creates lift to support the orbiter in the air*

ORBITER FLIGHT DECK ▲
The space shuttle is piloted from a two-person cockpit in the front of the orbiter's upper deck. The crew are surrounded on all sides by instruments and controls, and are seen here wearing orange launch and entry suits (LES). The LES protects the astronaut in the event of a pressure leak in the flight cabin. It also provides an emergency oxygen supply, a parachute, life raft, emergency water supply, and full survival and rescue kit.

Payload bay *carries payload (cargo), such as parts for the International Space Station*

Radiator panel *releases heat to stop shuttle overheating*

THE ORBITER ▶
Space Shuttle Endeavour is the most recent orbiter. It has completed 21 flights since it was first launched in 1992. The winged orbiter houses the crew and carries the payload (cargo). The crew fly the shuttle from the upper flight deck and have their living quarters on the mid-deck. There are three orbiters in use – Discovery, Atlantis, and Endeavour.

Pressurized cabin *houses the crew*

TO SPACE AND BACK

The shuttle operates from Complex 39 at the Kennedy Space Center in Florida, USA. The different parts are put together in the huge Vehicle Assembly Building, originally built to house the gigantic Saturn V Moon rockets. The shuttle stack sits vertically on the launch pad on top of a mobile launcher. It stands about 56 m (184 ft) high. During launch, the orbiter discards first the twin boosters, then the external tank, before climbing into orbit.

Payload is released

1 The orbiter's main engines and the twin solid rocket boosters (SRBs) all fire together at lift-off.

2 About 2 minutes after lift-off, the SRBs separate and parachute back to Earth to be recovered from the sea.

3 About 6 minutes later, the external fuel tank is jettisoned. It is not recovered, but burns up in the atmosphere.

4 Orbiter enters orbit, circling Earth about every 90 minutes. It usually stays in orbit for about a week.

5 On its return, the orbiter re-enters the atmosphere. Its heat shield glows as it is heated by friction.

6 Once in the atmosphere, the orbiter flies like a glider and lands on an ordinary runway.

FIND OUT MORE ▶▶ Astronauts 192–193 • Flight 96 • Friction 68 • Interplanetary Missions 198–199 • Moon 177

ASTRONAUTS

Since Yuri Gagarin became the first
spaceman in 1961, hundreds of space travellers,
or astronauts (called cosmonauts if Russian),
have ventured into space. Experience has shown
that humans can work well in space both on board
their spacecraft and outside on **EVAs**. In space,
astronauts' bodies are constantly monitored, both to check
their health, and as part of the study of **SPACE MEDICINE** –
research into how the body is affected by weightless conditions.

*Gloves
contain their own
heating units*

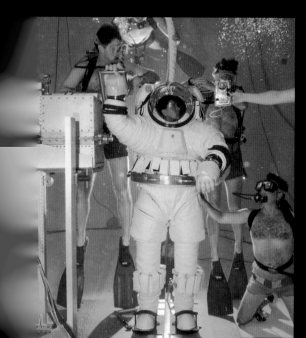

◄ IN TRAINING
On Earth, an astronaut practises for
spacewalking in a water tank. He wears a suit
like a spacesuit, which is weighted so that it
neither rises nor sinks. Such neutral buoyancy
(floating) conditions are similar to the state
of weightlessness that astronauts experience,
and have to work in, while in space.

*Red stripes
on one extravehicular
mobility unit (EMU), or
spacesuit, help observers to
identify the spacewalkers*

ALMONDS

*The backpack
provides oxygen, water for
cooling the spacesuit, and
electrical power*

◄ ON THE MENU
Astronauts on today's space flights eat a
variety of foods. Some are in their natural
state, such as nuts and biscuits, and some are
canned or frozen. Other foods are dehydrated
and need to be mixed with water before
eating. In the early days of space flight,
astronauts ate nutritious but unappetizing
food pastes out of toothpaste-type tubes.

TINNED PINEAPPLE   **GRANOLA**

*A metal arm
is used to secure tools that
astronauts needs in their work*

*The spacesuit
is a pressurized multilayer garment
that protects the astronaut from
extreme temperatures, dangerous
radiation, and meteorite particles*

EVA

Up in orbit, astronauts sometimes have to work outside their spacecraft. This extravehicular activity, or EVA, is popularly called spacewalking. Russian cosmonaut Alexei Leonov and US astronaut Edward White pioneered spacewalking in 1965. Today, astronauts go on EVAs to recover and repair satellites and carry out construction work on the International Space Station.

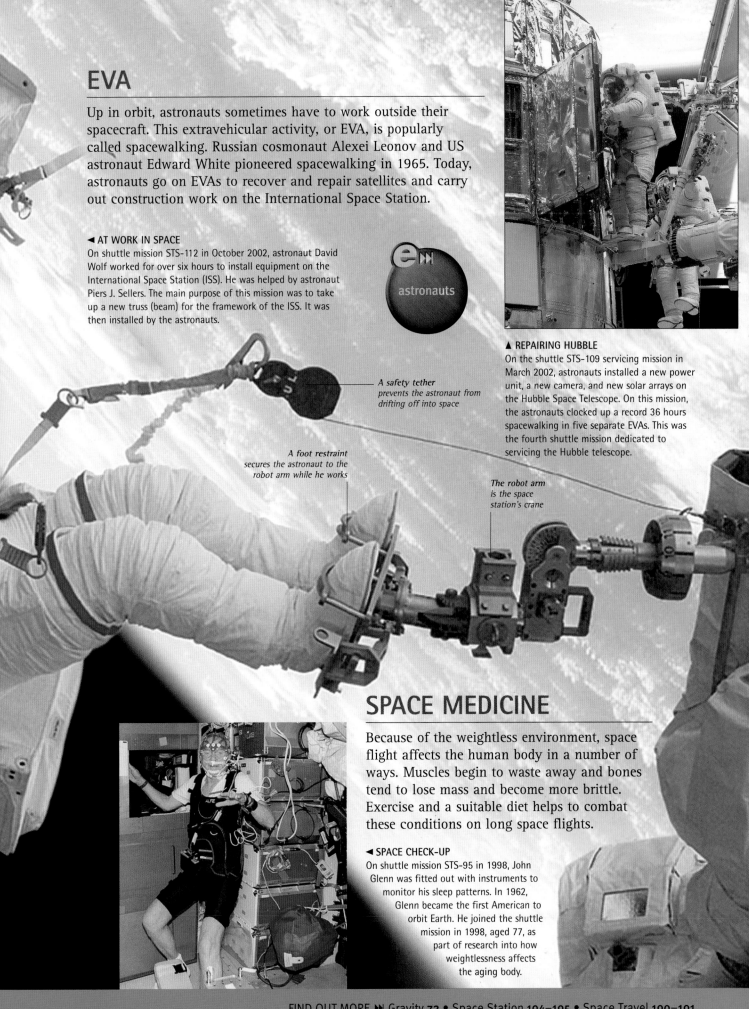

◄ AT WORK IN SPACE
On shuttle mission STS-112 in October 2002, astronaut David Wolf worked for over six hours to install equipment on the International Space Station (ISS). He was helped by astronaut Piers J. Sellers. The main purpose of this mission was to take up a new truss (beam) for the framework of the ISS. It was then installed by the astronauts.

astronauts

A safety tether prevents the astronaut from drifting off into space

A foot restraint secures the astronaut to the robot arm while he works

The robot arm is the space station's crane

▲ REPAIRING HUBBLE
On the shuttle STS-109 servicing mission in March 2002, astronauts installed a new power unit, a new camera, and new solar arrays on the Hubble Space Telescope. On this mission, the astronauts clocked up a record 36 hours spacewalking in five separate EVAs. This was the fourth shuttle mission dedicated to servicing the Hubble telescope.

SPACE MEDICINE

Because of the weightless environment, space flight affects the human body in a number of ways. Muscles begin to waste away and bones tend to lose mass and become more brittle. Exercise and a suitable diet helps to combat these conditions on long space flights.

◄ SPACE CHECK-UP
On shuttle mission STS-95 in 1998, John Glenn was fitted out with instruments to monitor his sleep patterns. In 1962, Glenn became the first American to orbit Earth. He joined the shuttle mission in 1998, aged 77, as part of research into how weightlessness affects the aging body.

FIND OUT MORE ►► Gravity 72 • Space Station 194–195 • Space Travel 190–191

▲ OFF TO SALYUT
Two cosmonauts look out of the capsule of the Soyuz 37 spacecraft before lifting off to visit the Russian space station Salyut 6 in 1980. Some cosmonauts in Salyut 6 remained in space for as long as six months, smashing all space endurance records. Salyut 6 was replaced in 1982 by Salyut 7.

SPACE STATIONS

A space station is a spacecraft designed to stay in orbit for many years. Conditions inside are carefully controlled: solar panels supply power, a comfortable atmosphere is maintained, and air and water are recycled. On board, astronaut-scientists conduct experiments, studying how the condition of weightlessness affects materials, people, plants, and animals. The **INTERNATIONAL SPACE STATION** is a project by many countries working together.

SPACE STATION			
NAME	COUNTRY	LAUNCH DATE	COMMENTS
Salyut 1	Soviet Union	1971	Visiting crew died on re-entry
Skylab	USA	1973	Visited by three crews
Salyut 7	Soviet Union	1982	3,216 days in orbit
Mir	Russia	1986	Date of first launch
ISS	International	1998	Date of first launch

The solar panels' 4,000 m² (43,000 sq ft) of solar cells convert sunlight into electricity to power the station

▲ SKYLAB IN ORBIT
The United States launched its first space station, Skylab, in 1973. One of its solar panels was ripped off at launch, damaging the heat shield over the crew's quarters. The first crew managed to repair the damage and stayed in orbit for 28 days. Later crews spent 59 and then 84 days on board.

▼ THE COMPLETED ISS
This picture shows what the ISS will look like when it is completed. It will measure 110 m (360 ft) end to end – as big as a football pitch. The units are carried up into space by Russian Proton rockets and US space shuttles. Once in orbit, they are put together by spacewalking astronauts and with the help of a travelling robot arm provided by Canada.

Truss assembly acts as a framework for the station

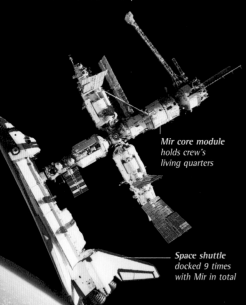

Mir core module holds crew's living quarters

Space shuttle docked 9 times with Mir in total

▲ MISSION TO MIR, 1995
In February 1986, Russia launched the first module of a space station that would remain in orbit for 15 years. It was named Mir (Peace). Five more modules were added over the years, carried into orbit by unmanned rockets. In 2001, Mir descended from orbit and broke up in the Earth's atmosphere.

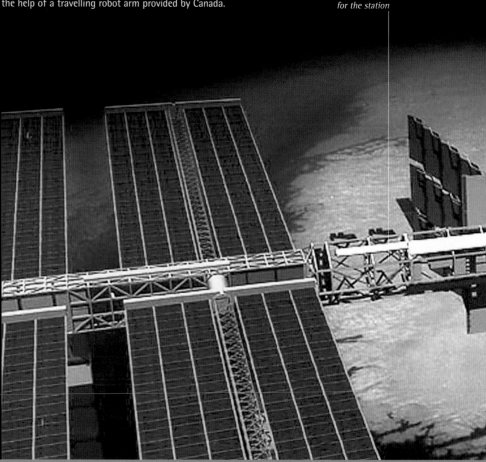

INTERNATIONAL SPACE STATION

Circling in orbit 390 km (240 miles) above Earth is the biggest
structure ever built in space, the International Space Station (ISS).
The nations involved include the United States, Russia, Canada,
Japan, Brazil, and 11 of the European Space Agency countries.
The ISS has been built by assembling its separate parts in orbit.

*A water droplet
forms a perfect
sphere in
weightless
conditions*

**◀ CONSTRUCTION
WORKERS**
Here in the shuttle orbiter
Endeavour, the crew posed
for a photo after delivering
the first US ISS element,
Unity, into orbit in December
1998. They had connected
Unity to the first ISS
element in orbit, the Russian
module Zarya. The poster
behind them displays the
flags of all the countries
involved in the ISS.

▲ SCIENTIFIC RESEARCH
Experiments in plant growth on the ISS are very important,
as plants could become a food supply for long interplanetary
flights of the future. Also, plants turn carbon dioxide into
oxygen, and could filter air. The laboratories of the ISS are
unique because experiments are carried out in conditions of
microgravity (near weightlessness). This is giving new insights
into many aspects of physics, chemistry, medicine, and biology.

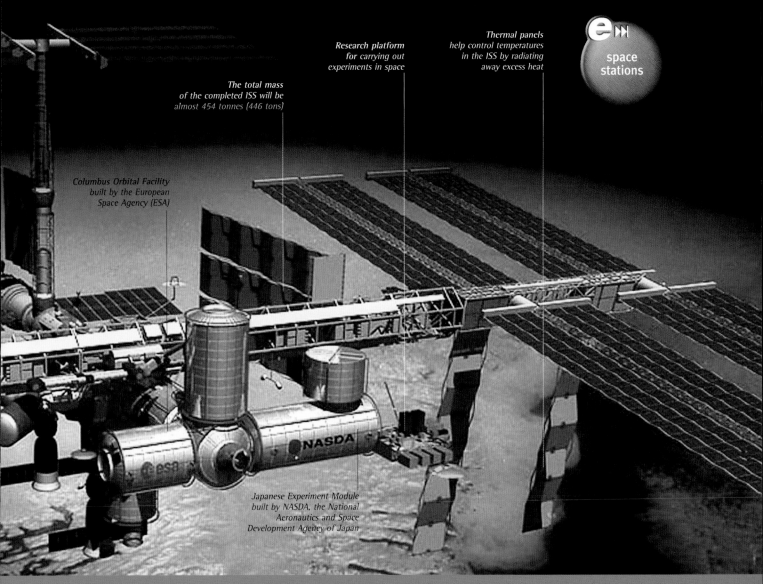

*Research platform
for carrying out
experiments in space*

*Thermal panels
help control temperatures
in the ISS by radiating
away excess heat*

**e ▸▸
space
stations**

*The total mass
of the completed ISS will be
almost 454 tonnes (446 tons)*

*Columbus Orbital Facility
built by the European
Space Agency (ESA)*

*Japanese Experiment Module
built by NASDA, the National
Aeronautics and Space
Development Agency of Japan*

FIND OUT MORE ▸▸ Astronauts **192–193** • Gravity **72** • Rocket **188** • Space Travel **190–191**

ELECTROMAGNETIC WAVES

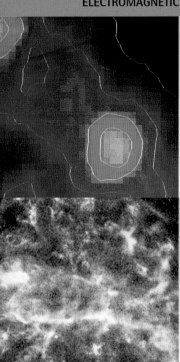

◄ GAMMA RAYS
This gamma ray image shows M1, the Crab Nebula. Gamma rays have the shortest wavelengths and the most energy. They are produced by some of the most violent events in the Universe, such as colliding galaxies.

◄ X-RAYS
High-energy X-rays are also emitted when violent events occur in the Universe. They carry more energy than visible light and have longer wavelengths than gamma rays. Here they are being emitted by the scattered debris from a supernova explosion.

◄ ULTRAVIOLET RAYS
This ultraviolet image of the Sun reveals different temperatures in the corona. Ultraviolet rays also have more energy than visible light, but less than X-rays. They come from very hot objects, such as the Sun. These are the rays that can burn the skin.

◄ INFRARED RAYS
M16, the Eagle Nebula, is seen here as an infrared image. Infrared rays have longer wavelengths than visible light. On Earth they are felt as heat, and are also known as heat rays. In space they penetrate interstellar dust and reveal what is behind it.

◄ RADIO WAVES
M51, the Whirlpool Galaxy, is seen here as a radio image. Radio waves have the longest wavelengths and the lowest energy. In space they are emitted from stars, galaxies, and gas clouds. Most can be picked up by ground-based radio telescopes.

SPACE OBSERVATORIES

High above the Earth's atmosphere, space observatories such as the **HUBBLE SPACE TELESCOPE** can view the Universe much more clearly than observatories on the ground. They can also pick up different, invisible forms of radiation that the atmosphere absorbs, such as gamma rays, X-rays, ultraviolet rays, and infrared rays. By looking at these different wavelengths, astronomers can better understand how the Universe works.

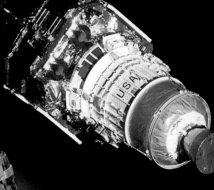

▲ COMPTON GAMMA RAY OBSERVATORY
When it was launched in 1991, the Compton Gamma Ray Observatory was the biggest space observatory ever. It mapped hundreds of gamma ray sources and recorded more than 2,500 gamma ray bursts, signs of the most violent happenings in the Universe.

CHANDRA X-RAY OBSERVATORY ►
The 13.7-m (45-ft) long Chandra X-ray Observatory is the world's most powerful X-ray telescope. In 1999, it was deployed from the Space Shuttle and then boosted into a highly elliptical orbit that took it 140,000 km (87,000 miles) above Earth.

◄ SOHO'S SUNWATCH
Since 1995, the Solar and Heliospheric Observatory (SOHO) has kept watch on the Sun at ultraviolet and visible light wavelengths. It is located in solar orbit 1.5 million km (930,000 miles) from Earth. SOHO investigates the Sun's surface, its interior, and its outer atmosphere, the corona.

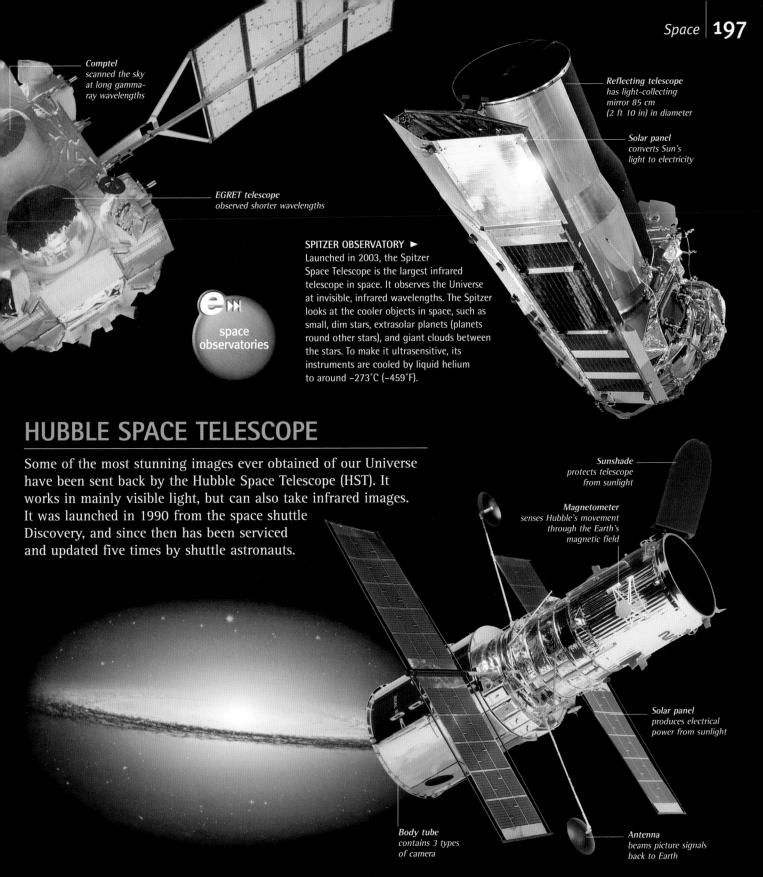

Comptel *scanned the sky at long gamma-ray wavelengths*

EGRET telescope *observed shorter wavelengths*

Reflecting telescope *has light-collecting mirror 85 cm (2 ft 10 in) in diameter*

Solar panel *converts Sun's light to electricity*

e ▶▶ **space observatories**

SPITZER OBSERVATORY ▶
Launched in 2003, the Spitzer Space Telescope is the largest infrared telescope in space. It observes the Universe at invisible, infrared wavelengths. The Spitzer looks at the cooler objects in space, such as small, dim stars, extrasolar planets (planets round other stars), and giant clouds between the stars. To make it ultrasensitive, its instruments are cooled by liquid helium to around –273°C (–459°F).

HUBBLE SPACE TELESCOPE

Some of the most stunning images ever obtained of our Universe have been sent back by the Hubble Space Telescope (HST). It works in mainly visible light, but can also take infrared images. It was launched in 1990 from the space shuttle Discovery, and since then has been serviced and updated five times by shuttle astronauts.

Sunshade *protects telescope from sunlight*

Magnetometer *senses Hubble's movement through the Earth's magnetic field*

Solar panel *produces electrical power from sunlight*

Body tube *contains 3 types of camera*

Antenna *beams picture signals back to Earth*

▲ SOMBRERO GALAXY
This HST image of the spiral galaxy M104, named the Sombrero Galaxy, taken in 2003, shows the galaxy more clearly than has ever been seen before. From the side, the galaxy's disc appears as a dark band against the bulge of bright stars in the centre of the galaxy. The HST does not take real colour pictures. Instead, it takes pictures through colour filters which are then combined together to reproduce the true colour. By computer processing, false-colour images can also be produced to emphasize certain features.

▲ HOW THE HST WORKS
Inside the body tube is a conventional reflecting telescope. A curved primary mirror 2.4 m (8 ft) across collects incoming light and reflects it onto a secondary mirror along the tube. This secondary mirror reflects and focuses the light through a hole in the primary mirror. The light is then fed to the cameras and other instruments. Hubble is controlled at NASA's Goddard Space Flight Center, Maryland, USA. Engineers there monitor the telescope and operate the telescope remotely, as instructed by astronomers based in Baltimore, USA.

FIND OUT MORE ▶▶ Astronomy **186** • Energy Waves **98–99** • Light **110–111** • Observatories **187** • Telescopes **117**

Camera platform
has wide- and
narrow-angle lenses

INTERPLANETARY MISSIONS

Scientists launch interplanetary missions to study the planets, asteroids, and comets close up. Mariner 2 was the first successful interplanetary craft flying past Venus in 1962. **MARS EXPLORATION** began with Mariner 4 in 1965. Since then all the planets have been visited by interplanetary craft. Some spacecraft study their targets as they fly by, some orbit their targets, and others even land on them.

Dish antenna
receives instructions from
Earth and transmits data

◀ VOYAGER FLY-BY

Two Voyager spacecraft are now winging their way out of the Solar System after highly successful missions to the outer planets that began in 1977. Both Voyager 1 and 2 flew past Jupiter and Saturn. Then Voyager 2 continued on to Uranus and Neptune, revealing their secrets for the first time.

RTG
(Radioisotope
thermoelectric
generator)
produces electrical
power

Long antenna
detects radio signals from
planets

▲ LANDING ON EROS

A spacecraft called NEAR-Shoemaker made an unexpected landing on the asteroid Eros in February 2001. It had spent the previous year in orbit around the 33-km (21-mile) rocky body, which sometimes comes within 22 million km (14 million miles) of Earth. At the end of the mission, the scientists decided to let the craft get closer and closer to the surface, taking pictures as it went. To their surprise, the craft survived quite a hard landing on the asteroid, and one instrument continued to work for several days afterwards.

▼ ORBITING VENUS

The Magellan spacecraft went into orbit around Venus in 1990. It used radar to find out what lay beneath the clouds that permanently cover the planet. The images it sent back showed that the landscape of Venus was covered in huge lava flows from hundreds of volcanoes. There were also spidery cracks, called arachnoids, in the planet's surface.

space
missions

Magellan
is deployed from the space
shuttle and sent into orbit

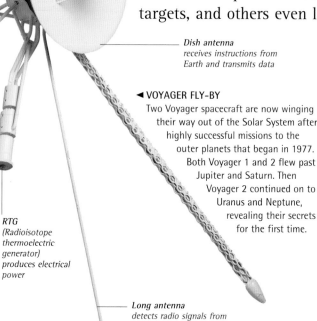

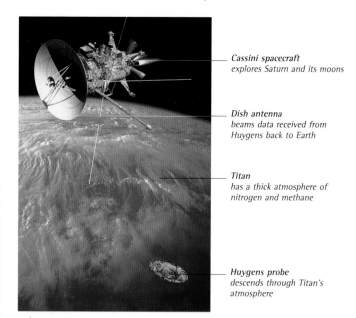

Cassini spacecraft
explores Saturn and its moons

Dish antenna
beams data received from
Huygens back to Earth

Titan
has a thick atmosphere of
nitrogen and methane

Huygens probe
descends through Titan's
atmosphere

▲ DESTINATION SATURN AND TITAN

In 2004 the Cassini spacecraft reached the beautiful ringed planet Saturn after a seven-year journey from Earth. The aim was to study the planet and many of its moons over a long period of time. Cassini released a probe called Huygens into the thick atmosphere of Saturn's largest moon, Titan. Huygens landed on Titan in January 2005 and sent back photographs of the moon's surface.

MARS EXPLORATION

Some of the most exciting interplanetary missions have been to Mars, our neighbouring planet. Mars is the only other planet where life may once have existed and where human beings could possibly settle in the future. Mars is being explored in depth by spacecraft on its surface and in orbit around it. These missions have found a lot of frozen water in the Martian rocks, perhaps all that is left of ancient Martian oceans.

▲ MARS EXPRESS

The European spacecraft Mars Express was launched in June 2003, and arrived at Mars seven months later. Since then, its instruments have made a detailed study of the world below, from imaging the entire surface, to mapping its mineral composition, to studying its atmosphere. Many of the images show the effects of the water that flowed across Mars 3–4 billion years ago.

▲ MARTIAN LANDSCAPE

This image, taken by the rover Spirit in 2004, shows the surface of Mars is rust-red and scattered with small rocks. Fine particles of dust blown up from the surface make the sky look pinkish-orange. Rover Opportunity landed in a crater, and its pictures of the crater rocks suggest that they were formed by water, from either a shallow, salty sea or pool, or under ice.

MAPPING MARS ▶

This is a topographical (three-dimensional) image of Mars taken by the Mars Global Surveyor in 1999. Blue areas are the lowest regions, red the highest. The blue region in the centre of the red areas is a deep basin known as Hellas, which is 1,600 km (1,000 miles) across.

▼ ROVING ACROSS MARS

In January 2004, two robot vehicles called Spirit and Opportunity touched down on Mars and began moving slowly over the surface. They carried instruments to study rocks, and cameras to take pictures of the surface, and for navigation. Before every stage of exploration, each rover took pictures of the area directly ahead. These were used to plan a route, avoiding hazards, to the next target area.

Pancam mast carries cameras

High-gain antenna transmits and receives radio signals

Wheel is powered by electric motor

Robot arm carries geological instruments and rock tools

FIND OUT MORE ▶▶ Asteroids **184** • Extraterrestrial Life **200–201** • Mars **178** • Saturn **180** • Venus **175**

EXTRATERRESTRIAL LIFE

Earth is the only world we know of that supports life. But is there extraterrestrial life – life beyond Earth – elsewhere in the Universe? It is possible that there is, or has been, other life in our Solar System, perhaps on Mars or Europa. But to find intelligent life, astronomers are looking much further away. In research programmes called SETI (Search for Extraterrestrial Intelligence), they monitor the skies for signals from intelligent life in deep space.

THE VOYAGER MESSAGE ▶
The two Voyager spacecraft travelling through the Solar System will soon leave it behind and enter interstellar space – space between the stars. Each spacecraft carries a disc on which images, natural sounds, speech, and music have been recorded. It is hoped that, one day, intelligent beings from another world may find a disc and get a picture of life on Earth.

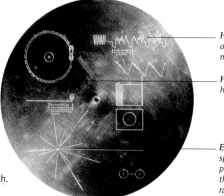

How to view the images of the natural and the man-made world

How to play the record to hear the sounds of Earth

Earth's position in space in relation to 14 pulsars (spinning stars that send beams of radiation across space)

▲ LIFE ON EUROPA?
Jupiter's large moon Europa, pictured here by the spacecraft Galileo, has a flat surface of pink ice. It is criss-crossed with cracks, which may have been caused by the movement of a liquid ocean beneath the surface, melted by energy caused by the powerful tidal effects of Jupiter's gravity. This has led to speculation that life may exist on Europa.

ET life

▼ SCANNING THE SKY
The Allen Telescope Array (ATA), in California, USA, listens for signals that may have come from other civilisations in space, while simultaneously observing stars and galaxies. The radio antennae below are part of the first batch of 42 that started work in 2007. When completed, the ATA will consist of 305 antennae studying the vicinities of one million nearby stars.

Each main dish measures 6.1 m (20 ft) across

PRIMITIVE LIFE ▶

The earliest forms of life on Earth appeared nearly four billion years ago in the form of cyanobacteria (blue-green algae). These organisms are still found today in Western Australia, where they form distinctive mounds called stromatolites. The evolution of life began from these single-celled organisms. On other worlds life may exist in a similar, primitive form.

▲ LOOKING FOR SIGNALS

A monitor screen displays signals received by the Arecibo radio telescope. Looking for a signal in space that may have been sent by intelligent extraterrestrials is like looking for a needle in a haystack. Huge quantities of data are received, which must then be processed. Through the SETI@Home project, ordinary computer users can help with this task.

Vertical lines make up the usual pattern – any variation may indicate a signal

EXTRASOLAR PLANETS ▶

More than 300 extrasolar planets, also called exoplanets, have been discovered. They orbit stars other than the Sun and may harbour life. A SETI project directs its radio telescopes at stars known to have planets, and stars like our Sun, as these are the most likely to have planets capable of supporting life.

THE DRAKE EQUATION

American radio astronomer Frank Drake, a SETI pioneer, drew up a list of key factors necessary for intelligent life to evolve on a planet. This led to an equation, named the Drake Equation, for calculating the number of possible civilizations in our galaxy. However, the factors are based on only one example of life on a planet, our own. We do not know if these factors will apply to life forms on other planets. Optimists estimate that there are millions of civilizations in the Milky Way. Pessimists estimate that there is just one — our own.

- How many stars in the galaxy are stable over the billions of years necessary for life to evolve?
- How many of these stars give birth to stable planetary systems around them?
- How many of these planets have suitable conditions for life?
- On how many of these planets does life begin and take hold?
- On how many of the planets does intelligent life evolve and become able to communicate?
- On how many of the planets with intelligent life are conditions right to create a technology suitable for communication across the Universe?
- How many potentially advanced civilizations are wiped out by natural or self-inflicted disasters?

FIND OUT MORE ▸▸ Algae **286–287** • Interplanetary Missions **198–199** • Jupiter **179** • Telescopes **117**

EARTH

PLANET EARTH

The rocky ball that forms our world is one of eight planets in the Solar System. Earth is a sphere, with a slight bulge in the middle at the Equator, and a diameter of 12,756 km (7,926 miles). It hurtles at speeds of 105,000 kph (65,000 mph) during its orbit around the Sun, turning on its **AXIS** once every 24 hours. This journey takes a year to complete. The Earth is the only planet that is known to support life, in a zone called the **BIOSPHERE**.

UNIQUE PLANET ▶
Water, oxygen, and energy from the Sun combine on Earth to help create suitable conditions for life. The planet's surface is mainly liquid water, which is why it looks blue from space. Earth is the only planet in the Solar System with an atmosphere that contains a large amount of oxygen. The Sun is 150 million km (93 million miles) away, producing heat that is bearable on Earth.

▲ ATMOSPHERE
The atmosphere is a layer of gas surrounding the Earth that is some 700 km (400 miles) thick. It is made up of nitrogen (78 per cent) and oxygen (21 per cent), plus traces of other gases. Tiny droplets of water vapour form the clouds we see.

▲ OCEANS
Oceans cover 70.8 per cent of the Earth's surface, to an average depth of 3.5 km (2 miles). The hydrosphere (watery zone) also includes freshwater rivers and lakes, but these make up less than 1 per cent of Earth's water.

▲ LAND
Dry land occupies 29.2 per cent of the Earth's surface, where the lithosphere (rocky crust) rises above sea level to form seven continents and countless smaller islands. Land can be categorised into biomes – major habitats such as forests, grasslands, and deserts.

BIOSPHERE

The biosphere is the part of Earth that contains what is needed for living things. This zone extends from the ocean floor to top of the troposphere (lower atmosphere). Tiny organisms can survive deep in the Earth's crust, but most forms of life are found from a few hundred metres below sea level to about 1,000 m (3,300 ft) above sea level.

THE LIFE ZONE ▶
Ozone is a gas spread thinly through the atmosphere. It filters harmful ultraviolet (UV) rays from sunlight, while allowing visible light (the light we can see) to pass through. Other gases in the atmosphere trap the Sun's heat when it is reflected from the Earth's surface, providing additional warmth for living things.

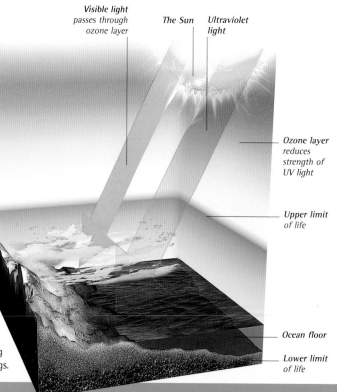

Visible light passes through ozone layer

The Sun

Ultraviolet light

Ozone layer reduces strength of UV light

Upper limit of life

Ocean floor

Lower limit of life

JAMES LOVELOCK
British, 1919–

Environmental scientist James Lovelock argues that the planet can be seen as a complete living organism, which he names Gaia, after the Greek goddess of Earth. Gaia theory states that Earth itself balances conditions to suit living things in the biosphere. This includes regulating the composition of the atmosphere, the chemistry of the oceans, and ground surface temperature.

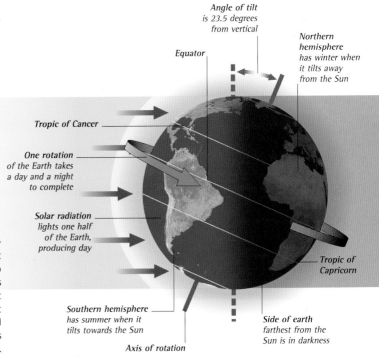

▲ ICE AND SNOW

The cryosphere (frozen zone) includes snow and glaciers on high mountains, sea ice, and the huge ice caps that cover the landmasses of Greenland and the Antarctic. In the past, during long cold eras called ice ages, ice covered much more of Earth's surface than it does today.

▲ EARTH SCIENCE

A meteorologist releases a weather balloon in Antarctica. Meteorology, the study of Earth's atmosphere, is one of the Earth sciences. Earth scientists study Earth's physical characteristics, from raindrops to rivers and the rocks beneath our feet. Other branches of study include geology (rocks), hydrology, (oceans and freshwater), and ecology (living things and the environment).

▲ STUDY TECHNIQUES

Satellite images allow scientists to monitor everything from ocean currents to minerals hidden below ground. Techniques such as radar and sonar have transformed our understanding of our planet. Some Earth scientists also spend time in the field, which means working outdoors, collecting data and samples from clouds, cliffs, craters, volcanic lava, and deep-buried ice.

AXIS

The ground beneath our feet may seem still, but in fact the Earth is spinning like a top as it orbits the Sun. The Earth takes 24 hours to rotate about its axis, an imaginary line running from the North Pole to the South Pole through the centre of the Earth. The Earth's axis is not at a right-angle to the path of its orbit, but tilts at an angle of 23.5°. The angle between each region of Earth and the Sun's rays alters through the year, producing seasonal changes in temperature and day length. These are most noticeable in regions next to the poles, which are most distant from the Equator.

planet Earth

DAY AND NIGHT ▶

As Earth turns about its axis, one half is bathed in sunlight and experiences day, while the other half is plunged into darkness and has night. The Earth always rotates eastward, so the Sun and stars appear to rise in the east and set in the west. The tilt of the planet means that at any time, one hemisphere (half of the Earth, as divided by the Equator) leans toward the Sun and experiences summer, while the other leans away and has winter.

Angle of tilt is 23.5 degrees from vertical

Equator

Northern hemisphere has winter when it tilts away from the Sun

Tropic of Cancer

One rotation of the Earth takes a day and a night to complete

Solar radiation lights one half of the Earth, producing day

Tropic of Capricorn

Southern hemisphere has summer when it tilts towards the Sun

Side of earth farthest from the Sun is in darkness

Axis of rotation

FIND OUT MORE ▶▶ Atmosphere **234–235** • Climate **236–237** • Earth **176** • Habitats **246–247** • Oceans **228–229**

EARTH'S STRUCTURE

The Earth is a giant, spinning ball of rock and metal. The rocky surface we live on is the Earth's thin outer layer, called the crust. In places the crust is just a few kilometres thick. Underneath the crust are two more layers, called the mantle, and the core, which combine to reach a depth of 6,370 km (3,960 miles). Scientists discovered these layers by studying how shock waves from earthquakes change direction and speed as they travel through the Earth. It is thought that the core creates Earth's **MAGNETOSPHERE**.

ANDRIJA MOHOROVICIC
Croatian, 1857-1916

Geophysicist Andrija Mohorovicic found that earthquake shock waves sped up when they reached about 20 km (12 miles) below the surface. He suggested that happened at a boundary where two different layers of material met. This boundary is between the crust and the mantle, and is now known as the Mohorovicic discontinuity, or Moho.

▼ EARTH'S LIFE STORY
The Earth came into being about 4,600 million years ago. Along with the other planets and moons in our Solar System, it was made from material left over after the birth of the Sun. Earth's surface has gone through many changes since, with the formation of the continents, oceans, and atmosphere, and the appearance of life.

Hazy atmosphere is beginning to form

▲ ACCRETION
Small particles of rock, dust, and gas in space are gradually pulled together by the gravity between them. The process is called accretion. The young Earth was formed by accretion over millions of years.

HEATING AND COOLING ▶
Huge pressure in the centre of Earth created heat that melted the rocks inside. For hundreds of millions of years the surface was bombarded by meteorites from space. About 4,200 million years ago, Earth's surface had cooled and a crust of solid rock had formed.

Band of light sedimentary rock

Band of rock containing iron oxides

OCEANS AND ATMOSPHERE ▲
The early atmosphere consisted of volcanic gases, which formed rain. From about 3,500 million years ago, this began to collect in oceans. Continents were also developing. Simple organisms in the oceans gave out oxygen into the atmosphere.

▲ PROOF OF EARLY OCEANS
Banded rocks found in the Hamersley Range National Park, Western Australia, are about 2,000 million years old. These ancient rocks provide evidence of Earth's history. They were laid down in layers under water, showing that oceans must have existed 2,000 million years ago. The red rock contains iron deposits, produced by residues of microorganisms reacting with seawater.

THE EARTH TODAY ▶
An imaginary slice out of the Earth shows that scientists believe it has a core made mostly of solid and molten iron, a mantle of solid and half-molten rock, and a crust of solid rock. The inside of the Earth is still extremely hot. Plate tectonics, mountain building, and erosion are constantly changing the appearance of the Earth's surface.

MAGNETOSPHERE

The Earth has a magnetic field around it, and the magnetosphere is the region in which this field can be felt. It stretches more than 60,000 km (37,000 miles) into space, like an invisible magnetic bubble, and protects the Earth from harmful solar radiation. The solar wind, particles which stream from the Sun, pull the magnetosphere into a teardrop shape.

THE CRUST CLOSE-UP ▼
Under the oceans, the crust is about 7 km (4 miles) thick, and made of young rocks. The continental crust is between 25 km (16 miles) and 90 km (56 miles) thick, and made of young and ancient rocks. The crust floats on the upper, semi-molten part of the mantle, and is cracked into giant plates that move slowly about.

Earth's structure

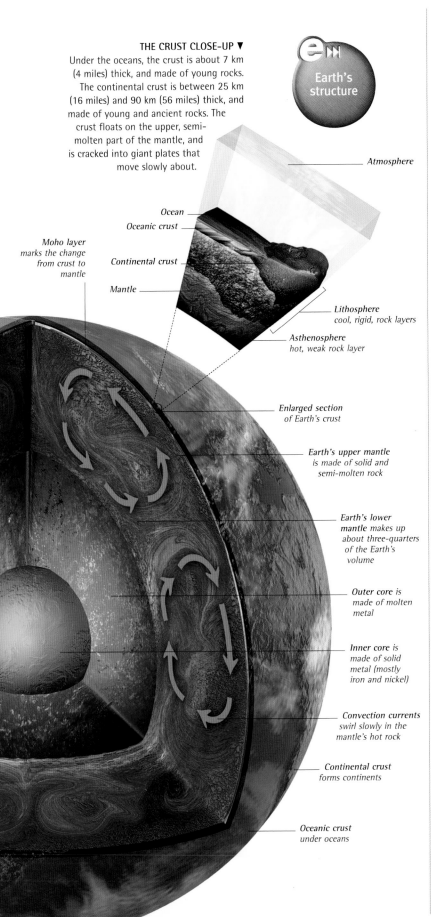

Atmosphere

Ocean
Oceanic crust
Continental crust
Mantle

Moho layer marks the change from crust to mantle

Lithosphere *cool, rigid, rock layers*

Asthenosphere *hot, weak rock layer*

Enlarged section of Earth's crust

Earth's upper mantle *is made of solid and semi-molten rock*

Earth's lower mantle *makes up about three-quarters of the Earth's volume*

Outer core *is made of molten metal*

Inner core *is made of solid metal (mostly iron and nickel)*

Convection currents *swirl slowly in the mantle's hot rock*

Continental crust *forms continents*

Oceanic crust *under oceans*

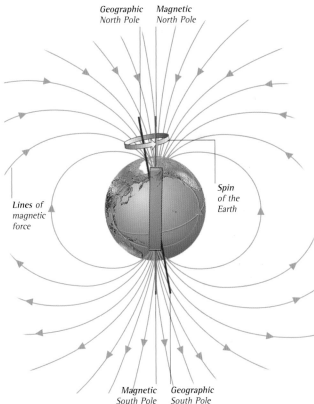

Geographic North Pole | Magnetic North Pole
Lines *of magnetic force*
Spin *of the Earth*
Magnetic South Pole | Geographic South Pole

▲ EARTH'S MAGNETIC FIELD
The Earth has a magnetic field that is the same shape as that of a bar magnet. It is as though the Earth contains a giant bar magnet with its poles located near the North Pole and South Pole. These magnetic Poles are tilted at a slight angle to the Earth's axis. Scientists think that the magnetic field is caused by currents of molten metal in the Earth's outer core. From time to time, these reverse, with north becoming south.

WILLIAM GILBERT
English, 1544-1603

William Gilbert was physician to Queen Elizabeth I of England. He was also the first person to realize that the Earth has a magnetic field similar to that of a bar magnet. He proved this by comparing the direction and tilt of a compass needle out in the open with its direction and tilt when held beside a model of the Earth containing a bar magnet.

FIND OUT MORE ▶▶ Erosion **222–223** • Plate Tectonics **208–209** • Rock Cycle **217** • Rocks **218–219**

PLATE TECTONICS

Scientists believe that the Earth's outer crust is made up of huge fragments, called tectonic plates, that fit together like a cracked eggshell. According to the theory of plate tectonics, devised in the 1970s, these plates ride like rafts on the softer, red-hot rock below and very move slowly over the globe, carrying the continents with them. Past arrangements of tectonic plates created one vast **SUPERCONTINENT**.

FRACTURED CRUST

Earth's crust is a giant jigsaw of seven enormous plates and about twelve smaller ones. Many scientists believe plate movement is driven by slow-churning currents deep in the mantle beneath. As the plates drift, they converge (move towards each other) and collide, or grind past one another at transform margins, or diverge (pull apart).

North America

Europe **7** Asia

Africa **6** India

South America

1 **2** **3** **4** **5** **8** **9**

Australasia

Antarctica

Plate names:
1 North American plate
2 Pacific plate
3 Nazca plate
4 South American plate
5 African plate
6 Arabian plate
7 Eurasian plate
8 Antarctic plate
9 Indo-Australian plate

Key to boundaries:
—— converging
—— diverging
—— transform
----- uncertain
☐ Boundary between African and Arabian plate

DIVERGING FAULT ▶
Plate boundary movement forms deep cracks known as faults. The Red Sea marks the boundary between plates bearing Africa and Arabia. As these two plates separate, molten rock wells up to fill the gap, creating new crust. The Red Sea is slowly widening because of this process. Now around 300 km (185 miles) across, it may one day be as wide as the Atlantic.

African/Arabian plate boundary runs through the Red Sea along the Gulf of Aqaba

GULF OF AQABA

AFRICAN PLATE MOVEMENT

ARABIAN PLATE MOVEMENT

Fault line

CORAL ISLANDS

THE RED SEA

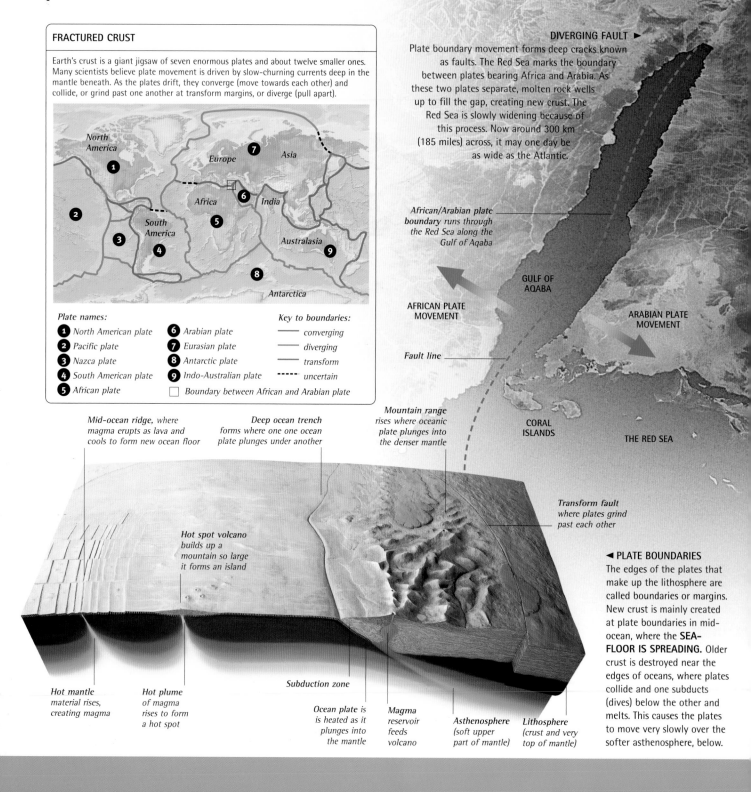

Mid-ocean ridge, where magma erupts as lava and cools to form new ocean floor

Deep ocean trench forms where one one ocean plate plunges under another

Mountain range rises where oceanic plate plunges into the denser mantle

Hot spot volcano builds up a mountain so large it forms an island

Transform fault where plates grind past each other

Hot mantle material rises, creating magma

Hot plume of magma rises to form a hot spot

Subduction zone

Ocean plate is is heated as it plunges into the mantle

Magma reservoir feeds volcano

Asthenosphere (soft upper part of mantle)

Lithosphere (crust and very top of mantle)

◀ PLATE BOUNDARIES
The edges of the plates that make up the lithosphere are called boundaries or margins. New crust is mainly created at plate boundaries in mid-ocean, where the **SEA-FLOOR IS SPREADING**. Older crust is destroyed near the edges of oceans, where plates collide and one subducts (dives) below the other and melts. This causes the plates to move very slowly over the softer asthenosphere, below.

SUPERCONTINENT

The shapes of continents such as eastern South America and western Africa would fit neatly if pushed together. The discovery of matching fossils and rock layers on land separated by wide oceans provided further evidence that landmasses were once united. Scientists call this supercontinent Pangaea. The slow movement of Earth's plates caused Pangaea to split apart.

250 million years ago continents formed part of Pangaea

Tethys

145 million years ago continents are slowly separating

Today's continents are still moving

◄ PANGAEA
Some 300 million years ago, plate movement drove Earth's landmasses together to form Pangaea (All-Earth). This was surrounded by the vast ocean Panthalassa. About 100 million years later Pangaea began to break up.

◄ MOVING CONTINENTS
An arm of the Tethys Sea, an ancient ocean, opened to split Pangaea in two. To the north lay Europe, North America, Greenland, and Asia, with South America, Africa, India, Australia, and Antarctica to the south.

◄ CONTINENTS TODAY
As plate movement continued, these large fragments split into smaller continents, which slowly came to their present positions. They continue to move at a rate of a few centimetres per year.

ALFRED WEGENER
German, 1880-1930

Climate expert and geophysicist Alfred Wegener pioneered the theory of continental drift in 1915. He became convinced that the continents were once joined, and put forward the idea of Pangaea. On the Arctic island of Spitzbergen, Wegener found fossils of tropical ferns, which suggested that the island had once lain in the tropics. His ideas were not taken seriously until the 1960s.

tectonics

SEA-FLOOR SPREADING

Mountain chains, longer and mightier than any on land, run down the centre of the oceans. At these mid-ocean ridges, where tectonic plates diverge, molten magma erupts to bridge the gap. Rock samples taken from the Atlantic floor in the 1960s showed that the youngest rocks lay in the centre of the ridges, with older rocks to either side. As the new rock forms, older rock is pushed aside, and the sea floor widens, or spreads.

MID-OCEAN FISSURE
As tectonic plates separate at mid-ocean ridges, the rocks split to form gaping fissures (cracks) like this one on the Pacific bed. Seawater entering the cracks is heated by upwelling lava and mixed with minerals, to belch in dark clouds from openings called black smokers. Crabs are among the sea creatures that feed on the micro-organisms that thrive here.

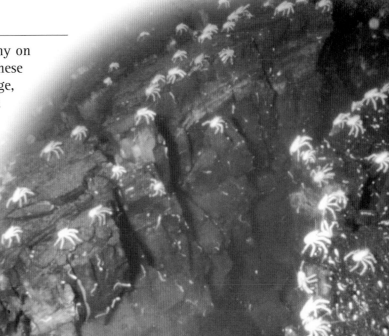

FIND OUT MORE ▸▸ Earth's Structure **206–207** • Earthquakes **210–211** • Ocean Floor **230** • Rock Cycle **217**

EARTHQUAKES

Earthquakes are caused by movements of the tectonic plates that form Earth's crust. **SEISMOLOGY** is the study of earthquakes. Most occur at cracks called **FAULTS**, at the boundaries where the plates meet. Every minute, the ground shakes somewhere in the world, but these vibrations are usually minor tremors that are barely noticed. When a major earthquake strikes, the ground shakes violently, and buildings and bridges topple.

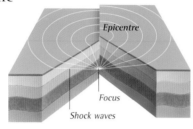

Epicentre

Focus

Shock waves

▲ SEISMIC WAVES

As the plates slowly shift, rocks are put under pressure. They stick, then stretch and, as the strain gets too great, they shatter and jolt into new positions. Seismic (shock) waves radiate from the earthquake's focus, underground. The epicentre, above the focus, suffers the worst damage.

◄ QUAKE-PROOF BUILDINGS

Major earthquakes can kill thousands of people. Most deaths occur when poorly constructed buildings collapse. Buildings in earthquake zones can be designed to withstand severe shaking using reinforced concrete and deep or flexible foundations. The Transamerica Building in quake-prone San Francisco, USA, has a sturdy triangular frame of concrete-clad steel columns.

EARTHQUAKE ZONES

Earthquakes can strike anywhere, but most occur along plate boundaries. This map shows that earthquake zones and tectonic plates are closely linked. The rim of the vast Pacific plate that lies below the Pacific Ocean is notorious for earthquakes. It is called the Ring of Fire because volcanoes are also common here.

North America

Europe

Asia

Africa

Pacific Ocean

South America

Australasia

Antarctica

◉ Earthquake zones

Roads and flyovers *are ripped apart*

Vehicles *are tipped sideways*

Trees *are uprooted*

▲ EARTHQUAKE DEVASTATION

Located on the edge of the Pacific plate, Japan is regularly hit by earthquakes. On 17 January 1995, a major quake devastated the city of Kobe, killing 5,400 people. The Hanshin expressway collapsed, scattering cars and trucks like skittles. Fires raged after shock waves in the ground fractured gas, oil, and electricity lines. Earthquakes can also trigger landslides, avalanches, volcanic eruptions, and giant waves at sea, called tsunamis.

FAULT

Faults are deep cracks in rocks, mostly caused by movement at plate margins. Deep earthquakes strike in subduction zones where two plates collide and one slides below the other. Shallow earthquakes occur mostly where two plates grind past one another. The rocks may be shifted only a few centimetres, but over millions of years, this can add up to hundreds of kilometres of movement sideways, and up to 30 km (19 miles) of vertical movement.

FAULTLINE IN ALGERIA ▶
The Oued Fodda faultline stretches across open countryside surrounding the Algerian town of El Asnam. Huge cracks opened up as rocks were shattered in two massive earthquakes in October 1980, which destroyed 80 per cent of buildings in the town. A block of land was being wrenched upwards in a movement called reverse faulting.

earthquakes

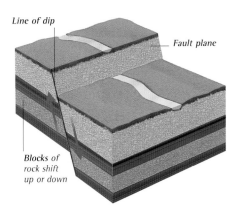

Line of dip

Fault plane

Blocks of rock shift up or down

▲ NORMAL DIP-SLIP FAULT
The rocks along a fault may move up or down, sideways or diagonally, depending on the angle of the fault plane. The angle of the fault plane to the horizontal is known as the dip. In a normal fault, also known as a dip-slip fault, the rocks shift straight down or up, following the line of dip.

Throw

One block slides up over the other

▲ REVERSE FAULT
The distance that the rocks slip up or down during a quake or tremor is called the throw. In a reverse fault, pressure causes one block of rock to overhang another. As the rocks shift, the block is forced farther up and over the other. A reverse fault with a fault plane of 45° or less is called a thrust fault.

Heave

Blocks slide past each other horizontally

Blocks slide past each other vertically

OBLIQUE-SLIP FAULT
In a strike-slip fault, rocks scrape sideways past one another. The amount of sideways slip is called the heave. The San Andreas Fault, which runs along the west coast of North America, is a famous example. The rocks in an oblique-slip fault slide past each other, and also up and down in a diagonal movement.

SEISMOLOGY

Seismologists study earthquakes. They also examine the behaviour of seismic waves passing through the Earth to find out about its structure. Instruments called seismographs measure the intensity of seismic waves. The magnitude (size) of earthquakes can be rated by measuring either these waves, on the Richter scale, or the damage caused – the Mercalli scale. Earthquakes cannot be prevented, but they can sometimes be accurately predicted.

SEISMOGRAM ▶
This seismogram records the size of earth tremors in Kobe, Japan. It is made by a seismograph, which contains a weighted pen hanging over a rotating roll of paper. As the seismograph shakes in an earthquake, the pen stays still and traces the vibrations as a series of jagged lines.

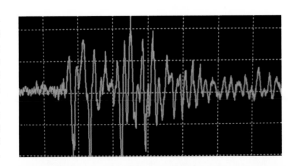

▲ EARTHQUAKE MONITORING
In Parkfield, California, USA, laser beams have been used to monitor minute plate movements along the San Andreas Fault. This is one of a variety of ultrasensitive instruments scientists have designed to monitor quakes and the tiny tremors that sometimes come before them. Devices called strainmeters and creepmeters measure rock shifts along faults.

FIND OUT MORE ▶▶ Earth's Structure **206–207** • Mountain Building **214–215** • Plate Tectonics **208–209**

VOLCANOES

Volcanoes are vents (openings) in the ground from which magma (molten rock), ash, gas, and rock fragments surge upwards, in an event called an eruption. They are often found at boundaries between the plates in Earth's crust. Volcanic eruptions produce volcanoes of different shapes, depending on the type of eruption and the region's geology. **HYDROTHERMAL ACTIVITY** occurs where underground water is heated by rising magma.

CLOUDS OF ASH, STEAM, AND GAS BLAST HIGH IN THE AIR

BLAST WAVE OF TOXIC GASES

VOLCANO VENT

COOLED AND HARDENED LAVA

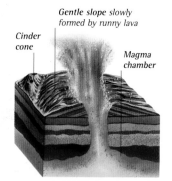

Gentle slope slowly formed by runny lava

Cinder cone

Magma chamber

◀ SHIELD VOLCANO
Magma that flows over the Earth's surface is called lava. A shield volcano produces lava that spreads over a wide area to form a broad mound. Magma collects underground in a space called a magma chamber, before erupting through vents to form low cones, and through fissures (long cracks).

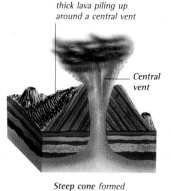

Dome formed by thick lava piling up around a central vent

Central vent

◀ DOME VOLCANO
A dome, or cone, volcano is formed when thick, sticky lava erupts from a volcano crater. The lava cools and solidifies quickly to form a dome. Further eruptions may add more layers. The collapse of a dome can produce dangerous pyroclastic flows – fast-moving flows of hot gas and volcanic fragments.

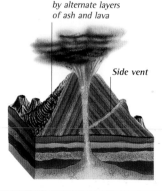

Steep cone formed by alternate layers of ash and lava

Side vent

◀ COMPOSITE VOLCANO
A steep-sided composite volcano is made of alternating layers of ash and lava, produced by a series of eruptions. Its thick magma does not flow far before solidifying. This type of volcano often has a main vent, fed by a chimney rising from its magma chamber, and additional side vents.

▲ ERUPTION OF MOUNT ST HELENS
Mount St Helens in Washington, USA, erupted with huge force on 18 May 1980. Before the eruption, gases and magma rose to fill the chamber beneath the volcano, forming an explosive mixture trapped by a plug of solidified lava. An earthquake triggered the explosion, and ash and rock avalanched down the mountain. The eruption continued for four days, killing 57 people.

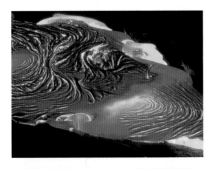

◄ PAHOEHOE LAVA

Magma forms when the rocks below the Earth's crust melt. A flow of erupted magma along the Earth's surface is called lava. When red-hot lava flowing from volcanoes cools, it solidifies into many different forms. One, pahoehoe lava, is fast-flowing and runny. As it cools, it forms a smooth, shiny skin, under which lava continues to flow. This sometimes wrinkles the smooth surface into ropelike coils.

◄ AA LAVA

Unlike smooth-skinned pahoehoe lava, aa lava has a rough surface, which is difficult to walk on and sharp enough to rip rubber shoes. This jagged material is formed when slow-moving, sticky lava cools and breaks up into sharp, blocky shapes. Flows of aa lava can be thick, reaching heights of up to 100 m (330 ft). The words for aa (pronounced ah-ah) and pahoehoe (pahow-ee-how-ee) lava come from Hawaii, where these lava types occur and were first studied.

HYDROTHERMAL ACTIVITY

The word "hydrothermal" comes from the Greek words for water and heat. In volcanic regions, the combination of heat and water below ground produces remarkable effects. In the oceans, openings called hydrothermal vents form when cracks containing red-hot magma fill with seawater. They spout black clouds of hot water mixed with gas and minerals. Hydrothermal activity on land produces hot springs, geyser, and pools of bubbling mud.

▲ BEFORE AN ERUPTION

Before 1995, the small Caribbean island of Montserrat had a thriving tourist industry. Its volcano, Soufrière Hills, had not erupted for 400 years. Volcanic eruptions can cause great destruction, reducing leafy landscapes to barren wastelands in hours. Despite this threat, people live in volcanic areas because the rock makes fertile soil for farming.

▲ AFTER AN ERUPTION

In 1995, Montserrat's volcano began a series of violent eruptions that continued for several years. When rain or snow mix with erupting ash and lava, the result is a fast-moving, deadly tide of mud. The capital Plymouth, shown here, and much of the island was buried beneath 2 m (6½ ft) of ash and mud. People had to leave the island and many have not returned.

volcanoes

◄ CRATER LAKE

The waters of this crater lake on the Kamchatka Peninsula, Russia, owe their bright blue colour to dissolved minerals. Most volcanoes have a hollow crater at the top, formed by eruption. A basin-shaped crater or caldera develops if part of the cone collapses into the empty magma chamber below. Rainwater fills it to make a crater lake.

▲ GEYSER

This geyser in Yellowstone Park, USA, shoots a hot jet of water and steam high into the air. A geyser is produced by water being heated underground by hot rocks. As the water boils deep underground, it makes steam that expands, driving the water above it towards the surface with tremendous force. The water and steam gush forth, releasing the pressure.

FIND OUT MORE ►► Earth's Structure **206–207** • Plate Tectonics **208–209** • Rock Cycle **217** • Rocks **218–219**

MOUNTAIN BUILDING

New mountains are built when rocks are pushed upwards by the movement of the giant rocky plates that make up the Earth's crust. The rocks are pushed upwards in two ways: **FOLD** mountains are formed when layers of rock become buckled, and **BLOCK** mountains are formed when giant lumps of rock rise or fall. Volcanic eruptions also create mountains. Many mountain ranges have been built up and eroded away since the Earth was formed.

Mountain peak

Glaciers (rivers of ice) erode mountain sides

Himalayan mountain *Annapurna II* is 7,937 m (26,030 ft) high

Sharp ridges and steep cliffs created by erosion

◄ THE ANDES
The Andes is the longest mountain range on land. It was formed along the western margin of South America, where two tectonic plates (rocky plates that make up the Earth's crust) collided. The mountains are still rising by about 10 cm (4 in) every century.

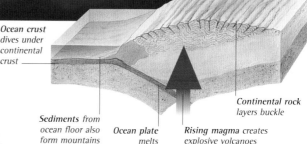

Ocean crust dives under continental crust

Sediments from ocean floor also form mountains

Ocean plate melts

Rising magma creates explosive volcanoes

Continental rock layers buckle

▲ COLLIDING PLATES
Fold mountains are pushed up at a boundary where two tectonic plates collide. The boundary between an ocean plate and a continental plate is called a subduction zone. Here, the thin ocean crust slides slowly under a thicker continental crust, making the rocks buckle and fold. The ocean plate also melts, creating magma (molten rock) that rises to form volcanoes.

WORLD MOUNTAIN RANGES

The world's major mountain ranges, such as the Andes, the Himalayas, and the Alps, are situated along the boundaries where tectonic plates collide. These ranges formed in the last few hundred million years, so are they quite young. The map also shows thin lines of volcanoes that erupt from the ocean floor, forming chains of mountainous islands.

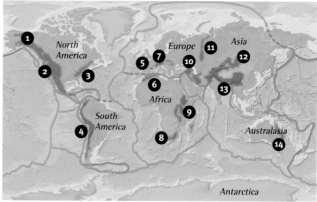

North America

Europe

Asia

Africa

South America

Australasia

Antarctica

Major mountain ranges:

1 Alaska Range
2 Rocky Mountains
3 Appalachians
4 Andes
5 Pyrenees

6 Atlas Mountains
7 Alps
8 Drakensberg
9 Ethiopian Highlands
10 Caucasus

11 Ural Mountains
12 Tien Shan
13 Himalayas
14 Great Dividing Range

◄ HIMALAYAN COLLISION
The Himalayas is a range of fold mountains formed by the collision between India and the rest of Asia. When the two tectonic plates collided, the southern edge of Asia buckled. The Indian plate continues to slide under Asia and, to date, has uplifted Tibet to a height of over 5 km (3 miles).

FOLD FORMATION

When layers of rock are pushed inwards from both ends, they crumple up into waves called folds. Rocks are too hard to be squashed into a smaller space. Instead they fold upwards and downwards. The immense forces that cause folding can crunch solid rocks into folds just a few metres across.

▼ FOLDING LAYERS
The rocks that buckle to form fold mountains are made up of layers of sedimentary rocks and igneous rocks. When the layers are folded, the rocks on the outside of a fold are stretched and the rocks on the inside of a fold are squashed. The folding also makes the layers of rock slide over each other.

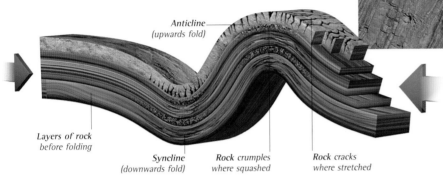

Anticline
(upwards fold)

Layers of rock
before folding

Syncline
(downwards fold)

Rock crumples
where squashed

Rock cracks
where stretched

FOLDED ROCK ▲
Folds in rocks are often visible where rocks are exposed by erosion or earth movement, such as here in Hamersley Gorge in Western Australia. These rocks are thought to be nearly three billion years old, making them some of the oldest rocks on Earth. The crust is full of folded rocks like this that were part of ancient mountain ranges.

BLOCK MOUNTAINS

Block mountains are mountains formed when layers of rock crack into giant blocks. Cracks in layers of rock are called faults. They form when the Earth's crust is stretched, squashed, or twisted. The blocks are free to slip up, down, or sideways, or to tip over. These movements are very slow, but over millions of years they form mountains thousands of metres high.

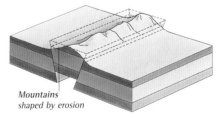

Block forced
upwards along faults

▲ BUILDING A BLOCK MOUNTAIN
Here, the layers of rock have been split into three blocks by two angled faults. Movements in the crust push the blocks together, forcing the centre block upwards. This has formed a block mountain called a horst, with a flat top, steep sides, and low, flat plains on each side.

Mountains
shaped by erosion

▲ ERODING A BLOCK MOUNTAIN
As soon as a new block mountain begins to rise, processes of erosion such as ice, wind, and water break the rocks down and remove the debris. Earthquakes, and the fault movements that cause them, speed erosion by breaking up rocks and causing landslides.

Range

Basin

▲ BASINS AND RANGES
The Basin and Range area in the southwestern USA is a typical block-mountain landscape. Here there is a mixture of low, flat areas called basins, and snow-capped mountain ranges. Millions of years ago, the crust was stretched, creating faults and blocks. Some blocks slipped down, leaving others sticking up to form mountains. The blocks have been eroded into jagged peaks, and the rock particles washed down have filled the basins with deep sediments.

mountains

FIND OUT MORE ▶▶ Earthquakes **210–211** • Erosion **222–223** • Plate Tectonics **208–209** • Rocks **218–219**

MINERALS

Minerals are the materials that make up the rocks of the Earth's crust. Among thousands of different minerals, only a few, including quartz, feldspar, and calcite, form most rocks. Native minerals, such as gold and copper, contain one chemical element. Compound minerals, such as quartz, contain two or more chemical elements. Most minerals are made up of **CRYSTALS** and can be described by their properties.

minerals

◄ CALCITE TERRACES
The peculiar rock terraces around the hot volcanic springs at Pamukkale in Turkey are made from the mineral calcite. This type of rock is called travertine. The terraces develop when water heated by hot rocks underground dissolves the rock and then flows to the surface to form pools. The travertine is deposited as the water drips slowly from pool to pool and evaporates. Limestone is also made of calcite type minerals.

MINERAL PROPERTIES
Colour, lustre, and habit   The colour of a mineral's crystal, its surface shine (lustre), and the form (habit) made by a group of its crystals.
Streak   The colour of a streak of powder left by the mineral when it is rubbed across an unglazed porcelain tile.
Cleavage   The lines of weakness along which the mineral breaks easily when it is hit with a hammer.
Hardness   Resistance to scratching, measured on Mohs scale, going from 1 (talc, very soft) to 10 (diamond, very hard).
Crystal system   The basic geometrical shape that a crystal of the mineral grows into. There are six crystal systems.

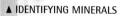

Colour darkens towards the tip of the amethyst crystal

Amethyst crystal forms in a trigonal (6-sided) structure

MINERALS IN ROCK ▼
In rock, minerals are normally found as tiny grains. It is harder to identify minerals in rock than when they are in the form of large crystals because the grains are often very small and do not have typical crystal shapes. One method used is to view a thin slice of rock under a microscope in polarized light. The minerals then show up in different colours, as in this piece of mica schist.

MICA SCHIST

MICA SCHIST UNDER MICROSCOPE

▲ IDENTIFYING MINERALS
A mineral is easy to identify when it is in the form of large crystals, such as these amethyst crystals. Then the colour, lustre, habit, and system of the crystal can be seen more clearly. Amethyst has a purple colour, a glassy lustre, a white streak, a hardness of 7 on Mohs scale, and hexagonal crystals that grow into flat six-sided columns.

CRYSTALS

Minerals come in the form of crystals. This means that the mineral's atoms (small particles) are arranged in a regular way, in neat rows and columns. The arrangement is not obvious when you look at mineral grains in rocks, but a free-growing crystal always forms a geometric shape, with flat faces. A crystal is normally symmetrical. All crystals form in one of six systems of symmetry, which help identify them.

CRYSTAL SYSTEMS

Cubic
Examples: diamond, fluorite, galena, garnet, gold, salt (halite), magnetite, pyrite

Orthorhombic
Examples: aragonite, barite, celestine, chrysoberyl, olivine, sulphur, topaz

Hexagonal/Trigonal
Examples: beryl, calcite, graphite, hematite, quartz (such as amethyst), ruby, sapphire, tourmaline

Tetragonal
Examples: cassiterite (a tin oxide), rutile (a titanium oxide), vesuvianite, zircon

Triclinic
Examples: turquoise, kaolinite, kyanite, labradorite, turquoise, wollastonite

Monoclinic
Examples: azurite, borax, gypsum, hornblende, malachite, mica, talc

▲ GEODES
A geode is a cavity in a piece of rock that is filled with concentric layers of minerals and crystals. This agate geode has been cut in two and polished, revealing a beautiful banded pattern. The layers of crystals grow from the outside inwards when water containing dissolved minerals flows into the rock cavity. Geodes are often found in lavas.

Layers of blue and white crystals

GARNET CRYSTALS

POLISHED GARNET GEMSTONE

▲ GEMSTONES
A gemstone, such as garnet, is a crystal found in rock, that is used for jewellery or other decoration. Raw (uncut) gemstones are cut and polished before they are used. Gemstones are also called precious stones because of their beautiful colours, hardness, and rarity. The most prized gemstones include diamond, ruby, sapphire, and emerald.

FIND OUT MORE ▶▶ Elements **22–23** • Rocks **218–219** • Solids **12–13** • Volcanoes **212–213**

ROCK CYCLE

The rocks under our feet seem permanent, but they are constantly being changed. This process is called the rock cycle. Rocks exposed on the Earth's surface are slowly broken down into sediments by water, ice, and wind. Meanwhile, new rocks are being created and recycled by forces in the Earth's crust and mantle, deep down under the surface.

HOW ROCKS ARE FORMED ▶
Older rocks on the surface are destroyed by erosion, or by being pushed down into the crust and melted. New rock is formed as sediments are compressed into sedimentary rocks, erupted magma cools and solidifies into igneous rocks, and heat and pressure changes rocks underground into metamorphic rocks.

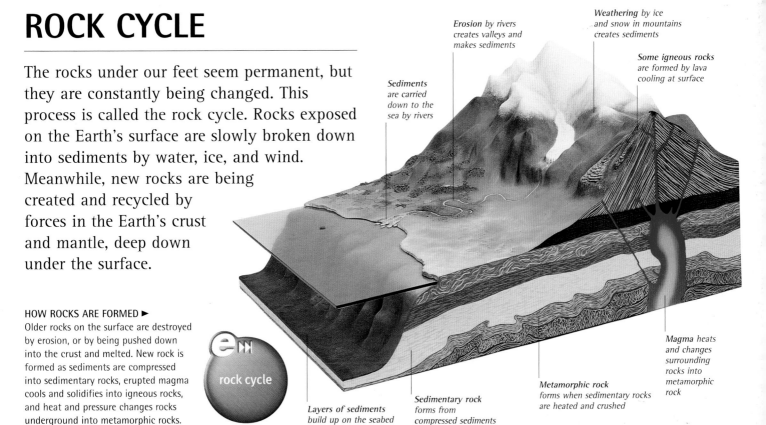

Erosion by rivers creates valleys and makes sediments

Weathering by ice and snow in mountains creates sediments

Some igneous rocks are formed by lava cooling at surface

Sediments are carried down to the sea by rivers

Magma heats and changes surrounding rocks into metamorphic rock

Metamorphic rock forms when sedimentary rocks are heated and crushed

Layers of sediments build up on the seabed

Sedimentary rock forms from compressed sediments

rock cycle

FIND OUT MORE ▶▶ Erosion **222–223** • Rivers **232** • Rocks **218–219** • Sediments **225** • Volcanoes **212–213**

Bedrock made up of beds (layers) of sedimentary rock

Sediments of sand and gravel made by erosion of cliffs by waves

ROCKS

The Earth is covered in a layer of solid rock called the crust. Rocks are either **SEDIMENTARY**, **IGNEOUS**, or **METAMORPHIC**. Almost all rocks are made of minerals, but different rocks contain different mixtures of minerals. Granite, for example, consists of quartz, feldspar, and mica. A rock can be identified by its overall colour, the minerals it contains, the size of the mineral grains, and its texture (mixture of grain sizes).

◄ BEDROCK
The solid rock that makes up the Earth's crust is called bedrock. It can be seen on coasts (as here at Burton Cliffs, Dorset, England) and in mountains, where it is being worn away by erosion. Erosion breaks the bedrock into small pieces, forming soil and sediments (such as mud, sand, and gravel), which cover up the bedrock in most places. The sediments may later turn into sedimentary rocks.

SEDIMENTARY ROCKS

Sedimentary rocks are made of particles of sediments such as sand and clay, or the skeletons and shells of sea creatures. When layers of loose sediment are buried and pressed down under more layers, the particles slowly cement together and lithify (form rock). Chemical sedimentary rocks, such as flint, form when minerals dissolved by water are deposited again.

▲ FLINT
Flint is hard and breaks into sharp pieces. It forms from silica in seabed sediments and grows into nodules with an irregular shape. It is often found as bands within chalk.

▲ CHALK
Chalk is soft, white, and fine-grained limestone. It is made from the remains of microscopic sea creatures, which were deposited at the bottom of ancient seas.

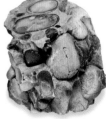

▲ CONGLOMERATE
Conglomerate is made up of rounded pebbles embedded fine-grained rock. It is formed when shingle is buried by other sediments.

▲ CLAY
Clay is a very fine-grained sedimentary rock. It is soft and crumbly when dry, and sticky when wet. Buried clay gradually turns to claystone and shale.

▲ LIMESTONE
Limestone is mostly made up of calcite, coloured by other minerals. It is one of the most common sedimentary rocks and often contains animal and plant fossils.

▲ SANDSTONE
Sandstone is a medium-grained sedimentary rock made from sand grains. It is formed when sandy beaches or river beds are buried by other sediments.

IGNEOUS ROCKS

Igneous rocks are created when magma (molten rock under the Earth's crust) cools and becomes solid. Magma loses heat when it moves upwards at weak spots, such as cracks, in the crust. Extrusive igneous rocks form when magma reaches the surface and cools quickly. Fast cooling produces fine-grained rocks. Intrusive igneous rocks form when magma cools slowly underground. This allows the minerals to grow into coarse grains.

BASALT COLUMNS ►
The Organ Pipes at Twyfelfontein in Namibia are made of basalt. This extrusive rock formation occurs when lava (volcanic magma) erupts and solidifies. The basalt cracks into flat-sided columns because it shrinks as it cools. Basalt is the Earth's most common igneous rock.

▲ OBSIDIAN
Dark, shiny obsidian is also known as volcanic glass. It is formed when volcanic lava cools so quickly that crystal grains do not have time to form. Prehistoric people used obsidian to make sharp tools.

▲ GRANITE
Granite's colour varies with the minerals it contains. This pink granite shows grains of pink feldspar, white quartz, and black mica. It is formed by slow cooling of molten rock deep in the Earth.

▲ PORPHYRITE
Porphyrite (also called microdiorite) is a grey or dark grey intrusive igneous rock. It takes its name from its texture of large grains (called porphyrites) set in a background of small grains.

▲ GABBRO
Gabbro is a coarse-grained grey or black intrusive igneous rock. It forms deep down in the Earth's crust and cools very slowly. It sometimes has obvious layers of colour.

METAMORPHIC ROCKS

Metamorphic rocks are formed when the minerals in rocks are changed underground by heat and pressure. Contact metamorphic rocks are produced when rocks are heated by magma rising through the crust. Rocks that are folded or crushed by immense pressure deep in the crust are called regional metamorphic rocks. The properties (characteristics) of a metamorphic rock depend on its parent rock (the original rock type) and how it was formed.

rocks

▲ MIGMATITE
Migmatite is a mixture of dark-coloured schist or gneiss and a lighter coloured rock similar to granite. This piece was found in the Highlands of Scotland.

▲ SLATE
Slate is fine-grained and dark grey or green. It splits easily into flat sheets, and is used to make roof and floor tiles. It is formed from the sedimentary rocks mudstone or shale.

▲ GARNET SCHIST
In a schist, lines of crystals can often be seen with the naked eye. This sample contains large crystals of garnet. Schists are mostly medium-grained and come from shales or granites.

▲ GNEISS
Gneiss (pronounced nice) is a coarse-grained, grey, or pink regional metamorphic rock formed from limestone or granite. Light and dark layers of minerals can be seen rippling across the rock.

▲ MARBLE
Heat and pressure in Earth's crust transform limestone into marble, one of the most popular stones for building and sculpture. Its colour ranges from white to pink, green, and black.

FIND OUT MORE ►► Earth's Structure **206–207** • Minerals **216** • Rock Cycle **217** • Sediments **225** • Volcanoes **212–213**

FOSSILS

Fossils are the only record we have of past life. Most often preserved in rock, they are sometimes also found in ice or amber (fossilized tree resin). Fossils come in many forms, from footprints and faint impressions, to mineralized shells and bones. They can tell us when species evolved and when they died out, and some can be used to work out the age of the rocks in which they are found. Animals and plants are **FOSSILIZED** after burial in sedimentary deposits that later become sedimentary rocks.

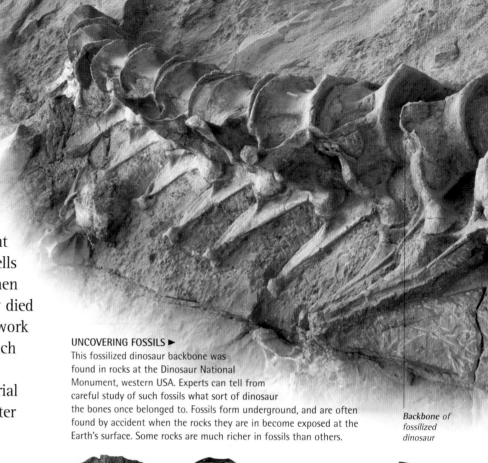

UNCOVERING FOSSILS ▶
This fossilized dinosaur backbone was found in rocks at the Dinosaur National Monument, western USA. Experts can tell from careful study of such fossils what sort of dinosaur the bones once belonged to. Fossils form underground, and are often found by accident when the rocks they are in become exposed at the Earth's surface. Some rocks are much richer in fossils than others.

Backbone of fossilized dinosaur

fossils

TYPES OF FOSSIL ▶
It is rare for the actual body parts of an animal or plant to remain intact when it is fossilized. Sometimes the parts are mineralized (replaced by minerals), or they may rot away, leaving a cavity called a mould. Sometimes the mould fills with rock to form a cast. A trace fossil is a fossil of a mark made by an animal, such as a footprint.

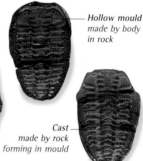

TRACE FOSSIL OF A DINOSAUR FOOTPRINT

Hollow mould made by body in rock

Cast made by rock forming in mould

TRILOBITE MOULD AND CAST

MINERALIZED AMMONITE

FOSSILIZATION

The process that turns the remains of an animal or plant into rock is called fossilization. It takes many thousands or millions of years. The remains become fossils only if they are buried by sediments before they rot away. This can happen at the bottom of lakes and seas or when dead animals beside a muddy river are quickly covered by floodwater.

▲ FISH TO SKELETON
When a fish dies, its body falls to the bottom of the sea or a lake. Its soft parts are eaten by other animals, leaving just a skeleton.

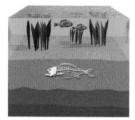

▲ SKELETON BURIED
A layer of sediment settles over the skeleton, which becomes buried. More sediment settles on top, burying the skeleton deeper.

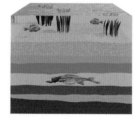

▲ FOSSIL FORMS
The buried sediment slowly turns to sedimentary rock. The fish skeleton is partly or wholly replaced by minerals that preserve its shape.

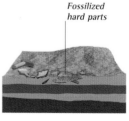

Fossilized hard parts

▲ FOSSIL EXPOSED
Millions of years later, earth movements thrust the rock with the fossil upwards. The fossil is revealed when the rocks above are eroded away.

WILLIAM SMITH
British, 1769-1839

William Smith is known as the father of British geology. While working as a surveyor, Smith noticed that layers of sedimentary rocks (known as strata) were laid down in a particular order and that the layers could be matched by the fossils they contained. This enabled him to draw the first proper geological map of England.

FIND OUT MORE ▶▶ Palaeontology 332–333 • Prehistoric Life 330–331 • Rocks 218–219 • Sediments 225

GEOLOGICAL TIME

Geologists (scientists who study rocks) divide the time since the Earth was formed until today into chunks called periods. During the various periods, different species of animals and plants lived on the Earth. For example, the Cretaceous period, which lasted from 146 million years ago to 65 million years ago, was the final period of the dinosaurs. Some rocks can be given a relative age by identifying the fossils they contain. The date of formation of some rocks can be found by using **RADIOMETRIC DATING.**

▲ STRATIGRAPHY
In the Grand Canyon, southwestern USA, the layers in the sedimentary rocks that make up the canyon are exposed. The rocks at top are the youngest and those at the bottom are the oldest. The study of the different rock layers is called stratigraphy. It can reveal how the rocks were formed over millions of years and what was happening where they were forming – for example, whether the area was under the sea or part of a desert.

geological time

RADIOMETRIC DATING

Radiometric dating is a way of measuring the age of a piece of rock. Igneous rocks contain tiny amounts of radioactive chemicals. As the rocks age, these elements gradually break down into elements that are not radioactive. By knowing the rate at which the elements break down and measuring their level of radioactivity, the age of a rock sample can be calculated.

ARGON-ARGON DATING ▶
In argon-argon dating, the amounts of two forms of the element argon in a rock sample are measured using a device called a mass spectrometer. One form only is produced by radioactive decay as the rock ages, so comparing how much there is of this in relation to the other form of argon reveals the age of the rock.

THE FOSSIL RECORD

Over many decades, palaeontologists have built up a database of fossils called the fossil record. This shows when different species of animals and plants lived in the history of the Earth. A fossil can be matched against the fossil record to find the age of the rock it was found in. The information in the fossil record is based on using stratigraphic layers to tell the age of a rock and the characteristic life forms within it. Here are some fossils and the periods they belong to.

Quaternary period
1.8 million years ago (mya) to present day

Mammoth tooth

Tertiary period
65.5–1.8 mya

Sabre-toothed cat's skull

Cretaceous period
145.5–65.5 mya

Tyrannosaur (meat-eating dinosaur)

Jurassic period
199.6–145.5 mya

Archaeopteryx (ancient form of bird)

Triassic period
251–199.6 mya

Skull of a therapsid (mammal ancestor)

Permian period
299–251 mya

Mesosaurs (reptiles that lived in water)

Carboniferous period
359.2–299 mya
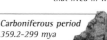
Fronds of an early fern

Devonian period
416–359.2 mya

Jawless fish

Silurian period
443.7–416 mya

Sea scorpion

Ordovician period
488.3–443.7 mya

Nautiloid (shellfish)

Cambrian period
542–488.3 mya

Trilobite

Precambrian period
before 542 mya

Early soft-bodied organism

FIND OUT MORE ▸▸ • Fossils **220** • Palaeontology **332–333** • Prehistoric Life **330–331** • Rocks **218–219**

EROSION

The process that breaks up and carries away the rocks and soils that make up the Earth's surface is called erosion. It is caused by flowing water, waves, glaciers, and the wind, and it constantly changes the shape of the landscape. Erosion happens more quickly on bare rock, which is unprotected by soil. It often begins with weathering, where rocks are weakened by the weather's elements, such as sunshine, frost, and rain. Rocks can be eroded by physical weathering through heat, cold and frost, and **CHEMICAL WEATHERING**. Erosion may lead to the **MASS MOVEMENT** of rock and soil.

◄ RIVER EROSION

In Marble Canyon, Arizona, USA, the Colorado River has worn through 600 m (2000 ft) of rock, revealing many layers. Flowing water, charged with stones and sand, is a powerful erosive force. Over millions of years, large rivers can carve out deep valleys. Rivers erode hills and mountains, and carry the debris down to the lowlands and sea. A river's erosive power is greatest in the mountains, where its flow is steep, fast and rough, and it carries boulders that bump along its bed.

High canyon sides were formed when the river cut down through rock layers

Softer rocks wear away faster than the harder rocks below them

Colorado River flows at the base of the deep canyon

▲ COASTAL EROSION

Waves erode the base of cliffs, undermining them and making them collapse. This can create coastal features such as the Twelve Apostles in Victoria, Australia. The stacks (rock towers) are left when headlands are worn away from both sides until they crumble. The broken rocks form shingle and sand beaches. Erosion happens faster when shingle is thrown against the cliffs by the waves.

▲ GLACIAL EROSION

Mountain ranges contain deep valleys that have been carved out by glaciers. A glacier has melted away on the slopes of Mount Kailas in Tibet, revealing this glacial valley. The glacier is like a slow-moving river of ice that flows downhill, carried forwards by its huge weight. The rocks dragged along underneath it gouge deep into the ground, creating U-shaped valleys with steep sides and flat bottoms.

▲ WIND EROSION

Sand blown by strong winds has sculpted the slender sandstone pillars of Bryce Canyon, Utah, USA. Their rugged outlines are caused by the softer layers of rock are being eroded more quickly than the harder layers. Wind erosion is common in deserts, where sand is blown about because there are few plants to hold the soil in place and there is no rain to bind the soil particles together.

CHEMICAL WEATHERING

Some rocks are broken down by chemical action, in a process called chemical weathering. The minerals they contain are changed chemically by the effects of sunlight, air, and especially water. The rocks are weakened and wear away more easily. Limestone, for example, is dissolved by rainwater, because the water contains carbon dioxide from the atmosphere, making it slightly acidic.

KARST LANDSCAPE ▶

The Guangxi province of China has spectacular scenery made by chemical weathering. Karst scenery is formed by rainwater dissolving cavities in limestone. These grow into caves and eventually collapse, leaving deep gorges and rocky pinnacles.

MASS MOVEMENT

erosion

Erosion normally breaks down the landscape a tiny piece at a time, but sometimes rocks and soil move downhill in large volumes. These movements, which include landslides, mudflows, and rock falls, are called mass movements. They happen when rock, debris, or soil on a slope becomes unstable and can no longer resist the downward force of gravity.

SOIL CREEP	SLUMPING	VOLCANIC MUDFLOW	DEBRIS SLIDE	ROCK MOVEMENTS
Soil creep is the extremely slow movement of soil down a steep hillside. It is caused by soil expanding and contracting, when it goes from wet to dry or frozen to unfrozen. The top layers of the soil move faster than the layers underneath. The movement is far too slow to see, but bent trees, leaning fence posts and telegraph poles, and small terraces in fields are all evidence of soil creep. Soil may also build up against a wall or at the bottom of the hillside.	A slump is a mass movement that happens when a large section of soil or soft rock breaks away from a slope and slides downwards. Short cliffs called scarps are left at the top of the slope. Slumps often happen where the base of a slope is eroded by a river or by waves, or when soil or soft rock becomes waterlogged.	A lahar is a mudflow of water mixed with volcanic ash. This forms when ash mixes with melting ice during an eruption, or with torrential rain. The mud flows down river valleys and sets hard when it comes to a stop. Lahars can cause destruction on a massive scale.	Debris is made up of broken rock, sometimes mixed with soil. These pieces of debris may collect on a slope and begin to roll or slide downwards. Debris slides often happen where people have cleared hillsides of trees and other vegetation, which causes the soil and rock to be eroded quickly.	Rock movements are the fastest type of mass movement. They happen when chunks of rock topple over or break away from cliffs and tumble or roll downhill. Many pieces of falling, tumbling rock make up a rock avalanche.

ROCK FALL

DEBRIS SLIDE

SOIL CREEP

SLUMPING

MUDFLOW

FIND OUT MORE ▶▶ Ice **226** • Coasts **227** • Rivers **232** • Rocks **218–219** • Soil **224**

SOIL

Much of the solid bedrock of the Earth's crust is covered in soil. This loose, soft material is a mixture of organic matter and particles of rock, made by weathering and erosion. The organic matter is made up of dead and living plants, animals, and other organisms. Many of the living organisms are **DECOMPOSERS** that live on the dead plants and animals. Plants get the water and nutrients they need from the soil they grow in.

Surface horizon is made up of dead and decaying plant material

A horizon contains plenty of humus (decaying organic material) and organisms – together with the surface horizon it makes up topsoil that plants grow in

Some minerals from decaying plants and animals are washed downwards by water

B horizon, made up of subsoil, contains less organic material than A horizon, and minerals washed down from above

Pressure downwards washes some minerals down and out of soil

C horizon is made up of broken rock with no organic material

Movement upwards brings minerals from weathered rock into soil

D horizon (sometimes also called bedrock) is solid rock. The rock particles in the soil come from here

▼ SOIL TYPES

The texture of a soil depends on the size of the rock particles it contains. Clay soil feels very smooth because it is made mostly of tiny particles. Sandy soil feels gritty because it is made of larger particles of up to 2 mm (¹/₁₀ in) across. Sandy soils are dry, while clay soils tend to be wet and sticky. Loam contains a mixture of sand, clay, and silt, and is a a good soil for growing crops.

CLAY SOIL   LOAMY SOIL   SANDY SOIL

NATURAL LAYERS ▲

A soil profile is a vertical slice through the ground showing the layers of the soil. Each layer is called a soil horizon. Most soils contain three major layers, called the A, B, and C horizons. The thickness of soil varies greatly. On mountains it can be a few centimetres thick. In valleys it can be many metres.

DECOMPOSERS

Many of the millions of organisms that live in the soil, including bacteria, fungi, insects, and earthworms, are known as decomposers. They live on the remains of dead plants and animals and break down these organic remains into simple chemicals that are released into the soil. Some of these chemicals provide nutrients for new plants to grow, so decomposers recycle plant material.

MICROSCOPIC VIEW ▶

Soil seen through a microscope reveals microorganisms called bacteria. A handful of soil contains millions of bacteria and fungi, which cling to particles of rock and decaying matter. Bacteria and fungi continue the decomposition started by larger organisms such as earthworms, woodlice, and slugs.

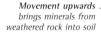

◀ SOIL ENRICHMENT

Earthworms do two important jobs to keep soil fertile, or good for plants to grow in. First, they feed on dead plant matter, helping to decompose it. Second, as they burrow, they mix and loosen the soil, which spreads organic matter and nutrients, allows air in, and improves drainage.

FIND OUT MORE ▶▶ Bacteria **284** • Ecology **326–327** • Erosion **222–223** • Fungi **282–283** • Nitrogen **42–43**

SEDIMENTS

The rocky material that is transported and **DEPOSITED** by rivers, seas, glaciers, and the wind is called sediment. Clay, sand, and gravel are all types of sediment. Sediments build up to form features such as mud banks along rivers or dunes in deserts. Sediments deposited on the seabed often build up over millions of years to form sedimentary rocks.

RIVER-MOUTH SEDIMENTS ▶
Sediments in a river are deposited in the sea at the river's mouth. In Spencer Gulf, South Australia, they have built up on the seabed to form long fingers of sand or shingle called spits and sand bars (low islands). The sea moves sediments along the coast to form beaches, and wind-blown sand forms dunes inland.

DEPOSITION

The laying down of sediments in water or on the ground is called deposition. Sediments are picked up by fast-flowing water, by strong, swirling winds, or by the ice in glaciers. Sediments are deposited when flowing water, wind, or glaciers cannot carry it any further – for example, when the water or wind slows down or stops, or when the glacier's ice melts.

FORMATION OF DUNES

The shape of a sand dune depends on the strength and direction of the wind, how much sand there is, and whether plants grow on the dune or on the ground. Barchan (crescent-shaped) dunes form when the wind blows from one direction most of the time. They move forwards at up to 30 m (100 ft) a year as the wind blows sand over the crest.

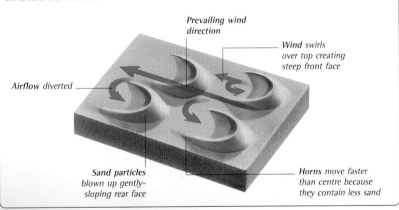

Prevailing wind direction

Wind swirls over top creating steep front face

Airflow diverted

Sand particles blown up gently-sloping rear face

Horns move faster than centre because they contain less sand

Crescent-shaped dune has gentle slope on one side and steep slope on the other

Crest is lower at the ends than in centre

▼ SAND DUNES
A dune is a heap or hill made of sand or other small particles. Dunes are formed in sandy deserts and on coasts when loose, dry sand is picked by the wind and then deposited. In the Sahara Desert, North Africa, dunes grow up to 100 km (60 miles) long and 200 m (650 ft) high, forming an ever-changing landscape like giant ocean waves.

FIND OUT MORE ▶▶ Erosion **222–223** • Ice **226** • Rivers **232** • Rock Cycle **217** • Rocks **218–219** • Wind **240–241**

ICE

About one-tenth of Earth's dry land and one-eighth of its oceans are covered with ice. This ice is made of snow that collects and becomes compacted (pressed down). Most ice occurs in thick **ICE SHEETS** that cap the land in the polar regions. In the past, during long cold eras called Ice Ages, ice covered much more of the Earth's surface than it does today. Scientists estimate that there have been over 15 Ice Ages in the last 2 million years.

ice

◄ POLAR ICE
An Inuit drives his team of huskies across a glacier in Greenland. There is relatively little land in the Arctic region. Most of it is covered by a huge ocean whose centre (near the North Pole) is permanently capped by salty sea ice.

ICE SHEETS

Ninety per cent of the world's ice is found in Antarctica. The ice cap here is 4,200 m (13,000 ft) deep in places. Over thousands of years, a thick ice sheet builds up over land when more snow falls during the winter months than melts each summer. The enormous weight of the ice pushes much of this vast, high landmass down below sea level.

▼ ICEBERG, ROTHERN POINT, ANTARCTICA
Icebergs are not formed from salty sea ice, but from land ice that calves (breaks off) from ice sheets or glaciers on the coast. Only 12 per cent of the iceberg's mass appears above the sea surface. The rest is hidden below. A fringe of sea ice also edges the Antarctic landmass, expanding in winter and melting in summer.

◄ RETREATING GLACIER
Glaciers are slow-moving rivers of ice that begin high on mountains. Fallen snow pressed down by new snow forms a dense ice called firn. When enough ice builds up, gravity and the glacier's own weight set it sliding downhill at a rate of 1–2 m (3–6½ ft) per day.

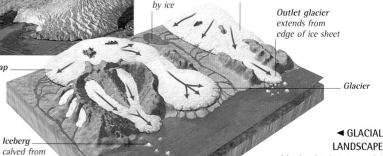

Meltwater lake left in hollow scooped out by ice

Continental ice sheet

Outlet glacier extends from edge of ice sheet

Ice cap

Glacier

Iceberg calved from glacier

Ocean

◄ GLACIAL LANDSCAPE
Moving ice is a powerful erosive force. As glaciers slip downhill they carve deep, U-shaped valleys, sharp peaks, and steep ridges. The gouging power of the ice is increased by rocks and boulders carried along at the front, sides, and beneath the glacier. When the glacier reaches the warmer lowlands, it melts.

FIND OUT MORE ▶▶ Coasts • **227** • Erosion **222–223** • Mountain Building **214–215** • Planet Earth **204–205**

COASTS

Coasts, which form the boundary between land and ocean, receive a constant battering from the wind and waves. In calm weather, the water merely laps at the shore, but on windy days, towering, foam-capped breakers smash onto coasts. It's no wonder that the shapes and even location of coasts are constantly shifting, as waves erode the land and as **SEA LEVELS CHANGE**. In some places, coasts are retreating inland by several metres each year.

Collapse of an arch leaves an isolated column called a stack

Sheer cliffs are caused by waves undercutting hard rock

Arch is formed by waves wearing away the rock

COASTAL EROSION ▶
Coastal features, such as the cliffs and arches seen here in Pembrokeshire, Wales, are formed by wave erosion. As the sea beats on rocky headlands, softer rocks are eroded (worn away) to form hollow caves. Twin caves on either side of a headland may eventually wear right through to form an arch. As the battering continues, the top of the arch collapses to leave an isolated stack.

SEA-LEVEL CHANGE

In the last few million years, sea levels have risen and fallen by up to 200 m (660 ft). Scientists believe these are caused by temperature changes, as Ice Ages come and go. During Ice Ages, sea levels are low because large amounts of water are frozen. When the climate warms, the ice melts and sea levels rise. Today, sea levels look set to rise because of global warming. This will bring a risk of flooding to coasts.

coasts

◀ DROWNED COAST
This Norwegian fjord is a deep coastal inlet, where sheer cliffs plunge into waters which can reach depths of up to 1,000 m (3,300 ft). A fjord starts to form during an Ice Age, when ice carves out a U-shaped valley near the sea. When the climate thaws, the glacier melts and the ocean rises to drown the valley.

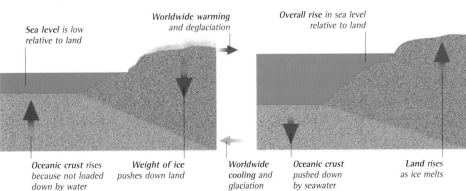

Sea level is low relative to land

Worldwide warming and deglaciation

Overall rise in sea level relative to land

GLACIAL CYCLES ▶
During an Ice Age, the weight of the ice depresses (pushes down) the land. Sea levels are low, so the crust beneath the ocean is not depressed. When the weather warms, melting ice causes sea levels to rise. This effect is partly offset by the land springing up when released from the ice's weight, while the ocean bed sinks beneath the weight of water.

Oceanic crust rises because not loaded down by water

Weight of ice pushes down land

Worldwide cooling and glaciation

Oceanic crust pushed down by seawater

Land rises as ice melts

FIND OUT MORE ▶▶ Climate **236–237** • Erosion **222–223** • Ice **226** • Oceans **228–229** • Pollution **250**

OCEANS

About 71 per cent of our planet's surface is covered by oceans and seas. In order of size, the five great oceans are the Pacific, Atlantic, Indian, Southern, and Arctic oceans. Seawater contains dissolved minerals, mostly sodium chloride (table salt), which make the oceans salty. The oceans are never still, but are stirred by powerful currents, **WAVES**, and **TIDES**.

oceans

COASTAL (NERITIC) ZONE ►
Coral reefs in warm, coastal waters are the ocean's richest habitat. These reefs are built by organisms such as corals with mineral skeletons. The oceans can be divided into two main biomes – the deep open ocean and the coastal, or neritic, zone. The shallow waters of the coastal zones occupy just 10 per cent of the total ocean area, but are home to 98 per cent of marine life.

OPEN OCEAN ZONES

EUPHOTIC (UPPER) ZONE
Jellyfish, fish such as herring, mackerel, and sharks, and crustaceans such as this lobster all inhabit shallow and surface waters. The open ocean can be divided into several layers, each with different levels of light, oxygen, and water temperature. The upper waters, or euphotic zone, down to 200 m (660 ft) are warm (up to 25°C or 77°F), sunlit, and rich in oxygen.

BATHYAL ZONE
Creatures such as this brittlestar, squid, and hatchetfish inhabit the bathyal zone, or mid-depths, between 200–2,000 m (660–6,600 ft). Some sunlight penetrates the upper bathyal zone down to about 1,000 m (3,300 ft). Here the water temperature may be about 5°C (41°F). No light reaches the dark zone beyond, where temperatures fall to -2°C (28°F).

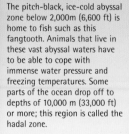

ABYSSAL ZONE
The pitch-black, ice-cold abyssal zone below 2,000m (6,600 ft) is home to fish such as this fangtooth. Animals that live in these vast abyssal waters have to be able to cope with immense water pressure and freezing temperatures. Some parts of the ocean drop off to depths of 10,000 m (33,000 ft) or more; this region is called the hadal zone.

OCEAN CURRENTS

The water in the oceans is never still, but moves continually in strong currents that flow both near the surface and at great depths. This helps to distribute the Sun's heat around the globe. Winds create surface currents, which are then bent by Earth's rotation and by land masses to flow in great circles, called gyres. Warm surface currents coming from the tropics warm the lands they flow past. Cool deep currents flowing from polar waters have the opposite effect.

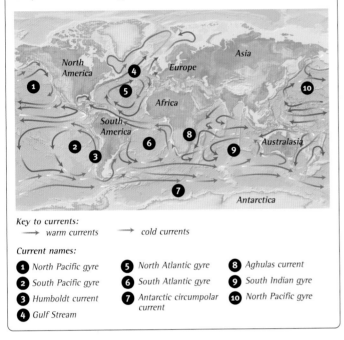

Key to currents:
→ *warm currents*   → *cold currents*

Current names:
1. *North Pacific gyre*
2. *South Pacific gyre*
3. *Humboldt current*
4. *Gulf Stream*
5. *North Atlantic gyre*
6. *South Atlantic gyre*
7. *Antarctic circumpolar current*
8. *Aghulas current*
9. *South Indian gyre*
10. *North Pacific gyre*

TIDES

The changes in the oceans' water levels, called tides, are caused by the tug of the Moon's gravity on Earth, the Sun's gravity, and Earth's spin. As the Moon orbits Earth, its gravity causes a bulge of water to build up on the ocean. The force of Earth's spin produces a matching bulge on the opposite side. Twice a day, these bulges form high tides.

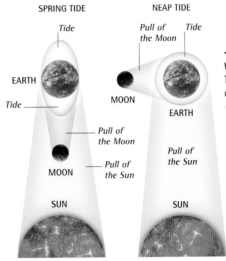

SPRING TIDE

Tide
EARTH
Tide
Pull of the Moon
MOON
Pull of the Sun
SUN

NEAP TIDE

Pull of the Moon   *Tide*
MOON
EARTH
Pull of the Sun
SUN

◄ **STRONG AND WEAK TIDES**
Tides vary from a few centimetres to 15 m (50 ft) or more. Very high tides, called spring tides, occur every two weeks, when the Sun and Moon line up so that their gravitational pulls combine. Weak tides, called neap tides, occur in the weeks in between, when the Sun and Moon lie at right angles to the Earth.

WAVES

Except in very calm weather, waves ruffle the surface of the oceans. They are caused by winds blowing over the surface, which creates friction. Winds blowing over vast expanses of open ocean form unbroken waves called rollers. Waves break into foamy crests as they reach shallow coastal waters, and finally smash onto coasts.

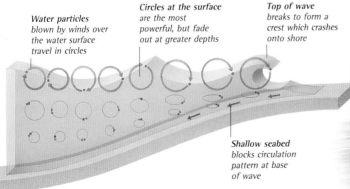

Water particles blown by winds over the water surface travel in circles

Circles at the surface are the most powerful, but fade out at greater depths

Top of wave breaks to form a crest which crashes onto shore

Shallow seabed blocks circulation pattern at base of wave

▲ **WAVE ACTION**
Waves may travel huge distances across oceans, yet surprisingly, the water in a wave stays in roughly the same place. As the wave passes, the water particles move around in a circle and return to their original position. As a wave moves in to the shore, its strength and size increases. The seabed disrupts the pattern, slowing the water in the lower part of the circle, causing the top of the wave to build up and break over it.

WAVE POWER ▲
Giant tidal waves, called tsunamis, are caused by earthquakes or volcanic eruptions on the ocean bed. As waves ripple outward from an undersea earthquake or volcano, they are barely noticeable. However in shallow waters they rear up to great heights and crash onto coasts with huge destructive force.

FIND OUT MORE ►► Coasts **227** • Habitats **246–247** • Islands **231** • Ocean Floor **230**

Just a century ago, the ocean floor was largely unknown. Now we know that the deep oceans have features such as mountains, deep valleys, and vast plains. Many of these are formed by the movement of the tectonic plates that make up Earth's crust. Far below the ocean's surface, volcanic mountain chains are rising in mid-ocean zones where plates pull apart. Elsewhere, deep trenches descend in subduction zones where plates collide and one dives below the other.

MARIE THARP
American, 1920–

Marie Tharp, along with colleague Bruce Heezen, made the first detailed map of the ocean floor using sonar readings. In the late 1940s, Tharp discovered a rift valley running down the centre of the Mid-Atlantic Ridge, and came to realize that a chain of mid-ocean ridges circles the globe.

◄ HYDROTHERMAL VENTS
In 1977, scientists used submersible vehicles to explore the seabed and discovered vents gushing dark plumes of superhot, mineral-rich water. These black smokers, are caused by volcanic activity at mid-ocean ridges. Water entering cracks in the crust is heated by magma and mixed with mineral sulphides, then belched forth in dark clouds.

Black smokers give off clouds of hot water containing sulphides

Chimneys are formed by minerals condensing from the vents

Tube worms and blind crabs thrive on bacteria that use the minerals to make energy

ocean floor

◄ SONAR
Oceanographers use sonar to map the ocean floor. The research ship directs sound waves at the bottom, and charts the echoes that bounce back to create a detailed map. Sonar has revealed features such as seamounts (submerged volcanic peaks), which rise 1,000 m (3,300 ft) from the sea floor, and guyots (flat-topped seamounts).

▲ SONAR IMAGE OF PACIFIC SEABED
This computer-generated map of the ocean floor off California, USA, was made using sonar. The floor is shown in different colours for different depths. The wide, flat ledge of the continental shelf which edges the land is shown in orange. On the seaward side, the continental slope drops away steeply to the abyssal plain below, shown in blue.

FIND OUT MORE ▶▶ Acoustics 106–107 • Mountain Building 214–215 • Oceans 228–229 • Plate Tectonics 208–209

ISLANDS

Islands are land masses entirely surrounded by water. They are found in oceans, seas, rivers, and lakes. Islands vary in size from tiny rock outcrops to vast areas such as Greenland, which covers 2.2 million sq km (840,000 sq miles). There are two main types of island: oceanic islands which are remote from land; and continental islands, which often lie close to the mainland. Many oceanic islands are volcanoes. Continental islands are often formed by changes in sea level.

◄ CONTINENTAL ISLANDS
Continental islands, such as the British Isles, rise from the shallow waters of continental shelves, which fringe the world's continents. Often these islands were once part of the mainland, but were cut off when sea levels rose to flood the land in between. Smaller islands, called barrier islands, sometimes form off coasts where ocean currents or rivers deposit sand or mud.

CORAL ISLANDS ▲
Coral islands, such as the Maldives in the Indian Ocean, are composed of the limey skeletons of coral polyps. Large colonies of these anemone-like creatures thrive in the warm, shallow waters off tropical coasts or around seamounts. The polyps' soft bodies are protected by cup-shaped shells, which grow on top of one another to form rocky reefs that eventually break the surface. If the seamount subsides, just a ring of coral, called an atoll, may be left.

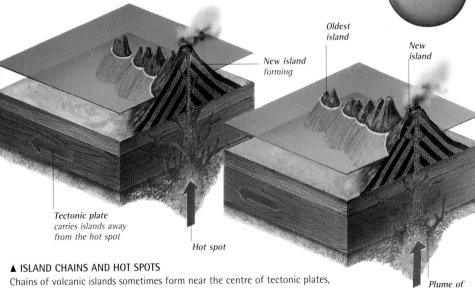

Oldest island

New island forming

New island

Tectonic plate carries islands away from the hot spot

Hot spot

Plume of magma rising from the mantle

▲ ISLAND CHAINS AND HOT SPOTS
Chains of volcanic islands sometimes form near the centre of tectonic plates, in zones called hot spots. Some scientists believe that hot spots occur where magma plumes surge up from the mantle below. The magma bursts through a weak point in the crust to form an island. Over millions of years, the hot spot stays in the same place as the crustal plate drifts over it, forming new islands.

▲ ISLAND ARCS
Oceanic islands are often formed by volcanic eruptions when plates collide. As one plate is forced below another, its crust melts in the red-hot mantle below. This molten rock rises up again to burn through the crust and erupt on the sea floor. Over time, the erupted rock forms a tall seamount and eventually breaks the surface as an island.

RIVERS

Rivers drain the surrounding land, carrying water that falls as rain and snow down to the sea. As rivers flow, they erode (wear away) rock, breaking it into fragments, called sediments, that are carried downstream. Most erosion occurs when rivers flood after heavy rain or as mountain snows melt in spring. Over time, erosion creates valleys and waterfalls, and sediments form land areas called flood plains and deltas.

◄ MOUNTAIN RIVER
The River Dora in northern Italy starts its journey on steep mountain slopes. The water in these rapids flows swiftly picking up rocks that bounce along the riverbed. The river has eroded a steep-sided, V-shaped valley.

Glacier

Rain and melting snow runs off mountains

Tributaries feed water into the main river

Waterfall and rapids

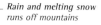

rivers

Meanders are bends where the river swings from side to side

SOUTH AMERICAN FALLS ▲
The Iguaçu Falls are on the border between Argentina, Brazil and Paraguay. Waterfalls form where water flows from hard to soft rock. Soft rock erodes faster, creating a vertical drop over the hard rock edge. Falling water undercuts the hard rock, making a plunge pool. The falls move upstream when this collapses.

LIMESTONE CAVES

When water flows over some rocks, such as limestone, caves may be formed by a process called chemical weathering. Water seeps into cracks and gradually dissolves the rock, widening the cracks until, over thousands of years, the limestone becomes riddled with caves and passageways. Water flowing through caves forms underground streams, rivers, and pools (such as this one in Mexico). Surface rivers disappear into sink holes and reappear many kilometres away. Eventually a cave roof may fall in, creating a gorge.

Sea   *Delta*

▲ RIVER'S COURSE
A river has three main stages. In the first, the river is steep and narrow and its flow is rapid and rough. In the second stage, it is wider, less steep and flows more smoothly through flat-bottomed valleys. In the third stage, the river is broad and flows placidly across flat coastal plains to the sea, where it drops its sediment.

RUSSIAN DELTA ►
This satellite image shows where the arctic River Lena (at the top in dark blue) flows into the sea. The green area is a delta, which is land formed by the river depositing (dropping) sediment as it slows down. As the river flows across the flat, marshy delta, it divides into many channels that fan out across the area.

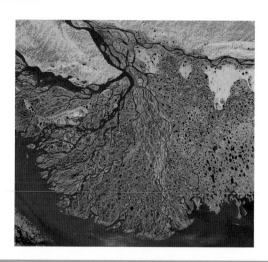

FIND OUT MORE ▸▸ Coasts **227** • Erosion **222–223** • Habitats **246–247** • Ice **226** • Rain **244–245** • Sediments **225**

GROUNDWATER

Groundwater is water under the Earth's surface. Most groundwater is found in porous rocks, which have tiny holes in them. If a hole is bored straight down through the rock, groundwater is eventually found at a certain level. This level is called the water table, and it usually rises when rainwater soaks into the ground. A spring is a place where groundwater emerges from a hillside.

▲ EGYPTIAN OASIS
Even in arid (dry) places such as deserts, groundwater sometimes comes to the surface. These lush green areas are called oases. The water at an oasis may have travelled underground from mountains hundreds of kilometres away. Oases are an important source of water, and towns often grow up around them.

▼ ARTESIAN BASIN
An aquifer is a layer of porous rock that can fill with water, like an underground reservoir. Sometimes, part of an aquifer is covered by rock that water cannot flow through, such as clay. This forms an Artesian basin. If a well is sunk through the basin to the aquifer, water flows into the well.

groundwater

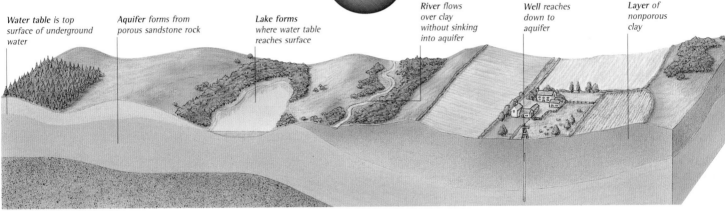

Water table is top surface of underground water

Aquifer forms from porous sandstone rock

Lake forms where water table reaches surface

River flows over clay without sinking into aquifer

Well reaches down to aquifer

Layer of nonporous clay

LAKES

lakes

Lakes form where water fills hollows in the landscape. Some of these hollows are formed by glaciers gouging into the ground, and some are created when river valleys are blocked by dams. Other lakes are formed in volcanic craters, or when land sinks during earth movements. Most lakes contain freshwater, but there are some saltwater lakes, such as the Dead Sea between Israel and Jordan.

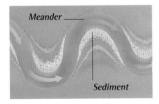

Meander

Sediment

▲ STAGE 1: BEND EROSION
Rivers can gradually produce lakes as they flow. The outside banks of the meanders (bends) erode and sediment builds up on the insides, making the meanders longer.

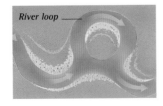

River loop

▲ STAGE 2: BREAKTHROUGH
Eventually two ends of a meander get so close to each other that the water breaks through. This often happens during a flood. Now most river water by-passes the bend.

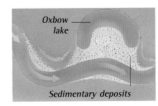

Oxbow lake

Sedimentary deposits

▲ STAGE 3: OXBOW LAKE
The water flowing into the bend slows down. It drops its sediment, which blocks the ends of the bend, leaving a crescent-shaped lake called an ox-bow lake.

▲ GLACIAL LAKE
These lakes on the Isle of Skye in Scotland were created by glaciers thousands of years ago. Glaciers begin to form high on mountain sides, from snow and ice which builds up and scours hollows in the rock called cirques. When the glaciers melt, these fill with melt water to form cirque lakes, which continue to be fed by rainwater flowing off the hillsides.

FIND OUT MORE ▶▶ Erosion 222–223 • Habitats 246–247 • Ice 226 • Rain 244–245 • Volcanoes 212–213

ATMOSPHERE

The Earth would be as lifeless as the Moon without the atmosphere – a blanket of gases surrounding the planet and extending about 700 km (430 miles) above its surface. This relatively thin layer, held in place by gravity, provides us with oxygen to breathe and separates us from the void of space. It includes the **OZONE LAYER**, which screens out harmful solar radiation. **PRESSURE SYSTEMS** in the atmosphere affect the Earth's weather.

◄ AURORAS

Auroras are shimmering curtains of light seen at night in the polar regions. They are known as the Northern Lights in the Arctic, and as the Southern Lights in the Antarctic. These spectacular displays are caused by charged particles from the Sun striking the upper atmosphere above the poles.

LAYERS OF THE ATMOSPHERE ►

The troposphere is the lowest layer of the atmosphere and contains 75 per cent of all its gases. Above is the stratosphere, which includes the ozone layer. Higher still is the thin air of the mesosphere, where meteors burn up. The thermosphere contains an electrically charged layer that radio waves bounce off. The exosphere is the top layer, fading off into space.

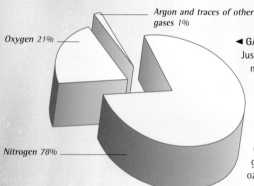

Argon and traces of other gases 1%

Oxygen 21%

Nitrogen 78%

◄ GASES IN THE ATMOSPHERE

Just two gases, nitrogen and oxygen, make up 99 per cent of the mixture of gases in the atmosphere. Nitrogen contributes 78 per cent and oxygen 21 per cent. The last 1 per cent is mainly argon (0.93 per cent), plus 0.03 per cent carbon dioxide and traces of other gases, including helium, neon, ozone, methane, and hydrogen.

OZONE LAYER

Ozone is a form of oxygen that gathers in the stratosphere to form a layer. This layer screens out harmful ultraviolet (UV) rays from the Sun, which can cause skin cancer. In the 1980s, scientists discovered that thin areas, or holes, were appearing in the ozone layer over the polar regions each spring. Ozone loss is caused by chemicals called chlorofluorocarbons (CFCs).

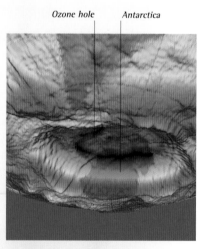

Ozone hole   Antarctica

▲ OZONE LOSS

This satellite image shows ozone loss over Antarctica. The thickest part of the ozone layer is shown in red, thinning through yellow to green and blue. Ozone loss is partly caused by chlorofluorocarbons (CFCs), which were used in refrigerators and aerosols. Today, many countries ban the use of CFCs.

PRESSURE SYSTEMS

Pressure systems in the atmosphere are masses of moving air that create winds and the weather. The air is set in motion by changes in temperature and air pressure (the pushing force of air from all directions). Air pressure is greatest at sea level, because there is a larger weight of air pushing down. The higher in the atmosphere you go, the less air and air pressure there is.

Exosphere
450–900 km
(280–560 miles)
above sea level

Thermosphere
80–450 km
(50–280 miles)
above sea level

Meteors burn up in the upper atmosphere, producing shooting stars

Auroras appear in the upper atmosphere in polar regions

Solar radiation is partly reflected and partly absorbed at various heights in the atmosphere

atmosphere

Mesosphere
50–80 km
(30–50 miles)
above sea level

Ozone layer

Stratosphere
12–50 km
(7–30 miles)
above sea level

Troposphere
12 km
(7 miles)
above sea level

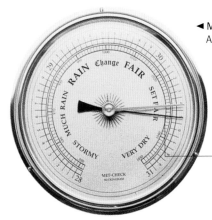

◄ MEASURING AIR PRESSURE
Air pressure is measured in units called millibars (mb), using an instrument called an aneroid barometer. Behind the barometer's dial is a chamber from which some air has been removed. Changes in air pressure cause the air in the chamber to expand or shrink, and this moves the needle around the dial.

Barometer needle shows air pressure and the likely weather

HIGH AND LOW PRESSURE ►
Warm air is lighter than cool air, so it rises above it. Rising warm air creates low-pressure areas called cyclones, or lows. Sinking cool air forms high-pressure anticyclones, or highs. Air moves from higher to lower pressure, bringing different weather systems. Highs usually bring dry, fine weather, and lows bring rain.

Rising warm air creates a low, or depression

Cold air flows towards areas of low pressure

Sinking cool air creates zone of high pressure

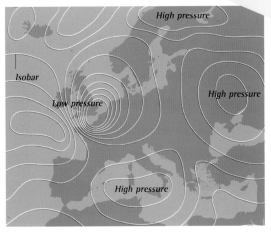

High pressure

Isobar

Low pressure

High pressure

High pressure

▲ ISOBARS
Air pressure is shown on weather maps using curved lines, called isobars, which link areas with equal air pressure. This map of western Europe shows an area of low pressure, over the UK, which is surrounded by several high-pressure areas. In regions where the isobars are packed tightly together, the air pressure is changing rapidly. These areas are said to have a steep pressure gradient and are characterized by strong winds.

EVANGELISTA TORRICELLI
Italian, 1608–1647

Torricelli discovered air pressure in the 1640s. He made the first barometer by upending a glass tube filled with mercury (a heavy liquid metal) in a bowl. The mercury remained in the tube near the closed top to a height of 76 cm (30 in). Torricelli concluded that air pressure prevented the liquid from falling further.

FIND OUT MORE ▶▶ Nitrogen **42–43** • Oxygen **39** • Pressure **74–75** • Weather **238–239** • Wind **240–241**

▼ EQUATORIAL FOREST
Dense tropical rainforests grow in a belt north and south of the Equator, where the climate is hot and wet. Temperatures vary between just 24–27°C (75–82°F), and it rains nearly every day. The subtropics on either side of the tropics are cooler. Some parts of the subtropics have an annual dry season and rainy season.

CLIMATE

Every part of Earth has its own climate – the typical pattern of weather over a long period of time. An area's climate is affected by its latitude (its distance north or south of the Equator), its height above sea level, and how far it is from the sea. In many parts of the world, conditions also vary with the **SEASONS**. A region's climate affects the types of plants and animals found there, and the kind of homes that the local people build.

◄ CLIMATE CHANGE
Scientists can find out about the Earth's past climate and atmosphere by examining samples of ice buried deep in the polar regions – here, in Antarctica. The bottom-most ice gives information about weather conditions when the snow fell, hundreds of thousands of years ago. Earth's climate does not stay the same, but changes quickly sometimes, as cold Ice Ages give way to warm periods like the present.

CLIMATE ZONES

Earth's landmasses can be divided into nine major climate zones, based on their usual temperature, rainfall, and the type of vegetation that grows there. Tropical areas are hot all year round, while polar regions and the tops of high mountains are always cold. Temperate zones in between the poles and the tropics, such as temperate forests and Mediterranean regions, have moderate, but seasonally changing, climates. Deserts are dry, receiving less than 25 cm (9 in) of rainfall every year.

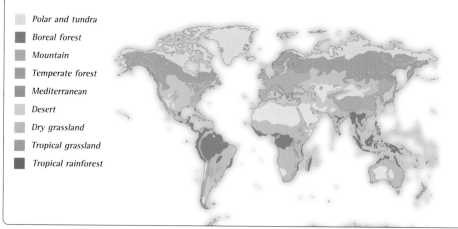

- *Polar and tundra*
- *Boreal forest*
- *Mountain*
- *Temperate forest*
- *Mediterranean*
- *Desert*
- *Dry grassland*
- *Tropical grassland*
- *Tropical rainforest*

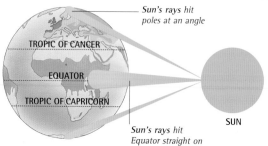

Sun's rays hit poles at an angle

TROPIC OF CANCER

EQUATOR

TROPIC OF CAPRICORN

SUN

Sun's rays hit Equator straight on

◄ HEAT FROM THE SUN
Earth's curving surface means that different regions receive different amounts of heat from the Sun. The midday Sun is directly overhead at the Equator, so the tropics are always hot. The Sun is low in the sky at the poles. Its rays are also spread over a wider area, and have further to travel through the atmosphere, so the poles are always cold.

Snow and ice cover the tops of high mountains – no plants can grow here, above the treeline

Conifer forests grow on lower mountain slopes, up to a point called the treeline

Flowers bloom in meadows below the treeline

Sun's heat on the sea causes warm air to rise, producing inshore winds

Pine trees shade the warm Mediterranean coastline

Succulent plants store water in their fleshy leaves for dry spells

climate

▲ MOUNTAIN CLIMATES

The thin air high on mountains cannot absorb as much of the Sun's heat as the air at sea level. The temperature therefore drops about 1°C (2°F) for every 150 m (500 ft) you climb. This results in various climate zones at different heights on mountains, each with its own characteristic vegetation. The snowline is at sea level near the poles and up to 5,000 m (16,500 ft) near the Equator.

▲ COASTAL CLIMATES

The Sun shines on a turquoise sea in Provence, France, where the summers are hot and dry. Regions near coasts are usually wetter and milder than those inland. The sea absorbs the Sun's heat more slowly than the land, but also releases heat more gradually. This gives coastal areas cooler summers and warm winters. Moist ocean winds blowing inshore bring rain, and help to cool coastal regions during the summer months.

SEASONS

Seasons are times of year characterized by certain weather conditions. In many parts of the world, temperatures and day length vary with the seasons. This affects plant growth, animal behaviour, and human life. The seasons occur because Earth is tilted on its axis (an imaginary line between the poles) as it travels around the Sun. Tropical regions have little seasonal variation: polar regions have the most.

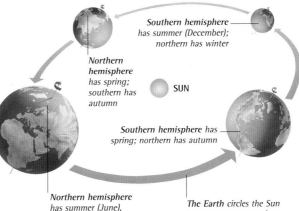

Southern hemisphere has summer (December); northern has winter

Northern hemisphere has spring; southern has autumn

SUN

Southern hemisphere has spring; northern has autumn

Northern hemisphere has summer (June), southern has winter

The Earth circles the Sun every year, or 365.2 days

◄ EARTH'S SEASONS

When the North Pole tilts towards the Sun, it is summer in the northern hemisphere (the half of the Earth above the Equator). Six months later, when the South Pole tilts towards the Sun, it is summer in the southern hemisphere (the half of the Earth below the Equator).

TEMPERATE SEASONS ►

Temperate lands located between the tropics and the polar regions experience four seasons: spring, summer, autumn, and winter. Many trees and plants in temperate regions reflect these seasonal changes. In spring, trees grow new leaves, which reach maturity in summer – the hottest season with the longest days. In autumn, trees shed their leaves in preparation for winter – the coldest season with the shortest days.

SPRING   SUMMER   AUTUMN   WINTER

◄ POLAR SEASONS

A time-lapse photograph in summer in northern Norway shows that it has sunlight 24 hours a day. Because of the Earth's tilt, the Sun dips low in the sky but never sets. In winter, the Sun never rises, bringing continual darkness. Polar temperatures are cool in summer and bitterly cold in winter.

12am   12pm

FIND OUT MORE ▶▶ Planet Earth 204–205 • Habitats 246–247 • Trees 268–269 • Weather 238–239

WEATHER

The weather is the day-to-day condition of the atmosphere at a particular place and time – whether the air is warm or cool, moist or dry, still or moving, and whether rain or snow is falling. **METEOROLOGY** is the study of the weather. The Sun is the driving force behind the weather. It heats air masses in different parts of the globe unevenly, creating differences in air pressure. This causes winds as air moves from zones of high pressure to low pressure. **WEATHER FRONTS** occur where moving masses of air collide.

CHANGING WEATHER ▲
Sunbeams highlight storm clouds sweeping across the sea and shedding heavy rain. In some parts of the world, such as the tropics, weather conditions remain fairly constant for weeks. In other places, the weather changes by the minute, as clouds drift over to obscure clear skies, or sunlight breaks through after rain.

WEATHER FRONTS

Weather fronts are border zones where masses of air of different temperatures and humidity (moisture) levels meet and push into one another. Warm air is less dense, or lighter, than cold air, and so it rises above the cold air. Rising warm air creates an area of low pressure or depression. Depressions are linked with unsettled weather conditions, including high winds and rainy spells.

weather

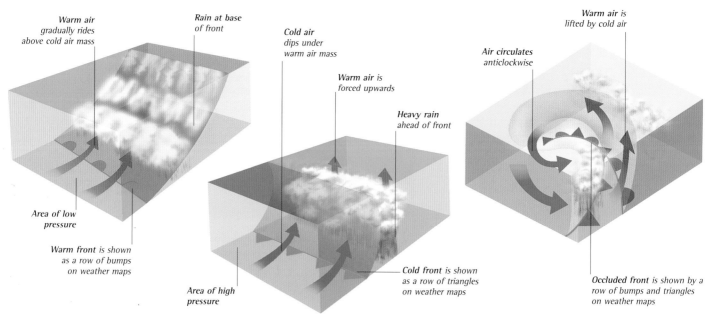

▲ WARM FRONT
A warm front occurs when a mass of warm air meets a mass of cold air. The warm air slowly rises above the cold air, forming a low pressure zone. As the rising warm air cools, the moisture in it condenses to form clouds, bringing drizzle or rain.

▲ COLD FRONT
A cold front occurs when a mass of cold air is driven towards a mass of warm air. As they collide, a steeply sloping front is formed and the warm air is forced to rise rapidly. This produces towering thunderclouds and brings torrential rain showers.

▲ OCCLUDED FRONT
Cold fronts often follow a few hours behind warm fronts. Earth's rotation bends the moving masses of air, causing the fronts to spiral around one another. The warm and cold air merge to form an occluded front, which brings cloudy skies and rain.

METEOROLOGY

Meteorology is the study of atmospheric conditions and weather systems. Meteorologists have the difficult task of predicting the weather for the next few days (short-term forecasts) and for a week or so ahead (long-term forecasts). We all rely on weather forecasts to help plan the day, but they are particularly important for farmers, shipping firms, and airlines, and also power stations, since the weather affects the amount of energy we use.

STUDYING THE WEATHER

WEATHER CENTRE ▶
Computer screens at a weather centre display information gathered by weather-sensing equipment. The WMO (World Meteorological Organization) has 13 main centres that coordinate data from weather stations across the globe. The data is fed into supercomputers, which predict how weather systems will develop.

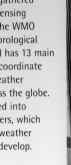

On-board equipment makes images of changing weather conditions

◀ WEATHER SATELLITES
Scientists use satellites orbiting high above Earth to track weather systems. Satellites provide images of clouds, storms, and hurricanes. They also monitor temperatures and humidity using sophisticated sensors.

Casing made of aluminium so satellite is strong and light

Clouds spiral anticlockwise around the centre of the depression

SATELLITE IMAGE ▶
Instruments called radiometers on board satellites provide pictures of clouds covering the Earth's surface. This allows meteorologists to track weather fronts. This satellite image shows a low, or depression, moving northeast of Japan, bringing wet and windy weather to the region.

WEATHER STATION ▶
A scientist checks sensing equipment at a weather station in Antarctica. There are over 10,000 land-based weather stations worldwide. Some are sited in remote places such as coasts, islands, and icy wastes. Others are in cities. Weather stations record air temperatures, pressures, and humidity, and note the speed and direction of winds.

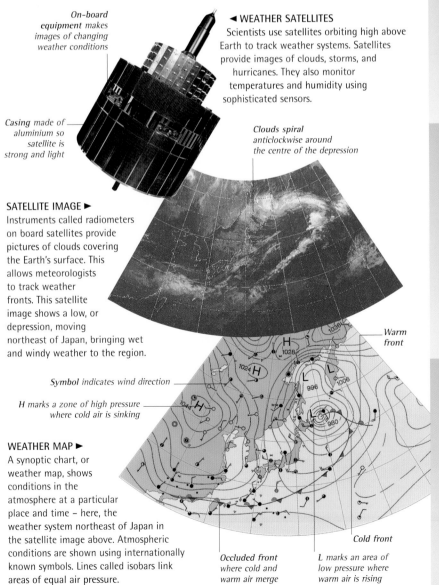

Symbol indicates wind direction

H marks a zone of high pressure where cold air is sinking

Warm front

WEATHER MAP ▶
A synoptic chart, or weather map, shows conditions in the atmosphere at a particular place and time – here, the weather system northeast of Japan in the satellite image above. Atmospheric conditions are shown using internationally known symbols. Lines called isobars link areas of equal air pressure.

Occluded front where cold and warm air merge

L marks an area of low pressure where warm air is rising

Cold front

WEATHER RADAR ▶
A radar dome in Kansas, USA, senses radio waves which have bounced back to Earth from moisture in the air. This technology allows forecasters to understand and predict changes in the weather. Data is not only collected by ground-based weather stations, but also by ships and buoys at sea, by aircraft, and by balloons that carry instruments high into the atmosphere.

FIND OUT MORE ▶▶ Atmosphere **234–235** • Climate **236–237** • Clouds **242–243** • Heat Transfer **82–83** • Wind **240–241**

WIND

Wind is the movement of the atmosphere. The atmosphere moves because the Sun heats the Earth's surface, causing air to increase in temperature, expand, and rise upwards. Cool air moves in to replace the warm air, and we feel this movement of air as wind. The air flows from areas of high pressure to areas of low pressure around the globe. The strongest winds occur during **HURRICANES** and **TORNADOES**.

▼ **LAND AND SEA BREEZES**
Local winds called land and sea breezes often blow in coastal areas. During the day, the land warms up faster than the sea. It heats the air above it, which rises. This is replaced by cool sea air flowing inland, creating a sea breeze. In the evening, the land cools faster than the sea. Now air over the sea rises, and a cool land breeze blows from land to sea.

DAYTIME SEA BREEZE

Air heats up overland

Warm airflow

Cold air drawn in

Cold air sinks

NIGHT-TIME LAND BREEZE

Cold air sinks

Air heats up and rises over sea

WHERE THE WIND BLOWS

At the Earth's poles, the air is at high pressure and low temperature; at the Equator, it is at low pressure but higher temperature. This, together with the spin of the Earth on its axis, creates a pattern of warm and cool winds around the globe. Continents and high mountains also produce wind patterns, such as the seasonal monsoon winds over southern Asia. Areas on the Equator where the winds are very light are called the doldrums. Sailing ships used to get stuck there for long periods.

🌀 *Tropical cyclones*

→ *Warm winds*

→ *Cool winds*

POLAR EASTERLIES

WESTERLIES

NE TRADES

NE TRADES

SW Monsoon

NE TRADES

Doldrums

SE TRADES

Doldrums

Doldrums

SE TRADES

WESTERLIES

WESTERLIES

POLAR EASTERLIES

PREVAILING WIND ▶
Winds often blow from one direction most of the time. If these prevailing winds are strong, trees grow lopsided. Wind direction is always given by the direction the wind is blowing from, rather than where it is blowing to. A southerly wind, for example, blows from the south towards the north. Wind speed is measured in kph or mph. At sea, it is measured on the Beaufort Scale, from force zero (no wind) to force 12 (hurricane).

Branches grow away from the direction of the wind

HURRICANE

A hurricane is a huge, spinning storm with very high-speed winds. A hurricane starts life as a group of thunderstorms near the Equator. If the storms begin to spin together, they form a tropical storm. If the storm's winds reach more than 119 kph (74 mph), it is called a hurricane. In the Pacific Ocean, hurricanes are called typhoons; in the Indian Ocean, they are called cyclones.

High-level winds carry dry, cool air away

Cool, dry air sinks through eye

INSIDE A HURRICANE ▶
A hurricane contains spiral bands of thunderstorms spinning around a still centre called the eye. Warm air within the bands flows around the eye and upwards. The air pressure in the eye is so low that, over sea, water bulges upwards. If the hurricane hits land, the bulge turns to a mass of water that floods the coast in a storm surge.

Eye wall contains strong, warm winds that spiral upwards

Winds and rain blow around the eye, spiralling in towards it

▲ EYE OF THE STORM
The central eye and swirling storm clouds of a hurricane can be seen in a satellite image taken high above the Earth. The hurricane's strongest winds are in the eye wall (the towering wall of cloud around the eye). These often reach speeds of 300 kph (190 mph). Inside the eye, however, there is almost no wind. The eye is usually between 8 and 25 km (5 and 15 miles) in diameter. The hurricane itself can be up to 800 km (500 miles) across.

TORNADO

A tornado is a spinning, funnel-shaped column of air. Inside, winds can blow at speeds of more than 480 kph (300 mph) – the fastest winds on Earth. These violent winds destroy buildings in their path. Tornadoes form underneath giant thunderstorms, and they can be anything from a few metres to 800 m (half a mile) across. Most tornadoes happen in the USA, especially in Tornado Alley in central USA.

SPINNING TWISTER ▶
A tornado in Kansas, USA, sucks up soil as it moves across the landscape. The super-strong winds rip up anything the tornado hits. Debris is hurled about, and even small objects, such as pots and pans, become deadly missiles. Tornado damage is very localized. One house can be wrecked but the next one left undamaged.

wind

FIND OUT MORE ▶▶ Atmosphere **234–235** • Climate **236–237** • Pressure **74–75** • Weather **238–239**

CLOUDS

Air always contains some water vapour from oceans, lakes, and the ground. Clouds form when the air cools below a certain temperature, so that some of the water vapour turns to liquid water or ice. Clouds are made up of millions of minute water droplets or ice crystals, which are so tiny that they float in air. The amount of water vapour that air contains is called its **HUMIDITY**. Warm, humid air often sets off **THUNDERSTORMS**.

▼ HOW CLOUDS FORM
For clouds to form, humid air must rise. It expands and cools as it rises, making its water vapour turn to liquid water or ice. Air rises in three different ways:

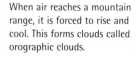

Convection currents

On warm days, the ground heats the air above it. The air expands and floats upwards, forming convection clouds.

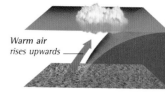

Air flows up mountain slopes

When air reaches a mountain range, it is forced to rise and cool. This forms clouds called orographic clouds.

Warm air rises upwards

At a weather front (where warm and cool air meet), the warm air rises over the cold air, forming frontal clouds.

CUMULUS

ALTOCUMULUS AND STRATUS

ALTOCUMULUS

CIRRUS

▲ TYPES OF CLOUD

Clouds are named according to their shape and height above sea level. Cumuliform clouds are heaped and stratiform clouds are layered; and "alto-" means medium-level and "cirro-" high-level. So cumulus clouds are low-level and heaped; stratus clouds are low and layered; and altocumulus are medium-level, heaped clouds. Cirris clouds are high and wispy.

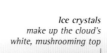

clouds

CUMULONIMBUS CLOUDS

Towering cumulonimbus clouds are formed when a cumulus cloud keeps billowing upwards. They can be more than 9,000 m (30,000 ft) in height. Cumulonimbus clouds produce heavy showers of rain or hail and gusty winds.

Ice crystals make up the cloud's white, mushrooming top

HAIL SHOWER

HUMIDITY

Humidity is the amount of water vapour in the air. The warmer the air is, the more water vapour it can contain. Saturated air is air that contains the maximum amount of water vapour for a particular temperature. Relative humidity is the actual amount of water vapour in the air, compared to the amount needed for the air to be saturated. Saturated air has a relative humidity of 100 per cent.

▲ MIST

Mist is a layer of cloud that lies close to the ground. It forms when warm, humid air comes into contact with an area of cold water or cold ground. This can happen when humid air touches ground that has cooled quickly on still, cloudless nights. Fog develops in the same way, but is thicker than mist.

▲ HOARFROST

If the air temperature falls below freezing (0°C or 32°F), hoarfrost may form. Surfaces on the ground become covered by ice crystals, which look like a light dusting of snow. Dew forms when some of the water vapour in humid air comes into contact with cold surfaces at ground level. The vapour then turns into tiny drops of liquid water instead of frost.

THUNDERSTORMS

A thunderstorm begins when a cumulonimbus cloud grows extremely large. The cloud produces lightning, thunder, heavy rain or hail, strong winds, and even tornadoes. About 40,000 thunderstorms happen in the world every day – mostly in the tropics, where the air is very warm and humid. A thundercloud can be recognized by its broad, flattened top.

LIGHTNING ▶

A flash of lightning is a giant spark of electricity. When ice crystals and water droplets move about and collide inside a thundercloud, static electricity builds up. Lightning is set off when the spark jumps through a cloud, or from one cloud to another, or from a cloud to the ground. A bolt of lightning heats the air to about 30,000°C (54,000°F) so the air expands suddenly and causes a clap of thunder.

HOW LIGHTNING STRIKES

Negative electric charge builds up in the base of a thundercloud, and positive charge in the top. The negative and positive charges are attracted to each other, so lightning can strike through the cloud. The negative charge in the cloud's base also attracts positive charges in the ground, so eventually a lightning spark leaps through the air between the cloud and the ground.

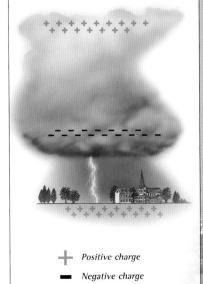

➕ Positive charge

➖ Negative charge

FIND OUT MORE ▶▶ Atmosphere 234–235 • Electricity 126–127 • Heat Transfer 82–83 • Rain 244–245

RAIN

The moisture gathered in clouds eventually falls to the ground as liquid rain or drizzle, or as solid, frozen **SNOW** or **HAIL**. Any kind of falling moisture is called precipitation. Rain forms when tiny droplets of water floating in clouds collide to form bigger drops. If the drops get large and heavy enough, the air can no longer support them, and they fall as rain. Rain also forms when falling snowflakes melt high in the air.

◄ RAINSTORM
A sudden cloudburst drenches city streets. The rain falls in streaks, but each raindrop has a rounded, flattened shape. Rain is vital to plants and animals, but torrential rain can bring floods. Some parts of the world have a higher rainfall than others. Coastal regions and the tropics are usually wet, while inland deserts may get almost no rain at all.

rain

▲ RAINBOW
The bright arch of a rainbow appears in the sky when the Sun shines through falling raindrops. Inside each raindrop, the sunlight is refracted (bent and split) into the seven colours that make it up – red, orange, yellow, green, blue, indigo, and violet.

THE WATER CYCLE ▼
Moisture rises from the Earth's surface and falls back in a never-ending cycle driven by the Sun's energy. As the Sun heats the surface of lakes, oceans, and icefields, moisture evaporates (turns into water vapour), rises into the air, and gathers in clouds. At cooler temperatures, the water vapour condenses (turns into a liquid) and falls back to Earth as rain, snow, or hail.

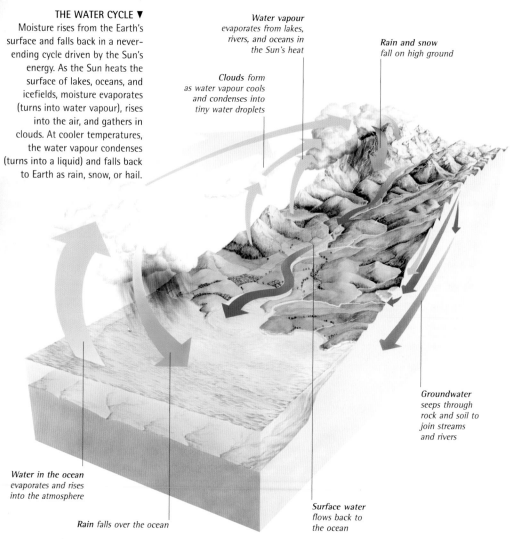

Water vapour evaporates from lakes, rivers, and oceans in the Sun's heat

Rain and snow fall on high ground

Clouds form as water vapour cools and condenses into tiny water droplets

Groundwater seeps through rock and soil to join streams and rivers

Water in the ocean evaporates and rises into the atmosphere

Surface water flows back to the ocean

Rain falls over the ocean

HAIL

Hailstones are balls of ice that grow from ice crystals. They form inside storm clouds that tower up to 10 km (6 miles) above the ground. The temperature at the base of a storm cloud is warmer than at the top. This causes powerful vertical air currents. Water droplets in the cloud freeze and are whirled up and down. A fresh layer of ice forms around a hailstone each time it is tossed up to the frozen cloud top. Eventually, it gets so heavy it falls to the ground.

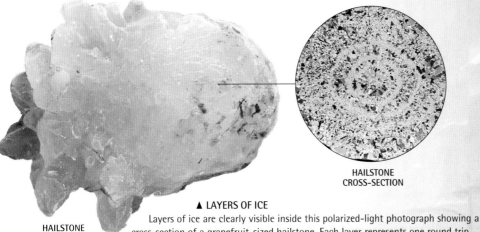

HAILSTONE

HAILSTONE CROSS-SECTION

▲ LAYERS OF ICE
Layers of ice are clearly visible inside this polarized-light photograph showing a cross-section of a grapefruit-sized hailstone. Each layer represents one round trip the hailstone has made to the top of a storm cloud and back down. Hailstones this big are rare, but many are often the size of marbles. Large chunks of falling ice are extremely dangerous, and may shatter glass, dent car roofs, ruin crops, or even kill people.

SNOW

Snow forms in clouds high in the atmosphere, in temperatures of -20 to -40°C (-4 to -40°F), as water vapour condenses to form ice crystals. The crystals collide and combine to form larger snowflakes, until they get too heavy to float, and drift to the ground. Sleet is a mixture of snow and rain, or partly melted snow.

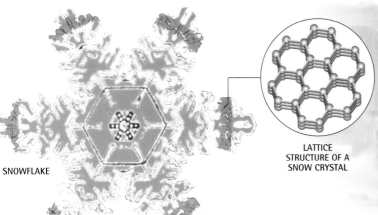

SNOWFLAKE

LATTICE STRUCTURE OF A SNOW CRYSTAL

▲ SNOW CRYSTAL STRUCTURE
A snow crystal is a single crystal of ice. A snowflake can be one snow crystal, or several stuck together. All snow crystals have a hexagonal (six-sided) structure. This is formed because the tiny water molecules inside ice line up in a regular hexagonal shape called a lattice.

BLIZZARDS ▶
A snowstorm hits Manhattan, New York, slowing traffic. A fresh fall of snow can make the countryside and city streets look beautiful, but severe blizzards can be dangerous. In January 1997, a heavy fall of snow in eastern Canada and north-eastern USA caused roofs, trees, and power lines to collapse.

FIND OUT MORE ▶▶ Climate **236–237** • Clouds **242–243** • Colour **122–123** • Water **40–41** • Weather **238–239**

◄ POLAR AND TUNDRA
The Svalbard Islands north of Norway combine a landscape of glaciers and arctic tundra – a boggy plain covered with mosses, lichens, and low-growing bushes. In polar habitats, temperatures may drop as low as -80°C (-112°F).

◄ BOREAL FOREST
In Siberia in northern Asia, and in northern North America and Europe, there are huge boreal forests, with many lakes, ponds, and rivers. Trees in this, the world's largest habitat, are conifers, such as firs, pines, and spruces. Their thin, needle-like leaves survive the cold climate. Plants grow only during the short, cool summer.

◄ TEMPERATE FOREST
The broad-leaved trees in this forest in Vermont, north-eastern USA, have produced spectacular autumn colours. Places halfway between the poles and the Equator have temperate climates, with cool winters and warm summers. Temperate forest habitats contain broad-leaved trees that lose leaves in autumn and grow new ones in spring.

◄ TEMPERATE GRASSLAND
Herds of horses graze on the open grassland of the Mongolian steppes. In temperate regions such as these, the summers are hot and dry and the winters are very cold and dry. Only grass grows in these areas, because it is too dry for trees. Temperate grasslands in North America are called the prairies; in South America they are known as the pampas.

HABITATS

A habitat is a place where plants and animals live, and provides them with food and shelter. It can be very small, such as a single tree or pond, or vast, such as a rainforest or desert. The physical conditions in a place and its vegetation are both part of the habitat. **HABITAT LOSS** is occurring in many parts of the world.

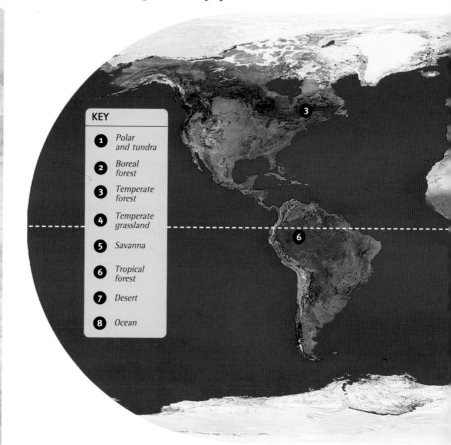

KEY

1. Polar and tundra
2. Boreal forest
3. Temperate forest
4. Temperate grassland
5. Savanna
6. Tropical forest
7. Desert
8. Ocean

HABITAT LOSS

Habitat loss is the destruction of habitats such as forests and marshes through human activities, especially forestry and farming. Many species of animals and plants live in one small habitat and cannot survive anywhere else. It is estimated that more than a hundred species become extinct every day through habitat loss.

DEFORESTATION ►
Cutting down natural forests (rather than forestry plantations) is called deforestation. The world's tropical rainforests have suffered most from serious deforestation. The trees are cut down for their valuable timber, or burned to make space for farming and ranching. Across the world, an area of rainforest larger than the city of New York is cut down every day.

▼ EARTH'S BIOMES

The colours of this satellite image show the vegetation on the different parts of Earth's surface, from polar ice sheets to steamy rainforests along the Equator. On land, a habitat is made up of the landscape of a place, the climate (the pattern of weather it experiences over a year), and the vegetation that grows there. Together, Earth's larger habitats, also known as biomes, form the biosphere – the zone that can support life.

habitats

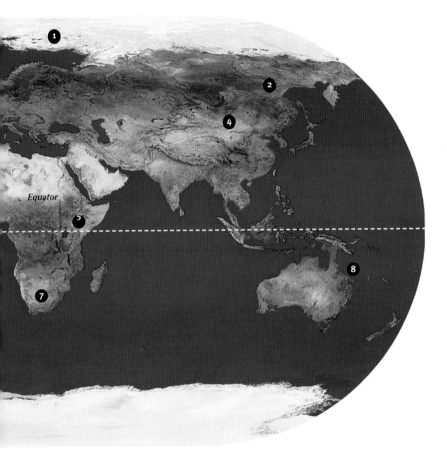

Equator

SAVANNA ►

In Kenya, East Africa, a cheetah looks out for grazing antelope on the open savanna. This grassland habitat is near the Equator, with weather that is warm all year round. There is just enough rain for grasses and a few trees, but not for a forest to grow.

TROPICAL FOREST ►

Tropical rainforests, such as the Amazon rainforest in South America, flourish in equatorial regions where it is hot all year round and it rains almost every day. A rainforest contains five main habitats: the forest floor, shrub layer, understorey, canopy, and the emergent layer at the top. A greater variety of species live here than in any other habitat.

DESERT ►

This quiver tree, in South Africa's Northern Cape, is adapted to survive its desert habitat by storing water in its stout trunk. Deserts are harsh, dry habitats, with sometimes no rain for years. Animals and plants that live there have to cope with daytime temperatures of up to 50°C (120°F) and nights that can be very cold.

▲ DAMAGED REEFS

The coral reefs that grow in shallow, tropical seas are one of The Earth's richest habitats. They support a huge variety of tropical fish and other marine animals. Here in the Maldives, in the Indian Ocean, coral reefs are being smashed to provide building materials and tourist souvenirs. This destroys the reef, and threatens the animals that depend on it for survival.

OCEAN ►

The corals that form the Great Barrier Reef off Australia's northeast coast teem with marine life. The three main habitats in the oceans are the sunlit surface waters; the cold, deep waters that extend to depths of more than 6,000 m (19,500 ft); and the sparsely inhabited ocean floor. There is life at every level, but it is most abundant in the sunlit zone, where there is a plentiful supply of food.

FIND OUT MORE ►► Conservation 335 • Oceans 228–229 • Planet Earth 204–205 • Trees 268–269

EARTH'S RESOURCES

The Earth has many natural resources that make life in the modern world possible. For example, rocks are used in their natural state to make buildings, but they can also be processed to provide the materials we need to make anything from bridges and cars to silicon chips and jewellery. **FOSSIL FUELS** provide us with energy, but so does water flowing down rivers, the wind, and even the Sun. Resources such as rocks and fossil fuels must often be extracted from the ground by **MINING**.

▲ WATER RESOURCES
Water collected by dams, such as the Ataturk Dam on the Euphrates River in Turkey, is piped to cities for use in homes and industries, and to fields for irrigation. The Ataturk Dam also contains a hydroelectricity station that provides Turkey with electricity. Water is Earth's most important natural resource. Without it there would be no life on our planet.

AGRICULTURE ▲
The prairies that stretch across the Midwest of the USA produce huge quantities of grain. Crops like these are grown on the soils that cover much of the Earth's surface. Farming uses up minerals in the soil, which farmers replace using fertilizers made from other resources, such as natural gas from underground and nitrogen from the atmosphere. Agriculture also requires large amounts of water.

◄ BUILDING MATERIALS
The marble being dug from this quarry near Carrara, Italy, is cut and shaped to make building stone, as are other rock types, such as granite, sandstone, and slate. Crushed rocks are used to make road surfaces. They are also used to make concrete, with sand, gravel, and ground limestone. Clay deposits are shaped and fired in kilns to produce bricks, tiles, and pipes.

FIND OUT MORE ▸▸ Energy Sources 86–87 • Minerals 216–217 • Nitrogen 42–43 • Rocks 218–219 • Soil 224

MINING

Rocks contain a great variety of useful minerals. Mining and quarrying involve blasting, drilling, and digging up rocks to extract the minerals. Most mines and quarries are worked for building materials, coal, metal ores, and gem-rich rocks and deposits. Mining is noisy, dusty, and can require the use of dangerous chemicals, all of which can cause environmental damage.

◄ UNDERGROUND MINING
A gold mine in Indonesia is an example of an underground mine, where rock is dug out by machinery deep under the surface. There are two main types of underground mine: shaft mines, which are normally deep, with vertical shafts leading to tunnels; and drift mines, which are near the surface. Underground mining is very dangerous because of possible flooding, explosive gases, and falling rocks.

OPENCAST MINING ▲
At the Bingham copper mine in Utah, USA, the ore deposit is close to the surface and is extracted by opencast mining. Opencast mining is cheaper and easier than underground mining because no shafts have to be dug, but it does more damage to the landscape. Once the ore is dug up, it is carried away by trucks, railway, or conveyor belts.

FOSSIL FUELS

Coal, oil, and gas are called fossil fuels because they were formed from the remains of animals and plants that were buried by layers of sediment millions of years ago. Most of the energy used today comes from burning fossil fuels. Fossil fuels are non-renewable sources of energy, which means that once they have been used they can never be replaced.

OIL DRILL

GAS

Earth's resources

OIL AND GAS ►
Many of the world's oil and gas supplies are found in rock under the sea, from where they are extracted through pipes drilled into the seabed from production platforms. Where oil and gas are found together, they were formed from the bodies of microscopic marine organisms. Oil is a source of chemicals as well as fuel.

YEAR 1   AT 90 MILLION YEARS   AFTER 360 MILLION YEARS

Plant matter   *Peat*   *Lignite*   *Bituminous coal*   *Anthracite*

◄ COAL
Coal is formed by the burial of plant remains before they rot completely. Surface deposits of vegetation form layers of peat that become lignite and coal as they are more deeply buried over time. Burial compresses the plant remains and squeezes out any water. Further pressure turns coal into anthracite.

POLLUTION

Pollution is waste that we put into the environment. It can harm plants and animals, including humans. Pollution comes from factories, and also from homes, farming, cars, ships, trucks, and aircraft. It includes smoke from fires, exhaust gases from engines, poisonous chemicals from industrial processes, rubbish such as plastic packaging, and sewage. These things pollute the landscape, rivers, lakes, seas, and the air. Noise and light can also be forms of pollution.

AIR POLLUTION ▲
The city of Bangkok in Thailand is affected by smog – a mix of smoke and fog. It causes difficulties for people with breathing problems, such s asthma and bronchitis. Smoke from car engines, power stations, and factories is made up of waste gases, dust, and tiny particles of unburned fuel. These harmful waste gases include sulphur dioxide, nitrogen oxides, carbon dioxide, and carbon monoxide.

pollution

◄ GREENHOUSE EFFECT
Some of the gases in the Earth's atmosphere trap heat from the Sun. This is called the greenhouse effect, because the atmosphere works like the glass in a greenhouse. The greenhouse effect is natural, but pollution causes an increase in the gases that produce it, including carbon dioxide, water vapour, and methane.

SUN

Greenhouse gases surround Earth, trapping some of the Sun's heat

HEAT FROM SUN WARMS EARTH

SOME HEAT FROM EARTH ESCAPES

SOME HEAT FROM EARTH TRAPPED BY ATMOSPHERE

◄ ACID RAIN
Many trees, such as these in Siberia, northern Russia, are killed by acidic rain formed from pollutants in the air, especially sulphur dioxide and nitrogen oxides. These gases mix with water in the atmosphere to form rain that is acidic. Acid rain also kills aquatic life when it runs into ponds and lakes.

◄ ACCIDENTAL POLLUTION
Pollution in this lake in southern Rio de Janiero, Brazil, has killed the fish by reducing oxygen levels in the water. Leaks from oil tankers into the sea create oil slicks on the water surface. Where the oil reaches the shore, it covers beaches and kills seabirds.

◄ GLOBAL WARMING
The snows in the Bolivian Andes are melting because of the gradual rise in temperature known as global warming. This is happening because carbon dioxide from burning fuels is increasing the greenhouse effect. If it continues, melting snow and ice will make sea levels rise.

FIND OUT MORE ▸▸ Acids 32 • Atmosphere 234–235 • Chemical Industry 50–51

SUSTAINABLE DEVELOPMENT

By using scarce resources such as oil, coal, and gas and producing pollution, we are storing up problems for future generations. We need to find ways of meeting our own needs now without spoiling the Earth for people in the future. The idea of sustainable development is to provide a good quality of life for everyone without causing pollution or using up resources that cannot be replaced. Using renewable forms of energy and reducing waste will help towards sustainable development.

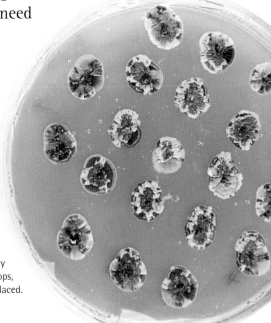

▼ TECHNOLOGICAL ECO-LIVING
An ecological village in Denmark is a good example of how technology can help sustainable development and reduce harm to the environment. Most of the electricity the villagers need for lighting and working machinery is generated by a wind turbine. The dome-shaped homes are mostly heated by the Sun and are well insulated to reduce heat loss. The villagers grow their own food organically, and their sewage is treated by their own waste-cleaning system.

ENERGY FROM BIO-FUELS ►
These bacteria are breaking down cellulose, a material from plants, into the chemical ethanol, which can be used instead of petrol. Ethanol is a bio-fuel. Bio-fuels are renewable energy sources because they are made from plants called energy crops, which can be grown and replaced.

sustainable development

SOLAR POWER ▼
At this solar-power station in the Mojave Desert, California, mirrors reflect the Sun to heat oil inside pipes. The oil heats water to make steam, and this drives the turbines in electricity generators. The Sun is also an important source of renewable energy in less developed countries, where smaller-scale solar power schemes can produce electricity and heating for remote villages.

FIND OUT MORE ▶▶ Earth's Resources **248–249** • Energy Sources **86–87** • Recycling Materials **60–61**

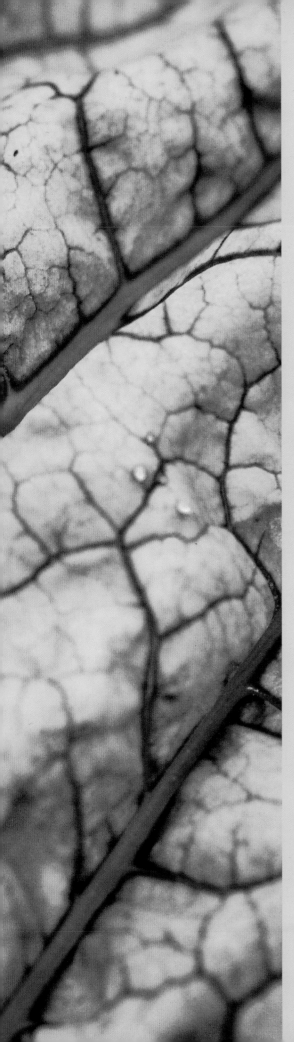

PLANTS

CLASSIFYING PLANTS

Plants belong to the **PLANT KINGDOM,** one of the five kingdoms of living things. Plants are classified into smaller groups, according to shared characteristics. All plants share certain features. They are made up of many cells. They also produce their own food by a chemical process called photosynthesis, using water, carbon dioxide, and the energy of sunlight. As a by-product, they release life-giving oxygen into the air.

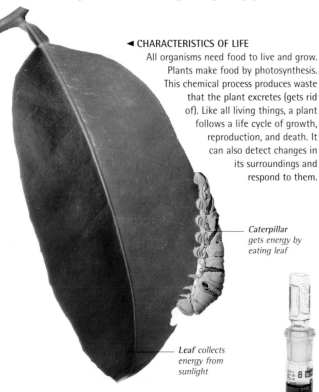

◄ CHARACTERISTICS OF LIFE

All organisms need food to live and grow. Plants make food by photosynthesis. This chemical process produces waste that the plant excretes (gets rid of). Like all living things, a plant follows a life cycle of growth, reproduction, and death. It can also detect changes in its surroundings and respond to them.

Caterpillar gets energy by eating leaf

Leaf collects energy from sunlight

CAROLUS LINNAEUS
Swedish, 1707–1778

Naturalist Linnaeus devised the first uniform, scientific way of defining and naming plants and animals. His system is still used as the basis for classification today. The first part of a Linnean name indicates the genus (group). The second part gives the particular species.

◄ CHLOROPHYLL

Plants are green because they contain a green pigment called chlorophyll. Chlorophyll captures some of the energy in sunlight and uses this to make food. This process is called photosynthesis. Most plants make food this way, but a small number also digest other living things.

Chlorophyll extracted from plant leaves

▼ THE FIVE KINGDOMS OF LIFE

The moneran kingdom includes the simplest, single-celled organisms. Protoctists are more complex single cells and include green algae, which contains the chlorophyll that is also found in plants. Fungi, plants, and animals are thought to have evolved from protoctist ancestors. The kingdom containing the most known species is the animal kingdom.

ANIMAL
LION

PLANT
SUNFLOWER

FUNGUS
MUSHROOM

PROTOCTIST
AMOEBA

MONERAN
BACTERIA

PLANT KINGDOM

Within the plant kingdom, plants are divided into two main groups. The largest group contains the plants that produce seeds. These are flowering plants (angiosperms) and conifers, Ginkgos, and cycads (gymnosperms). The other group contains the seedless plants that reproduce by spores. It includes mosses, liverworts, horsetails, and ferns. So far, scientists have named 400,000 separate plant species. Around 300,000 of these are flowering plants.

LIVERWORTS	FERNS	CONIFERS	FLOWERING PLANTS
475 MILLION YEARS AGO	362 MYA	290 MYA	130 MYA

EVOLUTION OF PLANTS ▲
From the fossilized remains of ancient plants, we know that the first plants probably developed from algae and lived in water. Mosses and liverworts appeared on land around 475 million years ago (mya). Next came clubmosses, horsetails, and ferns, between 390 and 350 mya. Cone-bearing cycads and conifers evolved much later, and flowering plants most recently of all.

Flowering plant produces seeds

plant kingdom

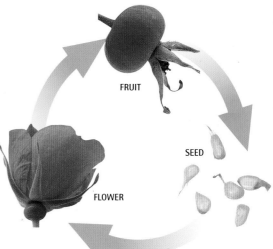

FRUIT

SEED

FLOWER

▲ CYCLE OF LIFE
A rose begins life as a seed that germinates (sprouts), growing roots and a shoot. The shoot puts out leaves to produce food and flowers to produce seeds. Seeds develop after pollination (when a male sex cell, carried in pollen, fertilizes a female one). Fruits called hips grow around the developing rose seeds. Birds eat hips and disperse seeds away from the parent plant.

◄ TROPICAL DIVERSITY
Plants are found in most of the world's habitats, including wetlands, grasslands, forests, and polar regions. By far the richest plant habitats are tropical rainforests. Tall, broad-leaved trees reach for the sunlight. Far below, ferns and mosses thrive in the warm, damp conditions. There are also flowering plants, often brightly coloured to attract animals that will help to pollinate them.

Fern is seedless, but produces spores

CLASSIFYING A LESSER CELANDINE

Class	Magnoliopsida (dicotyledons)
Order	Ranunculales
Family	Ranunculaceae
Genus	*Ranunculus*
Species	*Ranunculus ficaria*

PLANT ANATOMY

A plant's body has different structures designed for different tasks, such as making food and conserving water. **LEAVES**, usually broad and flat, take in energy from sunlight and carbon dioxide from the air. **ROOTS** snake through the soil to take in water and minerals. The **STEM** supports the part of the plant above ground. It contains a network of microscopic tubes that transport essential water, minerals, and food between the roots and the leaves.

PLANT CELL ▶

A plant is made up of microscopic living structures called cells. Like animal cells, plant cells have a nucleus and an oily membrane surrounding the whole cell. A plant cell is also encased in a tough cell wall and usually contains a large, fluid-filled bubble called a vacuole. Cells of green plant parts also contain chloroplasts.

Cell wall, made of tough cellulose fibres, maintains shape of cell

Cell membrane controls what enters and leaves the cell

Nucleus controls activity in the cell

Vacuole swells with water and keeps the cell firm

Chloroplast contains green chlorophyll for photosynthesis

Cytoplasm is the site of chemical reactions that release energy

Leaf node

Side shoot grows from bud at leaf node

Terminal bud contains actively dividing cells

TREE MALLOW

Petiole (leaf stalk)

LEAF

Leaves trap carbon dioxide and sunlight to make food by photosynthesis, but lose water from their surface. Larger, flatter leaves produce more food but lose more water. In dry habitats, leaves have specialized shapes and a waxy covering, to limit the amount of water loss.

TYPES OF LEAF ▶

The simplest type of leaf consists of a flat blade, but some leaves are divided into smaller "leaflets". These compound leaves may offer certain advantages, such as less resistance to the wind. Leaves of evergreen plants, such as camellias, are often extra waxy to protect against drought.

CAMELLIA LEAF (SIMPLE)

BITTERNUT LEAF (COMPOUND)

Transport vessels

Phloem   Xylem

Upper epidermis is waxy to help rainwater run off

Palisade cell contains chloroplasts

Spongy mesophyll contains air pockets

Lower epidermis

HELLEBORE LEAF ▶

This scanning electron micrograph (SEM) shows a cross-section through a leaf. Near the upper surface, where the light falls, are the palisade cells, containing chloroplasts for photosynthesis. The transport vessels, phloem and xylem, run through the next layer of cells, the spongy mesophyll. Gases are allowed in and out of the mesophyll through pores on the lower leaf surface called stomata.

STEM

The stem is the main support of the upright plant, and connects the roots and leaves. It contains bundles of microscopic tubes – xylem vessels, carrying water and minerals, and phloem vessels, carrying food. The stem also has cells with special thickened walls that help to provide strength. It may have layers of dividing cells, too, that allow the stem to grow thicker.

END OF A SYCAMORE STEM ▶
The bud at the tip of a shoot, known as the terminal bud, may develop into leaves or – in some plants – flowers. It also allows stems to grow, as the cells there divide and get bigger. Most cells of the stem are very long. As thousands of cells elongate (grow longer), the whole stem becomes taller.

Epidermis
has a thick, waxy cuticle (waterproof covering)

Air space
helps to protect stele

Stele, or vascular cylinder, contains transport vessels, xylem and phloem

Cortex (layer between epidermis and pith)

◀ MICROGRAPH THROUGH A MARE'S TAIL STEM
Different plants arrange their vascular tissue (bundles of transport vessels) differently. In the stem of the mare's tail plant, xylem and phloem run through an inner cylinder of large cells, called a stele. Around this is a light, protective cortex of air-filled spaces. The stem's outermost cells have a waxy coating to stop the stem from drying out.

Terminal bud

Elongating region

Root tip is where growth occurs

Side root

Woody lower stem is mostly made up of lignin

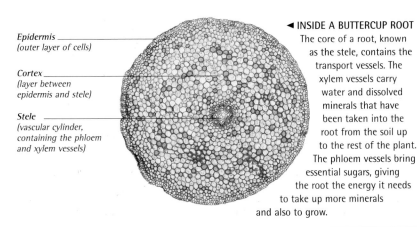

ROOT

A plant's roots hold it firmly in the soil and also take up water and minerals. Some types of plant have one main root, called a taproot, with smaller lateral (side) roots branching off. Other plants have root systems that form a dense tangle. Long taproots allow the plant to gain water from greater depths, but a thick network of roots may provide a more robust anchor in the soil.

Main root anchors the plant in the ground

plant anatomy

Epidermis (outer layer of cells)

Cortex (layer between epidermis and stele)

Stele (vascular cylinder, containing the phloem and xylem vessels)

◀ INSIDE A BUTTERCUP ROOT
The core of a root, known as the stele, contains the transport vessels. The xylem vessels carry water and dissolved minerals that have been taken into the root from the soil up to the rest of the plant. The phloem vessels bring essential sugars, giving the root the energy it needs to take up more minerals and also to grow.

HOW ROOTS GROW

Root tips produce substances that make roots grow down. The root cap secretes slime so that the growing root can slip through the soil. The region immediately behind the root cap is called the meristem and contains the actively dividing cells. The newly divided cells grow longer, lengthening the entire root as it grows downwards.

Elongating region

Meristem (region of actively dividing cells)

Root cap protects dividing cells

ROOT TIP OF A BROAD BEAN

FIND OUT MORE ▶▶ Animal Anatomy **292–293** • Photosynthesis **258** • Plant Survival **274–275** • Transpiration **259**

PHOTOSYNTHESIS

Unlike animals, most plants do not need to find food, because they can make it for themselves. Plants use energy from sunlight to turn water and carbon dioxide into an energy-rich sugar called glucose. This process is called photosynthesis, which means "making things with light". Photosynthesis takes place inside capsules in the leaf cells, called **CHLOROPLASTS**.

Light is absorbed by the leaf

MAKING FOOD AND OXYGEN ▶
Plants use their leaves to make food. Oxygen is created as a by-product. During photosynthesis, plant leaves take in carbon dioxide from the atmosphere. Using the energy from sunlight, this is combined with water drawn up from the roots to make glucose. Oxygen is also produced in this chemical reaction and exits the leaves into the surrounding air.

Glucose is carried from the leaf to other parts of the plant

Carbon dioxide enters the leaf

Oxygen leaves the leaf

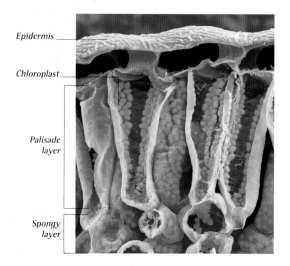

Epidermis
Chloroplast
Palisade layer
Spongy layer

◀ FOOD-PRODUCING CELLS
Different plant cells perform different tasks. Palisade cells and spongy cells are located just below the epidermis and are a plant's main food-producers. The tall palisade cells are packed with green chloroplasts, which carry out photosynthesis. The irregularly shaped spongy cells also have chloroplasts. Air spaces between the cells are filled with carbon dioxide, water vapour and other gases.

CHLOROPLAST

Many leaf cells contain tiny, lens-shaped organelles called chloroplasts. These can move around the cell towards the direction of sunlight. Chloroplasts contain a green, light-capturing pigment called chlorophyll. This chemical helps the chloroplasts to act like minute solar panels.

photosynthesis

Outer Membrane
Stroma
Grana

◀ INSIDE A CHLOROPLAST
Chloroplasts are made up of stacks of tiny disclike membranes called grana, held in a dense mass of material known as the stroma. The grana are where water is split into hydrogen and oxygen, using some of the light energy captured by the chlorophyll. The rest of the light energy is used in the stroma to combine the hydrogen with the carbon dioxide to make glucose.

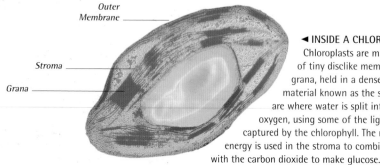

Nutrients are absorbed from the soil by roots

FIND OUT MORE ▶▶ Carbon 44–45 • Energy Sources 86–87 • Oxygen 39 • Plant Anatomy 256–257

TRANSPIRATION

In this process, water evaporates from the surface of a leaf through microscopic pores known as **STOMATA**. The loss of water creates a suction force that pulls up more water from the roots. Transpiration helps a plant to collect vital minerals from the soil. The amount of water lost from the leaves depends on how much water is in the soil, as well as other environmental conditions, such as temperature, humidity, and wind.

Water evaporates from the leaf

◄ UPWARD-FLOWING WATER
Water in plants is both pushed and pulled upwards inside transport pipes called xylem. This continuous flow of water is known as the transpiration stream and keeps the stem firm so that it can support the weight of the plant. The transpiration stream also transports water to the plant's leaves for photosynthesis and carries minerals around the plant.

Stem is cut in spring when lots of water is available

ROOT PRESSURE ►
The cut stem of this grape vine shows how water is forced out by pressure in the roots. This pushing pressure is created by water entering the roots from the soil and forcing water already in the roots upward. Water is also pulled through plants to replace water lost through transpiration from the leaves.

A well-watered plant stands upright – the water inside it keeps its cells and tissues firm

A dried out plant wilts and eventually falls over - if watered, it will gradually stand up again

◄ STANDING FIRM
Plants need a continuous supply of water to stand upright. Each plant cell holds water in swollen bags called vacuoles. This water pushes against the cell walls and keeps the cell firm. This pressure and firmness of plant cells is called turgor. The cells of the castor oil plant on the left are turgid. Those of the plant on the right are not, because they have lost too much water.

Water is pushed and pulled up through the stem

transpiration

STOMATA

The surface of a leaf has many tiny pores, called stomata. The stomata allow carbon dioxide into the leaf so that photosynthesis can occur. They also allow water to leave the leaf by transpiration. Plants that grow in full sunlight usually have most of their stomata on the shaded undersides of their leaves. This helps the plant to conserve water.

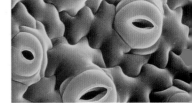

OPEN STOMATA

CLOSED STOMATA

Water is absorbed from the soil by roots

STOMATA BY DAY AND NIGHT ►
Seen in microscopic view, each individual stoma (stomata is the plural) in the leaf's surface is surrounded by two guard cells, which look a bit like lips. During the day, these guard cells swell with water and become bloated, opening the stoma. At night, the guard cells release their water and the stomata close.

FIND OUT MORE ►► Energy Sources 86–87 • Plant Anatomy 256–257 • Water 40–41

SEEDLESS PLANTS

Ferns, horsetails, mosses, and liverworts do not produce flowers or grow from seeds. The life cycles of these plants have two distinct stages – one in which **SPORES** are produced, and one in which sex cells (sperm and eggs) are produced. Most seedless plants live in damp and shady habitats. Certain types of mosses, called **PEAT MOSSES**, grow in vast expanses of wetlands in the northern parts of the world.

MODERN HORSETAIL

HORSETAIL FOSSIL

▲ ANCIENT HORSETAILS

The horsetails alive today are very similar to the types of horsetails that lived hundreds of millions of years ago, before there were any flowering plants. At this time, seedless plants dominated the land, and giant horsetails made up some of the earliest tall forests. Fossils of these prehistoric horsetails have been preserved in rocks from this period.

◄ FERN REPRODUCTION

Adult ferns produce spores in capsules inside chambers on the underside of their fronds. In dry conditions, the capsules release the spores into the air. When a spore lands on moist ground, it develops into a tiny, heart-shaped structure called a prothallus. This produces the sex cells. Fertilized by male sperm, the female egg of the prothallus develops into a new adult plant.

MODEL OF A SPORE CHAMBER

Fern frond
with dark clusters of
spore chambers

Spore capsule
in a tiny open
chamber

▲ FERN FRONDS

The leaf of a fern is known as a frond. At first, a young frond is curled up into a structure called a fiddlehead. The fiddlehead has this shape because its lower surface grows faster than the upper surface. As the plant matures, the frond unfurls. Fiddleheads of certain kinds of ferns have been used as a source of food, but some contain poisons.

Spore capsules sit on top of stalks. They release spores into the air

Leafy moss tufts produce male sperm and female eggs

▲ **MOSS REPRODUCTION**
Mosses and liverworts are known as bryophytes. Adult bryophytes produce the sex cells. Fertilized female eggs then grow into a stalked sporophyte, or spore capsule. Once they are released, the spores develop into the next generation of moss.

SPORE

Spores are minute, independent cells. Unlike sex cells, spores can divide on their own to make many-celled bodies. They have a simple structure, which consists of genetic material encased in a protective coat that can survive dry conditions. When spores land on damp ground, they grow into a plant that produces sex cells.

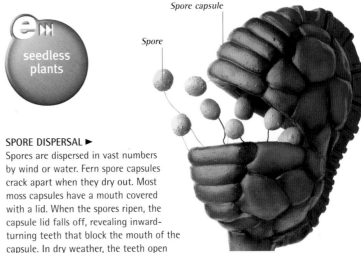

seedless plants

Spore capsule

Spore

SPORE DISPERSAL ▶
Spores are dispersed in vast numbers by wind or water. Fern spore capsules crack apart when they dry out. Most moss capsules have a mouth covered with a lid. When the spores ripen, the capsule lid falls off, revealing inward-turning teeth that block the mouth of the capsule. In dry weather, the teeth open outward, and the spores disperse.

FERN SPORES DISPERSING

PEAT MOSSES

Peat mosses, which are also called sphagnum mosses, grow in wetland areas known as peat bogs. These mosses have a spongy texture and can absorb large amounts of water. To get all the minerals they need, peat mosses use special chemical reactions that release acid by-products into the surrounding soil.

FROM PEAT MOSS TO PEAT ▶
Peat is dead peat moss and plant matter that has collected in peat bogs over hundreds of years. Peat forms in layers, which are compressed by the weight of water and living moss on the surface. Over time, the living layer dies down and is replaced by new moss. Peat is harvested and dried for use as fuel and fertilizer. The overuse of peat threatens peat bog habitats.

SPHAGNUM MOSS

▲ **HARDY COLONIZERS**
Bryophytes do not have true roots. They have hairy, rootlike growths called rhizoids that anchor the plants to the soil, but do not draw up water. Instead, their leaves absorb moisture in the air. Because they need little or no soil in which to root, bryophytes are often the first plants to colonize thin soil. Like the liverwort in this picture, bryophytes can also grow on bare rocks.

CUT PEAT

FIND OUT MORE ▶▶ Earth's Resources **248–249** • Pollination **266–267** • Seed Plants **262–263**

SEED PLANTS

Most plants grow from seeds. These seed plants fall into two groups, angiosperms and gymnosperms. Angiosperms are the flowering plants. Their seeds develop inside a female reproductive part of the flower, called the ovary, which usually ripens into a protective **FRUIT**. Gymnosperms (conifers, *Ginkgo*, and cycads) do not have flowers or ovaries. Their seeds mature inside cones. Seeds may be carried away from the parent plant by wind, water, or animals.

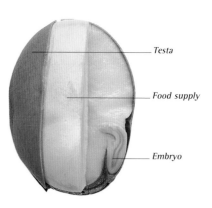

Testa

Food supply

Embryo

▲ INSIDE A SEED
A seed is the first stage in the life cycle of a plant. Protected inside the tough seed coat, or testa, is the baby plant, called an embryo. Food, which fuels germination and growth, is either packed around the embryo or stored in special seed leaves, called cotyledons.

Fertilized flowerhead closes and seeds develop inside

Flowerhead opens daily, and insects pollinate the florets

▲ FLYING FRUIT
Dandelion seeds have feathery parachutes to help them fly far from their parent plant. A dandelion flower is actually made up of many small flowers, called florets. Each develops a single fruit. The fruits form inside the closed-up seed head, after the yellow petals have withered away. When the weather is dry, the seed head opens, revealing a ball of parachutes. The slightest breeze lifts the parachutes into the air.

SPREADING WITHOUT SEEDS

Seeds are not the only means of reproduction. Some plants create offshoots of themselves – in the form of bulbs, tubers, corms, or rhizomes – that can grow into new plants. This type of reproduction is called vegetative reproduction. As only one parent plant is needed, the offspring is a clone of its parent.

Tulip bulb
A bulb is an underground bud with swollen leaf bases. Its food store allows flowers and leaves to grow quickly. New bulbs develop around the old one.

Jerusalem artichoke tuber
A tuber is a swollen stem or root with buds on its surface. When conditions are right, the tuber's food store allows the buds to grow.

Gladiolus corm
A corm is a swollen underground stem that provides energy for a growing bud. After the food in the old corm is used up, a new corm forms above it.

Iris rhizome
A rhizome is a horizontal stem that grows underground or on the surface. It divides and produces new buds and shoots along its branches.

seed plants

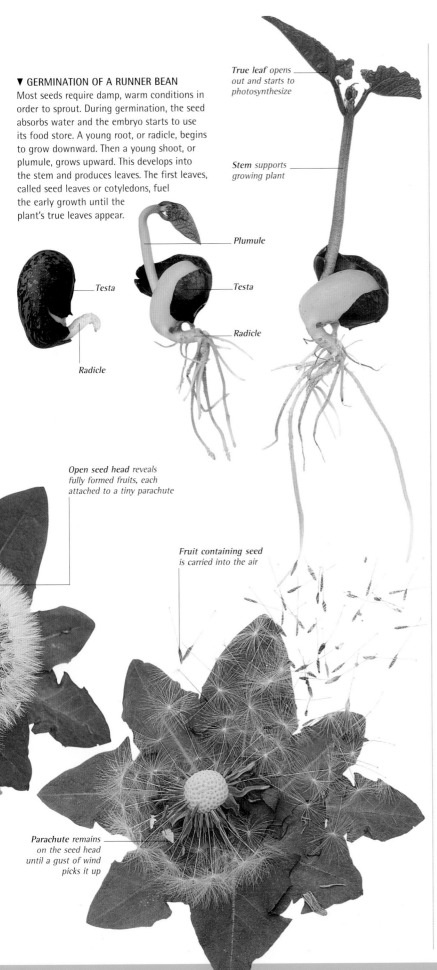

▼ GERMINATION OF A RUNNER BEAN

Most seeds require damp, warm conditions in order to sprout. During germination, the seed absorbs water and the embryo starts to use its food store. A young root, or radicle, begins to grow downward. Then a young shoot, or plumule, grows upward. This develops into the stem and produces leaves. The first leaves, called seed leaves or cotyledons, fuel the early growth until the plant's true leaves appear.

True leaf opens out and starts to photosynthesize

Stem supports growing plant

Plumule

Testa

Testa

Radicle

Radicle

Open seed head reveals fully formed fruits, each attached to a tiny parachute

Fruit containing seed is carried into the air

Parachute remains on the seed head until a gust of wind picks it up

FRUITS

A flower's ovary usually develops into a fruit to protect the seeds and help disperse them. A fruit may be succulent (fleshy) or dry. Fruit is often tasty and colourful to attract fruit-eating animals. Its seeds can pass through an animal unharmed, falling to the ground in droppings. Seeds may also be dispersed on animals' coats, by the wind, or by the fruit bursting open.

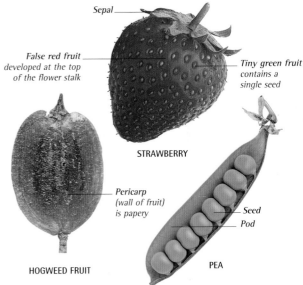

Sepal

False red fruit developed at the top of the flower stalk

Tiny green fruit contains a single seed

STRAWBERRY

Pericarp (wall of fruit) is papery

Seed

Pod

HOGWEED FRUIT

PEA

▲ DRY FRUITS

The seeds of dry fruits are dispersed in various ways. Peapods are dry fruits that split and shoot out their seeds by force. The hogweed fruit forms a papery wing around the seed, helping it to float on the breeze. The strawberry is a false fruit, but it is covered by tiny dry fruits, each with a seed.

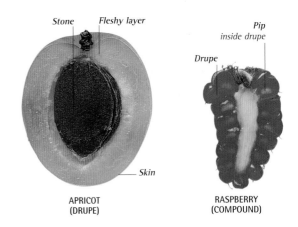

Stone

Fleshy layer

Pip inside drupe

Drupe

Skin

APRICOT (DRUPE)

RASPBERRY (COMPOUND)

▲ SUCCULENT FRUITS

Fleshy, brightly coloured, and often scented, succulent fruits are designed to attract the animals that eat and disperse them. Fleshy fruits such as apricots and cherries have a woody stone or pip that protects the seed. Called drupes, these fruits form from a single ovary. Many drupes, formed from many ovaries, may cluster to form a compound fruit, such as a raspberry.

FIND OUT MORE ▶▶ Flowering Plants **265** • Food Plants **276–277** • Pollination **266–267**

CONIFEROUS PLANTS

There are about 550 species of conifer, most of which are large, evergreen trees. Their leaves are often needle-shaped and usually have a thick, waxy coating that guards against water loss and freezing. Conifers produce their seeds on the woody scales of **CONES**, or in fleshy cups.

▲ SCALES
Cypresses produce scalelike leaves which are evergreen and aromatic. All conifers produce fragrant resins.

▲ DECIDUOUS NEEDLES
Larches are unusual because they are deciduous conifers. In the autumn, they shed their needlelike leaves.

▲ FLATTENED NEEDLES
Yews, firs, and some redwoods have small, flat leaves that grow on opposite sides of the stems.

CONES

The reproductive parts of coniferous plants are contained in cones. Most cones are woody, but some, such as those on yew trees, are soft and look like berries. The cones of pine and spruce trees usually fall to the ground in one piece, but the cones of cedars and most fir trees break up while still on the tree.

Inside the cone two pine seeds develop on each scale

Scales open and seeds are released when the cone matures

Male cone produces large amounts of pollen

Young female cone contains eggs inside sacs called ovules

Young shoots

◄ MALE AND FEMALE CONES
Coniferous plants have separate male and female cones. The male cones produce pollen (male sex cells). The female cones contain the eggs. When ripe, the male cone opens and releases pollen, which is carried on the wind to the open scales of the female cone. The female cone closes and the male cells fuse with the eggs to produce seeds.

◄ CONIFEROUS FOREST
Almost all coniferous trees and shrubs are evergreen, which means they keep their foliage throughout the year. This constant leaf cover prevents sunlight from penetrating down to the forest floor. Only shade-loving species, such as ferns and fungi, can cope with the dark conditions on the ground.

▲ WINGS FOR WIND DISPERSAL
When the weather is warm and dry, female cones open and release their seeds. Some seeds have papery brown wings that propel the seeds through the air for great distances. This ensures the seeds do not drop straight to the ground. Instead, they develop away from the shade of the parent tree.

FIND OUT MORE ▶▶ Classifying Plants 254–255 • Seed Plants 262–263 • Trees 268–269

FLOWERING PLANTS

Known as angiosperms (which means "seed cases"), flowering plants produce seeds inside the swollen base of the **FLOWER**, the ovary. Flowering plants make up over 80 per cent of all plant species. They are found in most parts of the world and range in size from tiny aquatic duckweed to gigantic eucalyptus trees. Flowering plants are divided into two groups, monocotyledons and dicotyledons.

▲ MONOCOTYLEDON
Flowering plants that have one seed leaf are monocotyledons. Their adult leaves have rows of parallel veins. Their petals and other flower parts are usually in multiples of three.

▲ DICOTYLEDON
The largest group of flowering plants, dicotyledons have two seed leaves. Veins branch out along the adult leaves. Their petals and other flower parts usually occur in fours or fives.

FLOWERS

Containing a plant's reproductive organs, flowers are the showy parts of flowering plants. Many depend on animal pollinators, which they attract with their colour, markings, or scent. Some plants, such as lilies, grow single flowers. Other plants produce large clusters of flowers. Daisies and sunflowers have many tiny flowers, or florets, that form a single flowerhead.

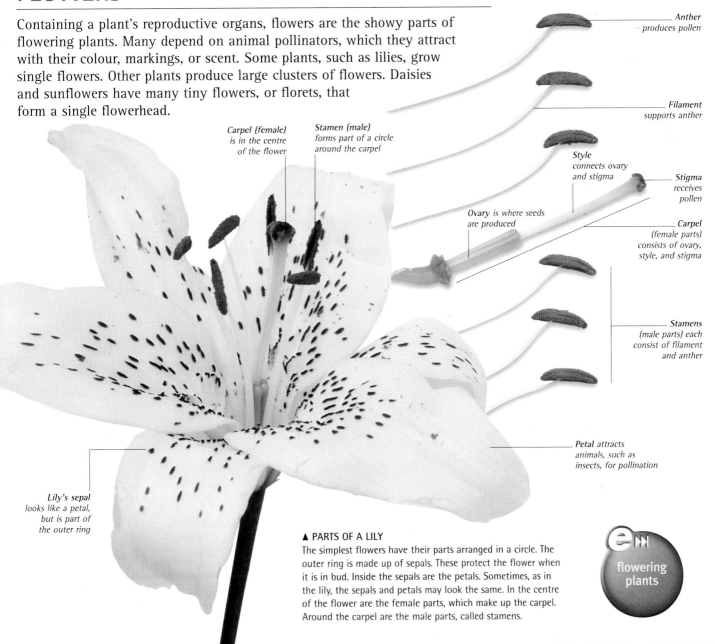

Carpel (female) is in the centre of the flower

Stamen (male) forms part of a circle around the carpel

Anther produces pollen

Filament supports anther

Style connects ovary and stigma

Ovary is where seeds are produced

Stigma receives pollen

Carpel (female parts) consists of ovary, style, and stigma

Stamens (male parts) each consist of filament and anther

Petal attracts animals, such as insects, for pollination

Lily's sepal looks like a petal, but is part of the outer ring

▲ PARTS OF A LILY
The simplest flowers have their parts arranged in a circle. The outer ring is made up of sepals. These protect the flower when it is in bud. Inside the sepals are the petals. Sometimes, as in the lily, the sepals and petals may look the same. In the centre of the flower are the female parts, which make up the carpel. Around the carpel are the male parts, called stamens.

flowering plants

FIND OUT MORE ▶▶ Pollination **266–267** • Seed Plants **262–263**

POLLINATION

The male sex cells of seed plants (flowering plants, conifers, and cycads) are contained in tough capsules called **POLLEN**. Pollen grains are produced by organs called anthers and must be transferred to the female parts of plants in order to form seeds. This process, called pollination, can be achieved in various ways. Some plants are assisted by animals that act as **POLLINATORS**. Others use the wind to take their pollen where it is needed.

pollination

◄ TRANSFERRING POLLEN
Pollen grains are released when the anthers of a flower split open. As this hummingbird probes a flower for nectar, it is dusted with pollen from the flower's anthers. When it feeds on another flower, some of that pollen will be rubbed off onto the stigma. Most plants have systems that prevent pollen produced by a particular flower fertilizing that same flower's eggs.

Stigma

Stigma
Pollen tube
Ovules

Pollen sends tube into stigma

POLLEN GERMINATING ON A POPPY STIGMA

Male sex cell *Polar nuclei* *Egg* *Male sex cell*

INSIDE AN OVULE

POPPY FLOWER

◄ GERMINATION OF POLLEN GRAINS
The female part of a flower has a special swelling, called a stigma. When pollen grains land on the stigma, they stick to it and begin to germinate. A microscopic tube sprouts from each pollen grain and starts to grow into the stigma. It then grows down through a stalk, called the style, towards the eggs in the ovules below. A flower's stigma is held up on a style so it can catch pollen.

▲ FERTILIZATION
A flower's egg is contained in a capsule called an ovule. The tip of the pollen tube penetrates the ovule, injecting the male sex cell. This fuses with the egg, fertilizing it. In flowering plants another male sex cell has to be injected before the ovule can become a seed. This second male cell fuses with nuclei, called polar nuclei, to form tissue that will make food for the baby plant, or embryo, as it develops inside the seed.

POLLEN

Unlike seedless plants, such as ferns, seed plants do not produce free-swimming sperm. Instead, their motionless male sex cell is completely enclosed by the tough casing of a pollen grain. Inside a pollen grain the male sex cell of a seed plant is protected from drying out. Instead of using water to swim to the egg, like the sperm of a seedless plant, it reaches the egg by being carried by the wind or animals.

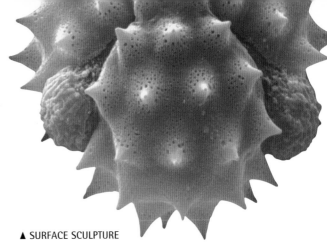

▲ SURFACE SCULPTURE
When viewed through a microscope, many pollen grains have incredibly intricate surface structures. Some grains are spiky, possibly to help them stick to the stigma of the female part of the plant. Scientists are often able to tell the species of plant that has produced a sample of pollen just by looking at the shapes of the grains. Pollen grains are produced by most flowering plants in huge numbers. Despite their tiny size, they affect many people, giving them hay fever.

▲ WIND DISPERSAL
Most plants that use wind to disperse their pollen have separate male and female flowers. The male flowers, such as these catkins, produce huge quantities of pollen and usually hang from twigs or are held up on stalks where they can catch the wind. Most female wind-pollinated flowers have long styles with exposed stigmas that increase their chances of catching pollen.

POLLINATORS

Many types of animals act as pollinators, transferring pollen between the anthers and stigmas of plants. Most plant species use insects as pollinators, since these flying animals are small enough to enter most flowers but can transfer pollen over large distances. However, some plants produce large flowers that are pollinated by animals such as birds or bats. Many plant species have flowers that are shaped in such a way that they can only be pollinated by a particular species of animal.

ATTRACTING POLLINATORS ▶
Different animal pollinators respond to different stimuli. Birds are attracted to the colour red but have a poor sense of smell, so many flowers pollinated by them are red but scentless. Insects are similarly attracted in different ways. Bees are drawn to blue or yellow, sweet-scented flowers. On the other hand, some flowers smell like rotting meat to attract flies.

Lipped-flower petals act as landing pads and support for the bee

FEMALE HOLLY

▲ CROSS FERTILIZATION
The healthiest seeds are produced by cross-fertilization, using pollen and eggs from different individual plants. Plants have many ways of ensuring that this happens. In some species, male and female flowers are never produced at the same time on the same plant. In others, the sexes are separate. Holly trees are either male or female, for example. Only the female plants produce berries.

◀ NECTAR GUIDES
Many flowers have patterns of lines on them. These guide insects towards the glands that produce nectar at the base of the petals. Some flowers display patterns in the ultraviolet light they reflect. Unlike us, the insects that pollinate these flowers have ultraviolet vision. These insects see the nectar guides, where we see plain-coloured petals.

FIND OUT MORE ▶▶ Flowering Plants **265** • Seed Plants **262–263**

TREES

These tall, seed-producing plants live for many years and do not die in winter. Trees have a single woody stem, called a trunk, which thickens as they get older and supports their increasing bulk. In some parts of the world, trees grow together in **FORESTS**. Trees are divided into two main groups: conifers and broad-leaved trees. Many broad-leaved trees are **DECIDUOUS**, which means they lose their leaves in autumn. Conifers are mainly **EVERGREEN TREES** and they keep their leaves throughout the year.

◄ WOODY TISSUE
Tree trunks are made up of different layers of cells. On the outside is a protective layer called bark. Just beneath the bark is a thin layer of phloem cells, which carry food from the leaves to the rest of the tree. Below the phloem is another thin layer of cells, the cambium, which constantly divide and make the trunk wider. Beneath them, the sapwood draws up water and minerals from the roots.

Heartwood is composed mostly of dead cells and gives strength to the tree

Sapwood is composed of dead and living cells

Bark is made up of tough, dead cells and protects the living tissue beneath

PROTECTIVE SKIN ►
A tree's bark is its skin. It shields the living wood inside, preventing it from drying out and protecting it from extreme cold and heat. Tiny slits in the bark, called lenticels, allow oxygen to enter the trunk and carbon dioxide to leave it. Different tree species have different bark textures. The bark of oak is cracked and rough. Beech bark is very thin and delicate. Birch bark is relatively smooth.

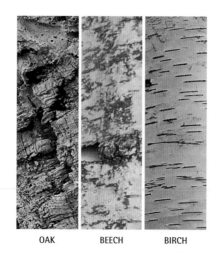

OAK   BEECH   BIRCH

FORESTS

Almost a third of the world's land surface is covered by forests, which are areas with dense tree cover. Forests differ according to the local climate – tropical, temperate, or boreal. Rainforests, thrive in warm and wet climates. Here there may be more than 200 tree species in one hectare (2½ acres) of land. The majority of trees in temperate forests are deciduous, such as oak and beech. In cold, northern regions, boreal forests contain hardy coniferous trees.

Emergent layer contains the crowns of the very tallest trees

Canopy is made up of the tops of most of the forest's fully grown trees

Understorey is filled with young trees growing towards the canopy

Shrub layer has plants with big leaves to catch what little sunlight filters down

Forest floor is home to ferns and other ground plants

▲ LAYERS OF LIFE
Rainforests grow in distinct layers, each with its own plant and animal species. Life is richest in the canopy, which contains most of the leaves, flowers, and fruit. A few very tall trees grow through this into the emergent layer. Beneath the canopy is an understorey of smaller trees and, below that, a shrub layer of big-leaved plants that can survive in low-light conditions. At the very bottom is the dark forest floor, where plants are smaller and fewer.

EVERGREEN TREES

Each leaf on an evergreen tree may last for years and only falls when it is ready to be replaced. As a result, these trees remain covered with leaves throughout harsh cold or dry seasons. Most conifers, such as pine trees, keep their leaves for between two and four years. In places where there is little seasonal change, such as the wet tropics, many other types of plant are also evergreen.

ANTIFREEZE CHEMICALS ▶
The leaves of many evergreen trees contain resins. These heal breaks in the surface and also prevent the leaves from freezing. The antifreeze properties of pine resins enables the cells in their needles to survive throughout the winter, even when they are completely covered with snow or ice.

Icicles hang from this conifer twig, but the needles beneath are undamaged

DECIDUOUS TREES

Trees that lose their leaves in the autumn are called deciduous trees. They grow in temperate regions – places that have warm summers and cold or cool winters. In autumn, temperatures start to fall and there are fewer hours of daylight for photosynthesis. To conserve energy during winter the trees shed their leaves and stop growing. This also helps them save water, which would otherwise evaporate from the surface of their leaves.

SEASONAL CHANGES ▶
The leaves of some deciduous trees turn brilliant colours just before they fall. For example, maple leaves change from green to yellow then turn orange and red. This colour change happens because the chlorophyll that made the leaves green is broken down in autumn and used by the tree. Other pigments that were previously hidden beneath the chlorophyll are then revealed.

Crown of the tree

Roots grow longer at their tips

▲ NETWORK OF ROOTS
Tree roots spread out in a huge network under the ground. This network is often as big as the crown of branches above. A tree's larger roots anchor it firmly in the ground. Most of the delicate feeding roots are nearer the surface. These are covered in root hairs which draw in water and nutrients from the soil.

trees

PARASITIC PLANTS

Most plants make all the food they need by photosynthesis, but some species are parasites. They steal food from other plants, known as host plants. Parasitic plants have special suckers that may invade the host plant's food channels and draw off sugars and minerals. Many parasitic plants are totally dependent on their host for food and no longer need green leaves. Others still have green leaves and make some of their own food through photosynthesis.

MISTLETOE SEED GERMINATING

MISTLETOE SHOOTS

MATURE MISTLETOE

GREEN PARASITE ▶
Mistletoe is a parasite that steals water and minerals from its host tree. However, mistletoe also has green leaves. It produces its own food through photosynthesis, using water stolen from the host tree.

TOTAL PARASITE ▶
Dodder is a parasitic plant that cannot photosynthesize at all. Its leaves are reduced to tiny brown scales. Since it has no green chlorophyll, it must obtain all its food from the host plant. Dodder is a climber related to bindweed. Its stems twine around the host, producing suckers, called haustoria, that invade the host and steal its food.

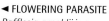
Vascular tissue of host plant

Dodder

Haustorium penetrates the host plant's vessels

▲ STEALING FOOD
This micrograph cross-section shows how a dodder's haustoria penetrate the stem of the host plant, pierce its vessels, and suck up sugars. As the dodder grows bigger and stronger, it puts out more haustoria. It steals more nourishment from the host, which weakens and eventually dies.

◀ FLOWERING PARASITE
Rafflesia arnoldii is a parasitic plant that produces the world's largest flower. It grows in the rainforests of Southeast Asia. The plant invades the underground roots of vines to take food. Sometimes the *Rafflesia* puts out a shoot from one of these roots. This develops into a giant, stinking flower that may be 1 m (3 ft) across. The flower's powerful smell of rotting flesh attracts pollinating flies.

parasitic plants

FIND OUT MORE ▶▶ Flowering Plants 265 • Fungi 282–283 • Photosynthesis 258 • Pollination 266–267

CARNIVOROUS PLANTS

Meat-eating, or carnivorous, plants can trap and digest insects and other small animals. They do this to obtain the vital nitrogen that they need to grow. Most plants absorb enough nitrogen from nitrates in the soil. Carnivorous plants live in bogs, where nitrates are in short supply, so they need to obtain their nitrogen by digesting prey instead. Carnivorous plants have developed unique ways to catch insects, such as fluid-filled **PITCHERS** and spring-loaded traps.

PITCHERS

The pitcher plant is named for the juglike traps that hang below its leaves or grow up from the ground. Each trap has its own lid to keep off the rain and contains special fluid at the bottom. Insects are attracted by the trap's red markings and the sweet nectar produced around its rim. If the insect lands to drink the nectar, it slips and falls into the trap. It drowns in the fluid at the bottom and its nutrients are slowly absorbed by the plant.

carnivorous plants

Insect
lands on open leaf
of Venus flytrap

Trigger hairs
detect movement
and shut the trap

▲ SPRING-LOADED TRAP
The Venus flytrap's leaves are hinged so that they can snap shut. Sensitive trigger hairs detect any insect that lands on the surface of an open leaf. At the slightest movement, the two halves of the leaf spring shut. As the sides of the trap close around the victim, the plant releases digestive juices. These break down the soft parts of the insect.

Insect
stuck to the hairs

STICKY TRAP ▶
Sundews are small bog plants that have hair-covered leaves. They produce a droplet of sticky "dew" at the tip of each hair. Insects are attracted to the fluid, but become stuck. Next, the hairs slowly bend inwards until the whole leaf has folded over the insect. Chemicals released from the hairs digest the insect's body, and nutrients are taken into the plant.

Drowned insect
is gradually
digested

▲ DIGESTIVE JUICES
An insect body has to be broken down before its nutrients can be absorbed into the plant. Carnivorous plants such as pitchers use enzymes, similar to the ones that break down food in an animal's gut. Acids help the enzymes to break down the body. A pitcher plant can digest a small insect within a few hours, but larger ones take days.

FIND OUT MORE ▶▶ Bacteria **284** • Feeding **312–313** • Food Plants **276–277** • Plant Survival **274–275**

PLANT SENSITIVITY

Like animals, plants sense changes in their surroundings and respond to them. Plants are able to detect and respond to light, gravity, changes in temperature, chemicals, and even touch. Unlike animals, plants do not have nerves or muscles, so they cannot move very fast. A plant usually responds to change by gradually altering its growth rate or its direction of growth. The slow movements that plants make towards or away from a stimulus, such as light, are known as tropisms. Tropisms are controlled with the help of special chemicals called **PLANT GROWTH REGULATORS**.

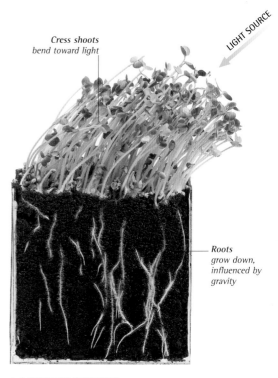

Cress shoots bend toward light

LIGHT SOURCE

Roots grow down, influenced by gravity

◄ TIME TO FLOWER
Many plants only bloom at certain times of year. They flower at the right time by responding to changes in light and temperature. A crocus plant is able to detect signs of spring, such as lengthening days (more light) and warmer soil (increased temperature). These changes cause chemical changes in the plant and the crocus starts to put out shoots and flowers.

▲ A SENSE OF DIRECTION
Light influences how shoots grow. They bend towards it, so that leaves will have the maximum amount for photosynthesis. Roots push down through soil because of the effect of gravity. They may also be drawn towards water, or away from bright light. Other factors, such as temperature and how wet the soil is, may affect when seeds germinate (sprout).

SURVIVING WINTER ►
Deciduous plants such as *Forsythia* respond to the lack of light and warmth in winter by entering a resting period. In preparation, the plant produces chemicals that weaken the leaf stalks, so the leaves fall. Over winter, the plant does not need to make food. Its shoots and buds are inactive. When spring comes, the plant produces chemicals that make buds and shoots start to grow again.

▼ FACING THE LIGHT
Over the course of a day, the flowerheads in a field of sunflowers gradually turn, tracking the Sun's path across the sky. The movement is almost too slow to notice. In the morning, the flowerheads all face east and by evening they face west. This is called phototropism, which means the movement of part of a body towards light. It happens as chemicals shift from one side of the stems to the other.

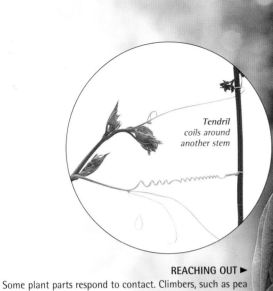

Tendril coils around another stem

REACHING OUT ►
Some plant parts respond to contact. Climbers, such as pea plants and this passion flower, put out long, reaching shoots called tendrils. When a tendril reaches something solid – such as a garden cane or the stem of another plant – it coils around it. By grasping at supports in this way, the plant is able to climb even higher.

▲ MANGROVE ROOTS UNDERWATER
Roots usually respond to light by growing away from it, but the roots of mangrove trees behave differently. Mangroves grow in coastal swamps where there is little oxygen in the waterlogged soil. Their roots compensate for this by growing upwards out of the mud. Each low tide, the mangrove roots are exposed to the air and can collect plenty of oxygen.

PLANT GROWTH REGULATORS

Certain chemicals influence different aspects of a plant's growth. These plant growth regulators may control how fast cells divide, or how they grow. Some are produced in the tips of shoots or roots. They can even change the direction the shoot or root takes as it grows. If cells on one side of the tip grow faster, the tip will start to curl in the opposite direction.

plant sensitivity

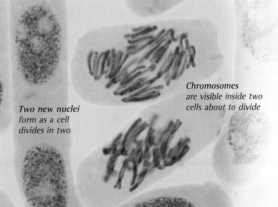

Two new nuclei form as a cell divides in two

Chromosomes are visible inside two cells about to divide

◄ CELLS DIVIDING
Some types of plant growth regulators encourage cell division, a complex process. Before dividing, the cell's genetic material (DNA) must be copied. Two cells here have copied their DNA strands, or chromosomes. The strands are separating, ready to be parcelled into two new cell nuclei. Then each cell will divide in two.

CELL DIVISION IN A GARLIC ROOT TIP

FIND OUT MORE ▶▶ Flowering Plants 265 • Photosynthesis 258 • Plant Anatomy 256–257 • Plant Survival 274–275

Nettle leaves are covered in stinging hairs

PLANT SURVIVAL

Some plants have special features that help them to repel predators. Other plants can survive and even thrive in hostile environments, such as cold and rocky mountains. In areas of little rainfall, plants known as **XEROPHYTES** have developed special methods for collecting and storing water. Another group of amazing plant survivors are known as **HALOPHYTES**. They can endure extremely salty regions, such as salt marshes, salt pans, and sand dunes.

Glassy tip caps poison-filled hair

▲ DEFENCES
Plants cannot move away from predators, so they must defend themselves in other ways. Some have thorns or spines. Others have foul-tasting poisons in their leaves. Stinging nettle leaves are covered in fine hairs that are filled with poison. Each hair ends in a swollen, glassy tip. When touched, the tip breaks off, leaving a jagged end that can pierce flesh and inject the poison from the hollow hair.

IN CAMOUFLAGE ▶
Some plants use disguise to hide from plant-eating animals. Blending in with the background like this is called camouflage. With its fleshy, grey leaves, the pebble plant is difficult to spot against the surrounding pebbles – only its flowers give it away. Most of the time, animals mistake the leaves for real stones, and do not try to eat them.

◀ TOUGH ALPINES
Known as alpines, mountain plants have to cope with strong sunshine, penetrating frost, and bitterly cold winds. Water may be scarce, too, as there is often low rainfall and thin, frozen soil. Alpines grow in dense cushions, which makes them less exposed. Fine hairs on their leaves reduce water loss and protect them from sun damage.

Alpines hug the ground to avoid the drying effect of the wind

Epiphyte leaf funnels the water to the centre of the plant

Flowers stand out to attract pollinating insects

◀ HIGH UP ON A HOST
The bromeliad lives in tropical rainforests. Seeking light, it grows high on the branches of a host tree, using its roots to anchor itself. The bromeliad's leaves direct any rainwater to the heart of the plant. Plants that fix themselves to other plants like this, but do not draw food from them, are called epiphytes.

▲ YELLOW WATER LILY
Aquatic (water) plants face their own survival problems. A water lily's flowers either float at the surface or are held high on long stems. The upper surface of each leaf is waxy and repels water. The broad, flat leaves float on the water and are supported by long stalks. The stalks are filled with air chambers supplying oxygen for respiration.

XEROPHYTES

Plants that have adapted to cope with dry desert conditions are called xerophytes. Many do not have regular leaves, which would lose water through evaporation in the heat. Cacti, for example, have defensive spines, which are essentially modified leaves. Some xerophytes have shallow roots that absorb water quickly after rain. Others have very long taproots that extract water from deep in the ground.

◄ WATER STORER

Succulents are plants that have swollen, fleshy parts in which they store water. The best-known succulent plants are cacti like this one. A cactus stores water in its stem and can cope with the driest climates. The thick green stem is also used for photosynthesis, as the leaves have been modified into spines.

DESERT BLOOM ►

Ephemerals are plants that carpet a desert after rare rainfall. In the space of a few days, they sprout, grow, flower, and produce seeds. The seeds of some ephemerals are coated in a chemical that prevents germination until rain has washed the chemical away.

HALOPHYTES

Plants that have adapted to live in salty environments are called halophytes. Salt draws water out of the roots of most plants, slowly drying them out. Some halophytes have ways to get rid of excess salt. Others need a salty environment in order to survive. Halophytes are able to grow in salt marshes, shallow coastal waters, dry salt pans, and on sand dunes.

Fleshy stem stores water absorbed by the roots

plant survival

Cactus spine is narrow so it loses little water

TROPICAL MANGROVE ►
Mangrove trees are halophytes that grow along tropical coasts. Their roots take in salt from the seawater. The salt is carried in the tree's sap up to old leaves, which are then shed, or to living leaves, which have glands that excrete the salt. Many mangroves have arching roots that are exposed at low tide. These roots have breathing pores for taking in oxygen from the air.

FIND OUT MORE ▸▸ Coasts **227** • Parasitic Plants **270** • Plant Sensitivity **272–273** • Transpiration **259**

NORMAN BORLAUG
American, 1914-

An agricultural scientist, Norman Borlaug was central to the 1960s' "Green Revolution", a major effort to reduce world hunger. He received the 1970 Nobel Peace Prize for developing high-yield, disease-resistant varieties of wheat. Today his wheat is grown in Asia, Africa, and South America.

food plants

FOOD PLANTS

Food chains begin with plants. The sugars they contain provide energy for plant-eating animals, which in turn feed meat-eaters. All plant parts are possible food sources: leaves, stems, roots, fruits, and seeds. Early people gathered wild plants but then, about 10,000 years ago, the first farmers began to cultivate food plants as crops. Insects, birds, and other animals eat plants too. When they feed on farmers' crops, they are treated as **PESTS**.

▼ PADDY FIELDS IN CHINA
Rice is the staple diet of more than half of the world's population. It is grown in flooded fields called paddies, because its roots need to be submerged in shallow water. Farmers have even found a way to grow rice on steep mountains, using terraces. Like wheat, maize, and barley, rice is a cereal (edible grass). Cereals are by far the most important crops, but farmers grow many other food plants, including vegetables, fruit, sugar, and tea.

▲ VERSATILE FOOD PLANT
The winged bean is a traditional crop in Southeast Asia. Its seeds, bean pods, leaves, and roots are all edible and they contain high levels of protein. This makes the bean a versatile food that could help to fight famine elsewhere in the world. Plants with useful characteristics are introduced to new regions of the world all the time.

Terracing allows field to be flooded without water flowing away

LIFE CYCLE OF RICE

Rice grains are the seeds of the rice plant. A rice plant's life cycle has two distinct growth phases. The first, the vegetative phase, is when the seed sprouts and the seedling develops. The plant puts out leaves and grows to about 1.2 m (4 ft) tall. The second phase is the reproductive phase. This is when the plant produces spikelets of flowers. Like most grass flowers, these must be wind-pollinated before they can develop into fruits.

Spikelets carry kernels of rice inside brown husks

▲ COMBINE HARVESTERS GATHER WHEAT
Machines such as combine harvesters help farmers to produce more crops from their land. Chemical fertilizers and pesticides also increase crop yields. However, intensive farming methods may remove natural nutrients and pest-killing predators. Some farmers produce crops organically, without using chemicals.

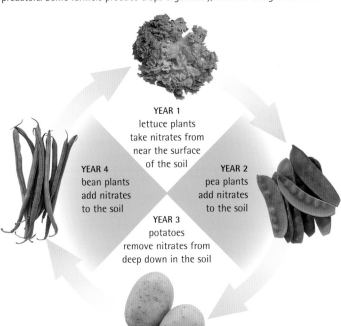

YEAR 1
lettuce plants
take nitrates from
near the surface
of the soil

YEAR 4
bean plants
add nitrates
to the soil

YEAR 2
pea plants
add nitrates
to the soil

YEAR 3
potatoes
remove nitrates from
deep down in the soil

▲ CROP ROTATION
Plants require nitrates, a usable form of nitrogen, in order to build the complex molecules they need to live and grow. To stop nitrates from being permanently removed from the soil, farmers alternate different types of crops. Legumes (peas and beans) have swellings in their roots, called root nodules, that return nitrates to the soil. They can replace the nitrates used up by last season's crop.

SELECTIVE BREEDING

PRIMITIVE
MAIZE COB

MODERN
MAIZE COB

Over thousands of years, people have improved the yield of crops through selective breeding. That means choosing the seeds of the best plants – the ones with the biggest seeds, tastiest leaves, or best resistance to disease – to grow into new plants in the next season.

Maize was first cultivated in Central America, and some primitive forms of the plant still grow there. As a result of selective breeding, modern maize is very different to its wild ancestor. It produces much larger cobs, with uniform rows of sweet-tasting corn kernels. All the basic food crops of the world were developed in this way.

*Plump kernels
of maize*

PESTS

Any organism that harms a crop plant is a pest. Many species of insect eat and damage crops. Some, such as aphids, may carry viruses that cause plant disease. Some fungi also cause disease. Weed plants are considered pests too, because they compete for the nutrients in the soil.

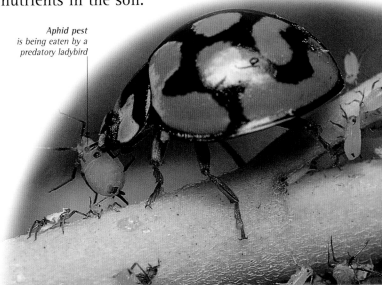

*Aphid pest
is being eaten by a
predatory ladybird*

▲ NATURAL PEST CONTROL
Ladybirds are important predators of aphids. Sometimes predators are deliberately introduced to keep aphids or other pests off crops, usually in an enclosed environment such as a greenhouse. Using nature to control pests is called biological pest control. Farmers can also control pests biologically by infecting them with diseases that are harmful only to them.

▲ SPRAYING WITH PESTICIDES
Chemicals called pesticides are often used to poison pests. They are usually very efficient, allowing farmers to grow fruits and other crops without wastage due to pest damage. However, pesticides can also kill harmless or beneficial organisms, such as honey bees. In apple orchards, honey bees are essential because they pollinate the flowers. Without them, there would be no fruit.

FIND OUT MORE ▶▶ Bacteria **284** • Ecology **326–327** • Genetically Modified Crops **278** • Seed Plants **262–263**

GENETICALLY MODIFIED CROPS

Plant crops provide us with food, clothing, and many other important products. Farmers select seeds from the best plants to grow next season. This selective breeding makes plants evolve features that people want them to have – such as heavier rice grains – but it is a slow process. To change faster, crops can be given new features directly from other life forms. Cells take their features from instructions they carry, called genes. Genetic modification (GM) allows us to move a gene from one life form to another.

◄ CLONING CROPS
These two plants are clones. They have been grown from cells taken from one "mother" plant, in a nutrient-rich sterile gel. This cloning technology, called micropropagation, can create thousands of young plants from a single parent. When any new plant has been created, it has to be multiplied millions of times before there are enough plants to sell to farmers around the world. This can be done through micropropagation or in other ways, for example growing plants from seed.

GM crops

▼ MOVING GENES
In genetic modification, scientists first identify the gene for the desired feature. They can cut this donor gene from its DNA strand, using chemicals called enzymes as scissors. The gene is kept in one piece and multiplied by inserting it into a bacterium. The bacterium is used to carry the new gene into the target plant, by "infecting" it.

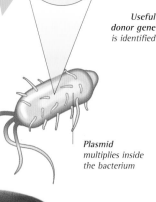

◄ CARRIERS OF GENES
This colour-coded micrograph shows a section of donor DNA in blue, attached to a tiny ring of genetic material called a plasmid, shown in red. Scientists use plasmids, which are found inside bacteria, to stop donor genes unravelling and to replicate (copy) them. The donor gene is zipped on to the plasmid using chemicals called enzymes.

Useful donor gene is identified

Piece of DNA carrying the gene is cut out using an enzyme

Donor DNA is slotted into plasmid (ring of DNA)

Plasmid multiplies inside the bacterium

Useful donor gene is identified

New plant, grown from infected cells, contains the extra gene

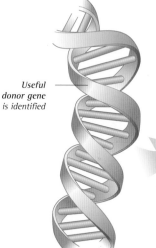

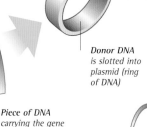

Mould grows on this non-GM tomato

GM tomato is mould-resistant

◄ IMPROVING SHELF LIFE
Tomatoes can be bruised when they are packed and stored. As they ripen, their skins become more delicate and are more easily damaged. Damaged tomatoes quickly begin to rot, because mould can grow on their skins. Scientists have genetically modified tomatoes by adding genes that stop their skins softening as they ripen. This means they are less likely to bruise in storage and be wasted.

FIND OUT MORE ▸▸ Bacteria **284** • Food Plants **276–277** • Genetics **364–365** • Plant Products **280–281**

MEDICINAL PLANTS

Many plants produce special substances in their roots, leaves, flowers, or seeds that help them to survive. For example, some plants make nasty-tasting substances to defend themselves against plant-eating animals. Since earliest times, people have gathered these substances to create herbal medicines to treat certain diseases. Many of the powerful drugs used in modern medicines originated in plants. Today's plant-based drugs treat a range of diseases, from headaches to cancer.

medicinal plants

Shaman cuts bark, which is an ingredient in traditional Amazon medicines

GATHERING BARK ▶
Rainforest people, such as this Yagua shaman in Peru, possess valuable knowledge about medicinal plants. With proper research, scientists believe they might find cures for some of the world's deadliest diseases among rainforest plants. Some of these plants have not even been discovered yet. Unfortunately, rainforests are being destroyed. As they disappear, so do thousands of possible life-saving drugs.

◀ MEDICINAL JUICE
Aloe vera is the thick juice of the aloe, a type of plant that comes from tropical Africa but is also cultivated elsewhere. The juice contains a chemical called alonin, that has been used in cosmetics and medicine. Its healing properties have made it especially useful as an ingredient for lotions and gels that soothe burns, including sunburn. It can also be used to repel biting insects.

Oil secreted by the aloe is a chemical defence against predators

FROM PLANTS TO MEDICINES

KHELLA
Also known as toothpickweed, this Mediterranean herb contains a chemical that opens up blood vessels, improving blood flow to the heart, and opens the breathing tubes of the lungs. The chemical has been used in medicines to treat asthma and angina (pain due to heart problems).

MADAGASCAR PERIWINKLE
The Madagascar periwinkle is the source of drugs used to treat diabetes and certain cancers, such as Hodgkin's disease and acute leukaemia. The drug for treating Hodgkin's disease has increased patients' chances of survival from one-in-five to nine-in-ten.

QUININE
The bark of this tropical tree contains a drug called quinine. Quinine is used in the prevention and treatment of malaria, a deadly disease carried by mosquitoes. Malaria is responsible for thousands of human deaths around the world every year.

MEADOW SAFFRON
This little plant contains a chemical called colchicine, which has been used to treat rheumatism and gout. As it tends to prevent cells from dividing too quickly, colchicine has also been used to suppress some types of cancer.

COCA PLANT
The coca plant grows naturally in South America and is the source of the drug cocaine. Although cocaine can be abused and is associated with addiction, it has also been used responsibly by doctors as a local anaesthetic and for pain relief.

OPIUM POPPY
Opium is a pain-killing drug extracted from the unripe seed pods of the opium poppy. In 1806, a German scientist isolated the drug morphine from opium. Morphine and its derivatives, such as heroin and codeine, remain important pain relievers.

MEADOWSWEET
Meadowsweet is a European wildflower that grows in wet soils and marshes. It has been used for pain relief in the treatment of many conditions, including headaches, arthritis, and rheumatism.

RAUVOLFIA
Rauvolfia is a small, woody plant that grows in tropical rainforests. It contains reserpine, a chemical that effectively relieves snake bites and scorpion stings. Reserpine was the first tranquilliser used to treat certain mental illnesses. It also lowers blood pressure.

FIND OUT MORE ▶▶ Disease **370–371** • Habitats **246–247** • Medicine **372–373** • Plant Survival **274–275**

PLANT PRODUCTS

As well as food and medicines, plants provide other useful products. Many plant cells form **NATURAL FIBRES** that strengthen and support the plant. The same properties make them perfect for textiles and paper. Timber from trees is used to build boats, houses, and furniture. Palm leaves are woven into baskets, hats, and mats. People also extract perfumed oils and natural dyes from the flowers and leaves of certain plants.

FIELD OF LAVENDER ▲

Vast farms of lavender are found around the Mediterranean, in Britain, and in the United States. The plant is grown for its scented oil, produced in oil glands on the stems, leaves, and flowers. The harvested flowers may be dried, or pressed to extract the oil. Sometimes the oil is distilled to create a purer, "essential" oil. Lavender oil is used in aromatherapy and as an ingredient for perfumes, soaps, and other cosmetics.

▲ MANUFACTURING PAPER

Most paper comes from softwood trees, such as pines. First, machines or chemicals break down the wood chips into fibres. This is called pulping. The fibres are soaked in chemicals, then pressed by heavy rollers into thin, flat sheets. Before pressing, the fibres may be bleached white or dyed different colours. Smoother paper is made by adding starch or clay.

WOOD CHIPS ARE USED
TO MAKE PAPER

◄ TAPPING RUBBER

The rubber tree grows naturally in South America, but there are also plantations in Asia. If its bark is cut, the tree produces a milky fluid called latex. People harvest the latex so that it can be turned into rubber, a useful, elastic material. Not all rubber comes from rubber trees. Most is made artificially from petroleum.

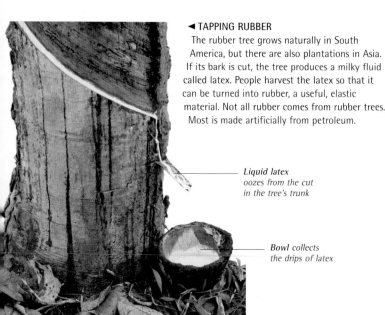

Liquid latex oozes from the cut in the tree's trunk

Bowl collects the drips of latex

▲ TIMBER FOR CONSTRUCTION

Harvested wood is called timber. Its strength makes it useful in the building trade, especially for creating supporting frameworks. Pine and other softwoods are the most widely used because they grow straight. They also grow fast, which makes their timber cheap and renewable (easily replaced). Hardwood, from flowering trees, grows slowly. It is more costly and is used for furniture.

HENNA FOR HANDS ▶
Henna is a shrub that grows in the Middle East and North Africa. Its leaves are harvested for their reddish-brown pigment. This is used to dye clothes, hair, and even people's skin. Greenish henna paste, made from powdered leaves, is used to paint the skin. When the paste dries and rubs off, the skin looks tattooed.

Henna leaves are pruned and harvested several times a year

Henna paste is used to paint intricate patterns

OIL GLAND ON LAVENDER LEAF

NATURAL FIBRES

Plants produce long groups of cells, called fibres. These can be used to make textiles, such as cotton, as well as other materials such as paper or felt. All plant fibres are strong, because their cell walls contain a tough molecule called cellulose, but to be useful fibres also need other properties, such as flexibility and length. Flax and hemp were two of the earliest fibres used by people.

TWISTED COTTON FIBRES

◀ SPINNING COTTON
The cellulose in cotton is arranged as interlocking, coiled strands of fibres. These can be spun into threads called yarn. Yarn is produced on an industrial scale and woven on looms to make textiles. Cotton textiles are hardwearing, "breathable", and take dyes well. They range from light gauzy fabrics to tough denims.

◀ COTTON PLANTS READY FOR HARVEST
The cotton shrub produces seedpods that burst open to reveal masses of fluffy cotton fibres. These fibres are harvested to produce cotton yarn and textiles. Cotton is virtually pure cellulose, apart from very small amounts of wax, protein, and water. The plant is cultivated in many parts of the world including China, the United States, and India.

plant products

FIND OUT MORE ▸▸ Food Plants 276–277 • Medicinal Plants 279 • Recycling Materials 60–61 • Synthetic Fabrics 56

FUNGI

Fungi grow without sunlight and feed on organic matter. A typical fungus is made of many threads growing on or in a food source. Each thread, called a hypha, oozes chemicals that break down the food. This releases nutrients that the hyphae can soak up. Fungi include **MOULDS**, mushrooms, toadstools, puffballs, and truffles. About one in four fungi lives in partnership with an alga – these partnerships are called **LICHENS**.

▼ EXPLODING PUFFBALL

Fungi produce fruiting bodies, such as puffballs, which we can see above ground. These fruiting bodies release tiny spores that are carried in the air and start to grow wherever they fall. A single fruiting body can produce millions of spores, so it is likely that some will land on a suitable food source.

◄ SOAKING UP FOOD

Hyphae branch to form a network called a mycelium. The combined surface area of all the hyphae allows the fungus to digest and absorb a lot of food. Many fungi play a vital role in food webs. Their digestive action is the first step in breaking down dead plant and animal matter, making it useful to other life forms.

YEAST-COVERED GRAPES ►

Grapes often have a fine coating of yeast. Yeasts are a type of fungus which grows in colonies of single cells. They thrive where there is a good supply of sugar, such as on the surface of fruit. As yeasts consume sugars they can create a toxic by-product that people value – alcohol. Yeasts are essential for producing alcoholic drinks, such as wine.

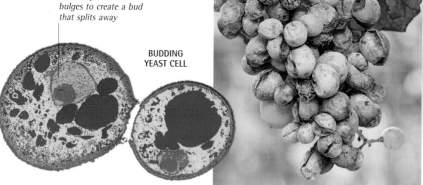

Mother yeast cell bulges to create a bud that splits away

BUDDING YEAST CELL

ANIMAL
LION

PLANT
SUNFLOWER

FUNGUS
MUSHROOM

PROTOCTIST
AMOEBA

MONERAN
BACTERIA

Cloud of spores drifts away from the puffball on the wind

MOULDS

Fungi called moulds do not produce large toadstools. Their tiny fruiting bodies look like peppery spots and are usually black or blue. Mould grows wherever spores land on suitable food, such as bread or fruit. The mould's threads, or hyphae, give it a woolly appearance.

▲ MOULDY LIFE SAVER
Greenish *Penicillium* mould grows outwards across the surface of a dish of nutrient gel. This mould releases a chemical called penicillin, which is an antibiotic. Antibiotics are used to kill bacteria that cause diseases, without causing harm to the organism or person infected with these bacteria. Moulds are now grown in huge vats to produce this medicine.

SIR ALEXANDER FLEMING
Scottish, 1881-1955

In 1928 Fleming discovered medicine's first antibiotic, penicillin, which has since saved millions of lives. He had noticed that one of his laboratory dishes of bacteria was infected with a mould. Around the mould, the bacteria had disappeared. Fleming realized that the mould produced a substance that killed bacteria.

LICHENS

Some fungi can combine with algae to form structures called lichens. Lichens can be flat or fluffy, living on rocks and tree trunks, and in environments too harsh for plants. They are often the first organisms to colonize a tough new habitat, such as a building's roof or walls.

▲ POISONOUS TOADSTOOL
Many fungi, including this fly agaric, use deadly toxins to deter animals from eating the fruit. The fly agaric fruit's bright red and white colouring also acts as a warning. Wild fungi must never be eaten unless an expert confirms they are safe.

LICHEN PARTNERSHIP ▶
Both the fungus and the alga benefit from living together as a lichen. The green alga photosynthesizes and makes sugar, some of which it gives to the fungus. In turn, the fungus gathers up nutrients and moisture and passes them to the alga. This type of two-way relationship between life forms is called symbiosis.

e ▶▶
fungi

FIND OUT MORE ▶▶ Algae **286–287** • Bacteria **284** • Medicinal Plants **279** • Medicine **372–373**

BACTERIA

Bacteria are monerans, the simplest single-celled organisms. They are the smallest of all cells, visible only through powerful microscopes. Bacteria are also the most abundant forms of life. They live in the air, on land, in water, and even inside the bodies of animals and plants. Some bacteria cause diseases, but others are useful. Bacteria recycle nutrients in the soil and aid the human digestive system.

INSIDE A BACTERIUM ▶
Most bacteria are surrounded by a tough cell wall. Inside, the genetic material is not contained in a nucleus, but is free in the cytoplasm. Some bacteria have fine hairs that enable them to stick to surfaces. Others have miniature tails that help them to swim.

◀ EARLY LIFE FORMS
Stromatolites, like these in Shark Bay, Australia, are huge communities of cyanobacteria (blue-green algae). They were among the earliest life forms, and used light energy to make food. As a by-product, they released oxygen into the air, making other life possible.

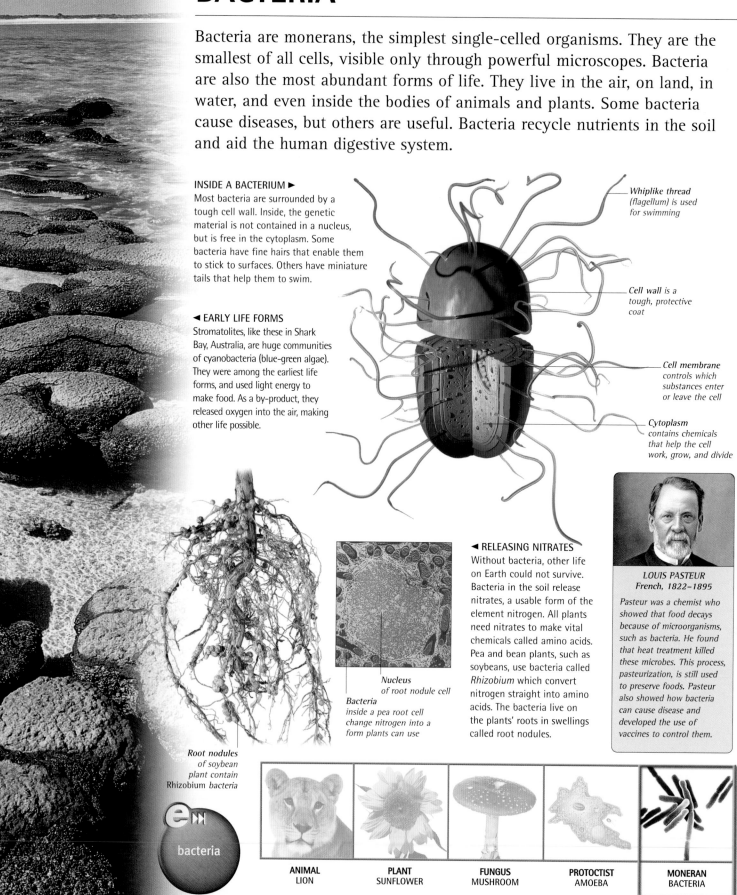

Whiplike thread (flagellum) is used for swimming

Cell wall is a tough, protective coat

Cell membrane controls which substances enter or leave the cell

Cytoplasm contains chemicals that help the cell work, grow, and divide

Nucleus of root nodule cell

Bacteria inside a pea root cell change nitrogen into a form plants can use

Root nodules of soybean plant contain Rhizobium bacteria

◀ RELEASING NITRATES
Without bacteria, other life on Earth could not survive. Bacteria in the soil release nitrates, a usable form of the element nitrogen. All plants need nitrates to make vital chemicals called amino acids. Pea and bean plants, such as soybeans, use bacteria called *Rhizobium* which convert nitrogen straight into amino acids. The bacteria live on the plants' roots in swellings called root nodules.

LOUIS PASTEUR
French, 1822–1895

Pasteur was a chemist who showed that food decays because of microorganisms, such as bacteria. He found that heat treatment killed these microbes. This process, pasteurization, is still used to preserve foods. Pasteur also showed how bacteria can cause disease and developed the use of vaccines to control them.

e ▶▶
bacteria

| ANIMAL | PLANT | FUNGUS | PROTOCTIST | MONERAN |
| LION | SUNFLOWER | MUSHROOM | AMOEBA | BACTERIA |

FIND OUT MORE ▶▶ Algae 286–287 • Classifying Plants 254–255 • Microscopes 116 • Nitrogen 42–43

SINGLE-CELLED ORGANISMS

Many life forms consist of a single cell. As well as simple bacteria, there are more complex organisms, known as protoctists. Unlike bacteria, they have complex internal structures, such as nuclei containing organized strands of genetic material called chromosomes. Most are single-celled, but some form colonies, with each cell usually remaining self-sufficient.

single cell

ANTONI VAN LEEUWENHOEK
Dutch, 1632–1725

Lens-maker Antoni van Leeuwenhoek made the first practical microscope in 1671. With it, he observed bacteria and protoctists, which he called "animalcules". Van Leeuwenhoek went on to study yeasts, plant structure, insect mouthparts, and the structure of red blood cells.

TWO AMOEBAE MEET ▶
An amoeba is a predatory single cell that does not have a fixed shape. It can project parts of its cell to create jellylike tentacles called pseudopodia. The amoeba uses these to move, touch, and grab prey. Amoebae live in water, where they creep along rotting vegetation. They hunt smaller single cells, such as bacteria.

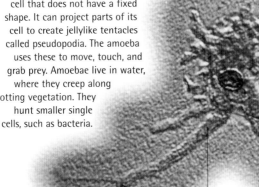

Nucleus

Pseudopod

Fruiting body contains spores that can grow into new cells

Chloroplasts develop when cell is exposed to light

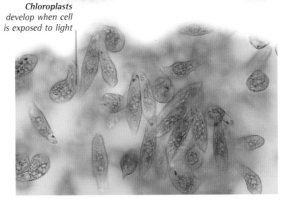

Red blood cell

Malaria

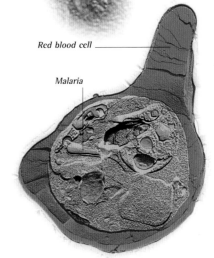

▲ SLIME MOULD FRUITING BODIES
This scanning electron micrograph (SEM) shows the mushroomlike fruiting bodies of a slime mould at x20 magnification. Slime moulds start out as amoebalike cells hunting for food in damp habitats. Later, the cells join together to build spore-producing structures.

▲ LIGHT MICROGRAPH OF *EUGLENA* ALGAE
Algae are now classed as protoctists, although scientists used to include them in the plant kingdom. Algae can make food by photosynthesis, as they contain green chloroplasts. *Euglena* algae live in ponds. They lose their chloroplasts in the dark and then feed like animals. Seaweeds are the best-known algae. They are made up of huge communities of algae cells.

▲ MALARIA PARASITE
Some protoctists obtain food by invading other organisms and living as parasites. This malaria parasite, shown in a transmission electron micrograph (TEM), has infected a human red blood cell. The malaria parasite first enters its human host through the bite of the female *Anopheles* mosquito. Once inside, it multiplies inside the blood and may infect the liver. The parasite causes malaria fever, a disease that may be fatal.

ANIMAL
LION

PLANT
SUNFLOWER

FUNGUS
MUSHROOM

PROTOCTIST
AMOEBA

MONERAN
BACTERIA

FIND OUT MORE ▸▸ Algae **286–287** • Bacteria **284** • Classifying Plants **254–255** • Disease **370–371**

BROWN SEAWEED

Air bladder
keeps fronds floating
at the water's surface

Red seaweed
can capture light energy at lower
depths than other seaweeds

RED SEAWEED

GREEN SEAWEED

Green seaweed often
grows in rock pools

▲ TYPES OF SEAWEED
All seaweeds contain green chlorophyll for photosynthesis, but some types have extra pigments which make them appear brown or red. Different seaweeds survive at different tidal zones on the seashore – the longer they can survive being exposed by the tides, the higher up the beach they can live.

▲ ALGAL OVERKILL
Lakes, ponds, and ditches can be smothered by algal growth when there are too many nutrients in the water. This is called eutrophication. Thick layers of algae cover the water's surface, cutting off sunlight to the plants and algae below and killing them. As these organisms rot, oxygen in the water is used up, and much of the life below the water's surface dies.

ALGAE

Algae are simple organisms that make food from sunlight by photosynthesis, but lack the roots, stems, and leaves of proper plants. Algae are found in all water environments, and some can live on land, forming a thin, greenish layer on damp surfaces. Algae make up most of the oceans' **PHYTOPLANKTON** – microscopic life forms photosynthesizing at the ocean surface. Larger marine algae called seaweeds are made of many cells, with structures called fronds that look similar to plant leaves.

FOREST OF KELP ►
Kelps are varieties of large, dark green or brown seaweeds that grow in cold seas around the world. One type, called giant kelp, can reach up to 60 m (196 ft) from seabed to surface. Giant kelp can form magnificent underwater forests. Kelp beds provide an important habitat for other marine life, including snails, crabs, sea urchins, seals, and sea otters.

Densely growing fronds
provide a habitat for
fish and sea urchins

Rootlike holdfasts
anchor the kelp to
the seabed

algae

PHYTOPLANKTON

Microscopic algae that float in the oceans, using the energy of sunlight to make food and grow, are called phytoplankton. Together with zooplankton – tiny animals and animal-like organisms – they float near the water's surface. In the right conditions, phytoplankton can multiply rapidly, turning water green or red. They are the ultimate source of food for almost all marine life.

Kelp fronds reach towards the sunlight they need in order to make food

◄ DIATOM

This single-celled alga has a hard, glassy case. Each case has two halves, which fit together like a lid on a box. When the diatom cell splits, each new cell keeps half of the case and makes a smaller new half to fit into it. Further generations are even smaller. When the diatoms become too tiny to split any more, they release spores that grow into new, full-size diatoms.

Phytoplankton is most dense in cooler water, away from the Equator

OXYGEN FACTORIES ►

This image from space shows where phytoplankton live. Red, yellow, and light blue mark the ocean's highest densities of chlorophyll, contained in phytoplankton and seaweeds. Dark blue and pink show the lowest density. Land plants are marked in green. Phytoplankton release more oxygen into the atmosphere than all land plants combined.

◄ POISONOUS RED TIDES

Algal blooms can happen in the sea when nutrients such as fertilizers or sewage make life too easy for phytoplankton. Algae such as *Noctiluca scintillans* can turn the sea red, poisoning animals such as shellfish with toxins which would normally be more dispersed. *Noctiluca* means "night light" – this phytoplankton can glow in the dark, creating flickers of light on the sea's surface.

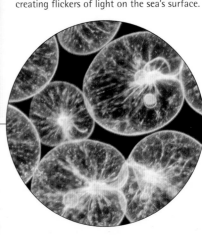

NOCTILUCA SCINTILLANS

FIND OUT MORE ►► Oceans **228–229** • Photosynthesis **258**

ANIMALS

ANIMAL KINGDOM

Animals belong to the largest and most diverse of the five kingdoms of living things. So far over two million animal species have been identified. All animals share certain features. Unlike plants, animals get the energy they need by eating food. They are all made up of many cells and many animals are highly mobile. Most reproduce sexually and have sense organs that allow them to react quickly to their surroundings. **CLASSIFICATION** uses these and other characteristics to group similar animals together.

CLASSIFYING A LION

Every animal species has a unique Latin name. The first word is the genus name, which is shared with closely related animals. The second word is the specific name, which, together with the genus, is unique to a particular species.

Species	*Panthera leo* ("panther-like" lion)
Genus	*Panthera* (large cat)
Family	Felidae (cat)
Order	Carnivora (flesh-eating)
Class	Mammalia (suckle young, endothermic)
Phylum	Chordata (rod-like backbone)

CLASSIFICATION

In order to make animals easier to study, scientists divide the animal kingdom into divisions and subdivisions. The first division is called a phylum. Each phylum breaks down into groups called classes. Classes are divided into orders, then families, and then genera. Each genus contains species, which are individual groups of animals that have the same characteristics and can breed together.

GIRAFFE

HARVEST MOUSE

VARIATIONS IN SIZE ▲
Animals are not classified by size. The giraffe and harvest mouse differ hugely in size but are both classified as mammals because they have fur, single-boned jaws, and suckle their young.

Eyes are the most prominent sense organs

Head is large and contains the brain

Feet are used to perch and walk, and to grasp food

EQUIPPED FOR FLIGHT ▶
Animals are the only living things to have conquered the air. Insects, birds, and bats are all capable of powered flight. Birds have strong muscles to power their flight, coordinated by a well-developed brain and nervous system.

ANIMAL
LIONESS

PLANT
SUNFLOWER

FUNGUS
TOADSTOOL

PROTOCTIST
AMOEBA

MONERAN
BACTERIA

Wings are powered by strong muscles

SIMPLE JELLYFISH ▶
Some animals, such as jellyfish, have a relatively simple structure. They have no skeleton, few muscles, and their movement is uncoordinated – they drift with ocean currents. Jellyfish are referred to as invertebrates because, like 98 per cent of animals, they have no backbone.

`◀ VERTEBRATES
Animals with backbones, like these zebras, are commonly referred to as vertebrates. Mammals, birds, fish, amphibians, and reptiles are all vertebrates. Zebras belong to the mammal order. Mammals, which also include humans, are the most complex animals in the animal kingdom.

animal kingdom

AQUATIC MAMMAL ▲
The blue whale is the largest living animal. It can reach 30 m (98 ft) in length. It can only grow to such a size because seawater supports its weight. Although the whale spends its entire life in water like a fish, it is classified as a mammal because it suckles its young.

Hard external skeleton prevents the mite from drying out

MICROSCOPIC MITE ▶
Animals can be very small. This mite (on a needle) is so small that it cannot be seen with the naked eye. Its size is limited because it can only grow by moulting (shedding its outer layer). Mites have a hard external skeleton and move on jointed legs. Other animals with these features include spiders and scorpions, many insects, and crabs.

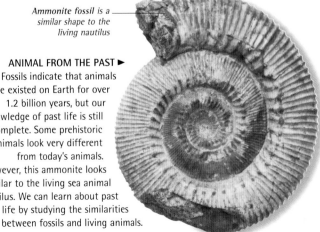

Ammonite fossil is a similar shape to the living nautilus

ANIMAL FROM THE PAST ▶
Fossils indicate that animals have existed on Earth for over 1.2 billion years, but our knowledge of past life is still incomplete. Some prehistoric animals look very different from today's animals. However, this ammonite looks similar to the living sea animal nautilus. We can learn about past life by studying the similarities between fossils and living animals.

FIND OUT MORE ▸▸ Classifying Plants **254–255** • Evolution **328–329**

ANIMAL ANATOMY

The study of the structure of living things is called anatomy. All animals are made up of **CELLS**, some of which are specialized to carry out different functions. Simple animals, such as sponges, are made up of only a few types of cell. In more complex animals, cells are organized into tissues, such as muscles and nerves that are necessary for movement. Tissues can form organs, such as the heart, which is used to pump blood around the **CIRCULATORY SYSTEM.**

SHARK ANATOMY ▼
Like all fish, sharks have a backbone, breathe through gills, manoeuvre using fins, and are ectothermic (cold-blooded). A shark's anatomy also bears the hallmarks of a predatory fish. They have a streamlined, torpedo-shaped body that allows them to cut easily through water to chase prey. They also have powerful jaws and sharp teeth.

◄ PERFECT SYMMETRY
Most animals, like this penguin, are bilaterally symmetrical. If the penguin were cut in half from head to toe, the two halves would be mirror images of each other. Other animals, such as sea anemones, are radially symmetrical. They have no head or tail and can be cut into identical halves along many lines. Of the two types, animals that are bilaterally symmetrical tend to move more quickly and precisely.

Ovary produces eggs which pass into a tube to be fertilized

Gall bladder releases substances into the intestine that help absorb fat

Gills take oxygen from water so the shark can breathe

Gill arches support the gills

Powerful jaw muscles for biting into prey

Eyes are well-developed

Nostrils help to detect the smell of prey from afar

Aorta carries blood to smaller arteries

Heart pumps blood all round the body

Teeth are numerous and sharp

Rib cage protects heart and lungs

Skull protects the brain, a vital organ

Backbone runs from head to tail

e ▸▸
animal anatomy

◄ INTERNAL SKELETON
All animals with backbones have an internal framework of support, called an endoskeleton. Bony skeletons, such as that of the squirrel, are light to aid movement. When an animal is young, bones in the skeleton can grow in length. Some bones protect vital organs, while limb bones are used in movement.

Many tail vertebrae make up a long tail for balance

Spiracle on the caterpillar's body

▲ RESPIRATORY SYSTEMS

Most animals need oxygen to survive. Simple animals exchange gases over the surface of their bodies. Insects, such as caterpillars, have openings along their bodies, called spiracles, through which air passes. Animals with lungs, such as birds and mammals, actively breathe.

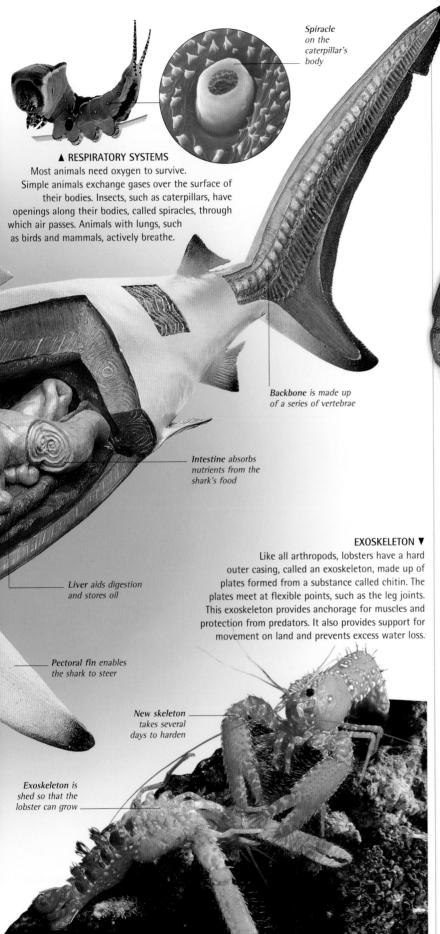

Backbone is made up of a series of vertebrae

Intestine absorbs nutrients from the shark's food

Liver aids digestion and stores oil

Pectoral fin enables the shark to steer

Exoskeleton is shed so that the lobster can grow

New skeleton takes several days to harden

EXOSKELETON ▼

Like all arthropods, lobsters have a hard outer casing, called an exoskeleton, made up of plates formed from a substance called chitin. The plates meet at flexible points, such as the leg joints. This exoskeleton provides anchorage for muscles and protection from predators. It also provides support for movement on land and prevents excess water loss.

CELLS

Animal cells are typically just 0.02 mm ($\frac{1}{1,250}$ in) across. Although they can be extremely varied, they share common features. Cells are surrounded by a skin called a membrane and contain a jelly-like fluid called cytoplasm. All the processes needed for life, such as producing energy from food, removing waste, and growth take place inside cells.

Cell membrane lets some substances pass through it, but not others

Nucleus contains DNA, which tells the cell how to grow and function

Cytoplasm is a jelly-like substance within the cell

Mitochondrion converts energy from simple substances

▲ CELL COMPONENTS

Inside an animal cell, the cytoplasm contains structures called organelles that have a variety of functions, from storing vital substances to destroying bacteria. The most important organelle is called the nucleus, which carries genetic information, controlling how the cell behaves. Another organelle, the mitochondrion, produces energy from food.

CIRCULATORY SYSTEM

The circulatory system carries blood around an animal's body, providing nourishment and oxygen to cells. In some animals it is open, in others it is closed. In an open system, blood flows freely around the body. In a closed system, blood is confined to a network of vessels. The circulatory system also helps distribute heat around the body.

ECTOTHERMIC LIZARDS ►

Many land animals, such as reptiles, are ectothermic – they rely on the Sun's heat to raise their body temperature to a level that allows them to be active. Birds and mammals are endothermic – they produce their own heat and maintain a constant body temperature.

FIND OUT MORE ►► Body Systems **338–339** • Movement **314–315** • Plant Anatomy **256–257**

SPONGES

The simplest of all animals, most sponges live in colonies (groups) that are little more than units of cells organized into two layers. Most live in the sea and are usually hermaphroditic – each sponge produces both eggs and sperm. The larvae are free-living, but adults are sessile – they remain anchored in one place.

CALCAREOUS SPONGES ▲
Sponges are classified by their spicules, the pointed structures that make up a sponge's framework. In a calcareous sponge these are made of calcium. There are about 150 species of calcareous sponge.

PHYLUM: PORIFERA

Sponges have a skeleton of spicules (pointed structures) but no distinct body parts. Many are essentially a tube, closed at one end. They are not symmetrical. There are about 10,000 species.

	Class: Calcarea (calcareous sponges) **Features:** often less than 10 cm (4 in) high, skeletal spicules of calcium carbonate
	Class: Hexactinellida (glass sponges) **Features:** skeletal, six-pointed, silica (glass-like) spicules
	Class: Demospongiae (demosponges) **Features:** some have three- or four-pointed silica spicules

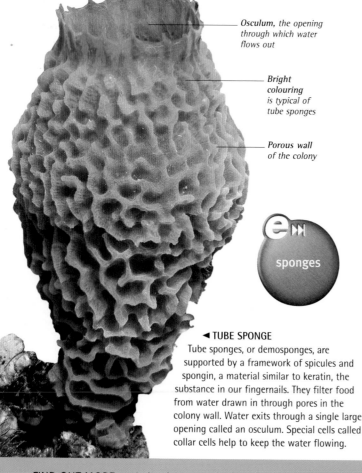

Osculum, the opening through which water flows out

Bright colouring is typical of tube sponges

Porous wall of the colony

sponges

◄ TUBE SPONGE
Tube sponges, or demosponges, are supported by a framework of spicules and spongin, a material similar to keratin, the substance in our fingernails. They filter food from water drawn in through pores in the colony wall. Water exits through a single large opening called an osculum. Special cells called collar cells help to keep the water flowing.

CNIDARIANS

Cnidarians are aquatic animals that have a simple, usually symmetrical, body with a mouth opening. Stinging cells on tentacles around the mouth catch prey. Cnidarians are either bell-shaped and mobile, like the jellyfish, or tubes anchored to one spot, like coral and sea anemones.

▲ BOX JELLYFISH
Jellyfish drift about in the ocean currents, trailing their tentacles through the water. They sting small animals with the cnidoblasts (stinging cells) on their tentacles and push the prey into their mouth. After digestion, waste passes out of their mouth.

▲ CORAL
Most corals live in colonies, but mushroom corals form a single polyp (anchored tube) that may grow 50 cm (20 in) wide. Their hard skeleton is made of chalk (calcium carbonate). Coral skeletons often build up to form a reef.

cnidarians

Sensory tentacles around mouth

PHYLUM: CNIDARIA

All cnidarians have stinging cells. Many are able to reproduce asexually (without mating) and sexually. There are 9,000 species.

	Class: Anthozoa (corals, sea fans, sea pens, sea anemones) **Features:** anchored polyp (tube-like) form, carnivorous (eat flesh), often in groups
	Class: Scyphozoa (jellyfish) **Features:** free-living, medusoid (bell-shaped) form, mouth on underside
	Class: Hydrozoa (hydrozoans) **Features:** some free-living, others anchored, most in colonies (large groups), mostly carnivorous
	Class: Cubozoa (box jellyfish) **Features:** free-living, box-shaped medusoid form, with long tentacles from corners

SEA ANEMONE ▲
Sea anemones are commonly found in coastal rock pools. They catch fish and other small animals in their stinging tentacles. When the tide goes out, they survive out of water by pulling in their tentacles. This helps them to conserve water.

FIND OUT MORE ▸▸ Animal Anatomy **292–293** • Communities **325** • Ecology **326–327**

WORMS

There are about one million species of worm, living in a wide range of habitats. They have a long body, and have no legs. Many worms are parasites that live on or in another animal and use strong mouthparts to feed off that animal. Others are predators, and can move quite quickly. The three main groups are **FLATWORMS**, **ROUNDWORMS**, and **SEGMENTED WORMS**.

e ▶▶ worms

TYPES OF WORM

There are many different phyla of worms. The following three are the best-known. Some worms live on land in burrows, feeding on plant matter; others live in the sea or fresh water, filtering food from water.

	Phylum: Platyhelminthes (flatworms) **Features:** about 20,000 species free-living forms have a mouth and no anus, many live in water
	Phylum: Annelida (segmented worms) **Features:** about 15,000 species segmented bodies, mostly burrowing, gut with mouth and anus, live on land and in water
	Phylum: Nematoda (roundworms) **Features:** about 25,000 species unsegmented bodies, gut with mouth and anus

FLATWORMS

There are about 20,000 species of flatworm. They have a solid, flat body that does not contain blood. Most flatworms are parasitic, but some are free-living.

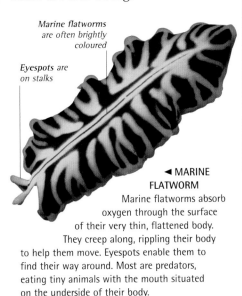

Marine flatworms are often brightly coloured

Eyespots are on stalks

◄ MARINE FLATWORM
Marine flatworms absorb oxygen through the surface of their very thin, flattened body. They creep along, rippling their body to help them move. Eyespots enable them to find their way around. Most are predators, eating tiny animals with the mouth situated on the underside of their body.

Hooks form a ring on the head

Sucker clamps onto animal's gut

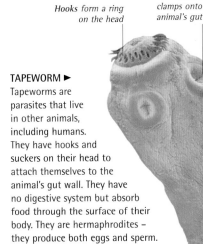

TAPEWORM ►
Tapeworms are parasites that live in other animals, including humans. They have hooks and suckers on their head to attach themselves to the animal's gut wall. They have no digestive system but absorb food through the surface of their body. They are hermaphrodites – they produce both eggs and sperm.

ROUNDWORMS

Roundworms, or nematodes, are found almost anywhere and exist in huge numbers. As many of the roundworms are transparent, few people are aware of them.

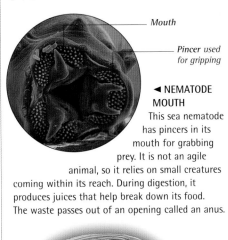

Mouth

Pincer used for gripping

◄ NEMATODE MOUTH
This sea nematode has pincers in its mouth for grabbing prey. It is not an agile animal, so it relies on small creatures coming within its reach. During digestion, it produces juices that help break down its food. The waste passes out of an opening called an anus.

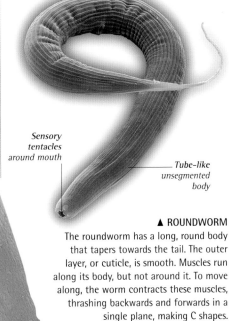

Sensory tentacles around mouth

Tube-like unsegmented body

▲ ROUNDWORM
The roundworm has a long, round body that tapers towards the tail. The outer layer, or cuticle, is smooth. Muscles run along its body, but not around it. To move along, the worm contracts these muscles, thrashing backwards and forwards in a single plane, making C shapes.

SEGMENTED WORMS

This group divides into earthworms, bristleworms, and leeches. All have segmented bodies. The worms' bodies are fluid-filled, but the leeches are solid.

Mouth

Saddle produces mucus

◄ EARTHWORM
Earthworms are formed from many segments. Only the gut runs through the whole body from head to tail. Worms have a circulatory system with blood vessels but no heart. The thickened area towards the front of their body secretes mucus, which binds mating worms together and forms a cocoon for eggs.

Body is made up of individual segments

Anus

LEECH ►
Leeches are parasites that attach themselves to the outside of an animal host. Their specialized cutting jaws bite through skin so that they can suck the animal's blood. Substances in their saliva prevent the blood from clotting and make the bite painless so that the animal is unaware it has been bitten. Leeches move by shifting one sucker forwards and then bringing the other one up behind it.

FIND OUT MORE ▶▶ Communities **325** • Soil **224**

CRUSTACEANS

Crustaceans have a hard, jointed external skeleton, called an exoskeleton, that protects them like armour. They have five pairs of jointed legs and in some species, the front pair of legs are modified to form strong pincers. Crustaceans have compound eyes (made up of lots of lenses) on stalks and two pairs of antennae, which help them to sense predators. Most crustaceans live in water, but some, such as woodlice, live in damp places on land.

PHYLUM: CRUSTACEA

Most crustaceans live in water. There are more than 45,000 species in seven classes, including:

	Class: Branchiopoda (fairy shrimps, water fleas) **Features:** small, free-living, filter feeders with bristled mouthparts
	Class: Cirripedia (barnacles) **Features:** box-like bodies, sessile (anchored to one spot) as adults
	Class: Malacostraca (crabs, lobsters, prawns, woodlice) **Features:** jointed legs, often pincers, eyes on stalks

JOINTED BLUE LOBSTER ▼
The common lobster is blue, and can be as much as 1 m (3¼ ft) long. Lobsters have a jointed body, a long abdomen, and a wide tail fan. As the lobster grows, it gets too big for its hard shell, or carapace. The shell splits, the lobster crawls out, and a new shell hardens. Lobsters feed at night, cracking open molluscs with their huge claws.

Compound eye *is on a stalk*

Long antenna *finds out about the lobster's surroundings*

Large claw *for feeding and defence*

Tail fan *works like an oar, enabling the lobster to move quickly*

Jointed leg *for walking along the seabed*

Cephalothorax *a single segment made up of the head and the thorax*

e⟫ crustaceans

Feeding legs *catch particles of food*

Barnacle larva *drifts until ready to change into the adult form*

◄ BARNACLE BABY
Some crustaceans, such as barnacles, can only move about as larvae. Barnacles release eggs that hatch into larvae and move away. These drift freely as they grow, before attaching themselves to a rock, the bottom of a ship, or a whale, and changing into the adult form. The adult barnacle cannot move around.

Adult barnacle *releases eggs that hatch into larvae*

Large claw *is brightly coloured*

Eye *on a long stalk to give a better view*

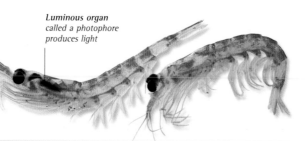

LOPSIDED CRAB ▲
Most crustaceans have two claws that are the same size, but the male fiddler crab has one enormous claw and one tiny one. It waves the giant claw about in order to attract a mate and also to frighten away competing males. The huge claw can make up half of the crab's total weight. Fiddler crabs live in mangrove swamps, where they make burrows in the mud.

KRILL ►
Krill use long hairs on their front legs to filter food particles from the seawater. They have a soft body and large eyes. They are sociable, often living in huge swarms, and are important in the marine food chain. Baleen whales feed mainly on krill.

Luminous organ *called a photophore produces light*

FIND OUT MORE ▶▶ Animal Anatomy 292–293 • Senses 316–317

INSECTS

An insect's body divides into three sections. The head holds the eyes, antennae, and mouthparts. The thorax bears three pairs of jointed legs and two pairs of wings. The abdomen contains the digestive system and the sex organs. Most insects undergo a complete change (called metamorphosis) between the larval stage and adult form.

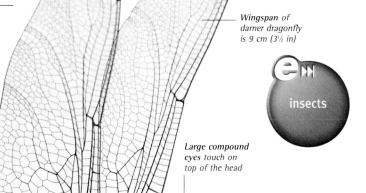

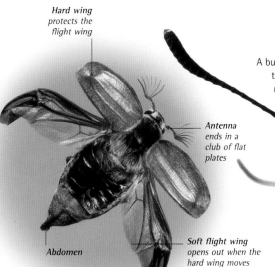

Wingspan of darner dragonfly is 9 cm (3½ in)

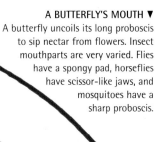

insects

Large compound eyes touch on top of the head

DRAGONFLY NYMPH ▼
A young dragonfly, known as a nymph, lives underwater. As it grows bigger, it moults – its skin splits and a new, larger one forms. Each time, the nymph begins to look more like the adult dragonfly. For the last moult, the nymph climbs out of the water, and the adult dragonfly emerges.

Long abdomen made up of 10 segments

AGILE DRAGONFLY ►
The dragonfly's slim body and long, thin wings make it one of the fastest fliers in the insect world. The front wings and the back wings beat alternately, giving the dragonfly excellent flight control. Dragonflies have large, compound eyes, each made up of about 30,000 lenses, and can see prey up to 12 m (39 ft) away. The legs form a basket to hold food.

Claw-like foot

Jointed leg

CLASS: INSECTA

There are more different types of insect than of any other kind of animal, with over one million identified species. All are terrestrial, most can fly, and some can walk on the surface of water.

	Order: Odonata (dragonflies, damselflies) **Features:** two pairs of matching wings, long abdomen, carnivorous when adult, dragonflies rest with wings open, damselflies with wings folded
	Order: Orthoptera (grasshoppers, crickets) **Features:** straight, tough forewings, short antennae, capable of huge leaps using powerful hindlegs, chewing mouthparts
	Order: Lepidoptera (butterflies, moths) **Features:** bodies covered with tiny sensory hairs and scale-covered wings, butterflies have club-ended antennae and fly by day, moths fly by night
	Order: Hemiptera (bugs) **Features:** two pairs of wings, protruding rostrum (mouthpiece) that is used for piercing and sucking
	Order: Coleoptera (beetles) **Features:** tough elytra (front wings) fold over membranous hindwings protecting them, can squeeze into small spaces
	Order: Diptera (flies) **Features:** most have a single pair of flight wings, some have a thin body and thread-like antennae, others a bigger body and short antennae
	Order: Hymenoptera (ants, bees, wasps, sawflies) **Features:** two pairs of membranous wings joined in flight by tiny hooks, many have a narrow "waist"

Hard wing protects the flight wing

Antenna ends in a club of flat plates

Abdomen

Soft flight wing opens out when the hard wing moves forward

▲ HARD WINGS AND SOFT WINGS
This cockchafer beetle has two pairs of wings but only uses the back pair to fly. The hardened front wings, called elytra, cover the back wings when they are not in use, protecting them. Some insects have muscles attached directly to the wings, others move their wings by changing their body shape. A small number of insects, such as silverfish, do not have wings.

A BUTTERFLY'S MOUTH ▼
A butterfly uncoils its long proboscis to sip nectar from flowers. Insect mouthparts are very varied. Flies have a spongy pad, horseflies have scissor-like jaws, and mosquitoes have a sharp proboscis.

Antenna helps the butterfly to smell flowers

Proboscis, coiled up when not in use

FIND OUT MORE ▸▸ Life Cycles **305** • Pollination **266–267** • Populations **324** • Senses **316–317**

ARACHNIDS

Spiders, scorpions, ticks, and mites have two body parts and four pairs of legs. They breathe using lung books (that look like an open book) in the abdomen. The front part of the body, known as the cephalothorax, bears the legs and two pairs of mouthparts: the chelicerae, which are like either pincers or fangs, and the pedipalps, which look like either legs or claws. Most arachnids live on land, but some live in water.

CLASS: ARACHNIDA

Most arachnids are predators, but some scavenge for food and a few mites are parasitic (live on another animal and feed off that animal). There are 75,500 species of arachnid, in 12 orders, including the following three main ones.

	Order: Scorpionida (scorpions) **Features:** predators, sting-bearing tails, large, claw-like pedipalps, bear live young
	Order: Acarina (mites, ticks) **Features:** body not distinctively segmented, many are pests and parasites
	Order: Araneae (spiders) **Features:** mostly eight-eyed, able to produce silk

MEXICAN RED-KNEED TARANTULA ▶
The red-kneed tarantula pounces on prey that comes close to its lair. Like most types of spider, the tarantula paralyses and kills its prey with venom, which it injects using fangs. The venom also breaks down the prey's flesh, so that the spider can suck it up as a liquid. Spiders are carnivorous, eating mainly insects.

Pedipalps, for holding and tearing food

Cephalothorax bears four pairs of legs and two pairs of mouthparts

Sharp fangs inject poison into the prey

Abdomen, covered in hairs that stand on end to put off enemies

arachnids

Jointed legs have red, hairy knees

SHEEP TICK ▲
A tick's soft, flexible abdomen can expand to 10 times its normal size as the tick sucks in blood with specialized piercing and sucking mouthparts. The tick fastens itself to a sheep while it drinks in blood, then drops to the ground. When it needs more food, it attaches itself to another animal that is passing by.

Spinneret produces silk

▼ SPINNING SPIDER
A spider's spinnerets produce liquid silk that hardens in the air. Many spiders spin a web with this silk, to catch prey. When an animal gets caught in the web, the spider wraps it in silk and kills it with venom. Spider silk is the strongest-known material – if a web were made with silk threads the diameter of a pencil, it would be strong enough to stop a plane in flight.

Soft young cling to the female's back

Large tail, bearing the scorpion's sting

◀ CARING SCORPION
The imperial scorpion is one of many arachnids that care for their young. A female scorpion carries about 30 young on its back until they have moulted (grown a new, larger skin) for the second time. The scorpion has a hard, black carapace (shell), large claws, and a poisonous sting.

Pedipalp, in the form of a large claw

FIND OUT MORE ▶▶ Animal Anatomy **292–293** • Feeding **312–313**

MOLLUSCS

Slugs, snails, oysters, squid, and octopuses are very different to look at, but they are all molluscs. They have a ribbon-like tongue, called the radula, covered in thousands of denticles (tiny teeth). Many have a calcium-carbonate shell. Most molluscs live in water, but many slugs and snails live on land.

Muscular foot secretes lubricating slime that helps the snail move

Coiled shell into which the snail withdraws for protection

Short tentacle feels things as the snail moves

Simple eye on antenna

GIANT SNAIL ▲
The African land snail can be 30 cm (12 in) long. Like all snails, it carries a coiled shell on its back and withdraws into the shell when threatened. It moves slowly on a large, muscular foot, using slime to ease the way. Its mouth, underneath its head, contains the denticle-covered radula.

Tentacle, covered in suckers for crawling over rocks and catching prey

▲ INTELLIGENT OCTOPUS
The blue-ringed octopus is a mollusc that does not have a shell. It has a large brain and big eyes. It uses its eight arms to crawl, but also squirts water from inside its body to move more quickly.

shell

PHYLUM: MOLLUSCA

Molluscs have unsegmented bodies generally protected by a shell. There are more than 50,000 species in seven classes, including the following five common ones.

	Class: Bivalvia (clams, mussels, oysters, scallops) **Features:** shell in two symmetrical halves with hinge
	Class: Polyplacophora (chitons) **Features:** shell made up of several overlapping plates
	Class: Gastropoda (snails, slugs) **Features:** suction foot, many have spiral shell
	Class: Scaphopoda (tusk shells, razor shells) **Features:** long, tubular shells, burrow into sediment
	Class: Cephalopoda (nautiloids, squid, octopus) **Features:** tentacles, suckers

ECHINODERMS

Echinoderms have a spiny body that usually divides into five equal parts. They walk on hundreds of tube feet that are full of water. If they lose part of their body, they can regrow it. They have a skeleton of calcium-carbonate plates.

Tube feet cover the underside of each tentacle

LONG-LEGGED STARFISH ▶
A starfish moves by pulling itself along on the sucker-like tube feet underneath its arms. These strong feet also enable the starfish to force open the shells of molluscs such as mussels or oysters. As the mollusc's muscles weaken and the shell opens, the starfish pushes its stomach out through its mouth and into the shell to digest the mollusc.

Calcium-carbonate plates cover the starfish's body

echinoderms

Tube feet, modified into feathery tentacles

SEA CUCUMBER ▶
The sea cucumber's tube feet are grouped around its mouth and filter food from the sand. If it is attacked, the cucumber pushes out its stomach and reproductive parts for the predator to take. The sea cucumber then grows replacement parts.

▲ SPIKY URCHIN
Sharp spines cover the sea urchin's hard skeleton and protect it from attack. Tube feet cover its body and spread between the spines. The sea urchin grazes on algae and small animals, with sharp jaws situated underneath its body.

PHYLUM: ECHINODERMATA

Most echinoderms are mobile, although sea lilies are stationary. They live on shores, reefs, and the seabed. There are about 6,000 species.

	Class: Asteroidea (starfish) **Features:** central mouth on underside, surrounded by arms (usually five)
	Class: Echinoidea (sea urchins) **Features:** spherical skeleton covered in spines
	Class: Crinoidea (feather stars, sea lilies) **Features:** mouth faces upwards, feed on plankton
	Class: Holothuroidea (sea cucumbers) **Features:** sausage-like body with tentacles around mouth

FIND OUT MORE ▶▶ Communication **318–319** • Defence **320–321**

FISH

Fish are aquatic animals that evolved about 500 million years ago. Most have a well-developed internal skeleton, scale-covered bodies with fins, and a tail for swimming. They breathe using gills to absorb oxygen from the water, although a few, such as the lungfish, can survive in air. The four classes of fish – jawless fish, sharks, lungfish, and bony fish – have common characteristics, but are only distantly related.

FISH	
The term fish is an informal grouping of chordate animals, living in water. They are ectothermic (cold-blooded) and move using fins. There are about 25,000 species.	
	Class: Cyclostomata (jawless fish) **Features:** sucker-like mouth, notochord (stiffened rod)
	Class: Chondrichthyes (cartilaginous fish, including sharks, skates, rays) **Features:** skeleton of cartilage, tooth-like scales
	Class: Osteichthyes (bony fish) **Features:** bony skeleton, flexible fins, swim bladder
	Class: Choanichthyes (lungfish) **Features:** lungs, internal nostrils

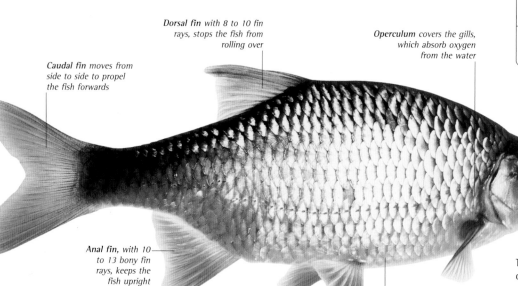

Dorsal fin with 8 to 10 fin rays, stops the fish from rolling over

Operculum covers the gills, which absorb oxygen from the water

Caudal fin moves from side to side to propel the fish forwards

Eye has no eyelid, as water keeps the eye wet and clean

Anal fin, with 10 to 13 bony fin rays, keeps the fish upright

Overlapping scales cover the flattened, streamlined body

Pelvic fin works with its pair to move the fish up and down

◄ **BONY SWIMMER**
Bony fish are good swimmers. Muscle blocks, called myotomes, contract in sequence as the fish moves. The tail fin provides thrust; other fins change position or direction. The lateral line (sensory organs along the side of the fish) detects movement in the water. The swim bladder contains the right amount of air, so that the fish does not float or sink.

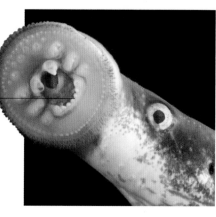

JAWLESS LAMPREY ▲
The sea lamprey has a circular sucker instead of a mouth. Around the edge of the sucker are rows of small teeth, while larger teeth surround the opening. The lamprey sucks blood with its rough tongue. Lampreys do not have jaws or scales, and their internal skeleton is a stiffened rod, or notochord.

Teeth bite into another fish so that the lamprey can suck out its blood

▼ **FILTER FEEDER**
The manta ray looks strange, but is a harmless filter feeder. Large flaps on each side of its head channel water into its mouth, and the gills filter out animal plankton and small fish for the ray to swallow. It moves through the water by beating its powerful, triangular pectoral fins. It can accelerate suddenly and leap out of the water, if threatened.

Flap guides water into the mouth

Large, pectoral fins

AFRICAN LUNGFISH ►
The African lungfish does not have gills. It lives in stagnant water and breathes air with its lungs. This African lungfish survives droughts by burrowing into mud and making a cocoon where it lies dormant until the rains fill the pool again.

Long, thin fin

FIND OUT MORE ►► Animal Anatomy **292–293** • Defence **320–321** • Movement **314–315**

AMPHIBIANS

Amphibians generally start life in water, but later change so that they can also live on land. Most return to the water to breed. The life cycle involves three stages: egg, larva, and adult. The change from larva to adult is known as metamorphosis. Amphibians are ectothermic (cold-blooded) animals that have a bony internal skeleton. They have small lungs and can also breathe through their smooth skin, which must be kept moist.

AMPHIBIA

Amphibians can lose water through their skin. There are no scales or hair protecting the skin. Most live in damp places. There are about 5,000 species in three orders.

	Class: Gymnophiona (caecilians) **Features:** limbless, worm-like, poor sight
	Class: Anura (frogs, toads) **Features:** wide head, no tail, powerful back legs
	Class: Caudata (newts, salamanders) **Features:** long tail, carnivorous larvae, good sense of smell

Eye positioned so that the frog can see when swimming

WET FROG ►
A frog looks wet all the time because glands in its skin produce mucus to keep it moist. The frog can then absorb oxygen through its thin skin directly into the bloodstream. Frogs are the only amphibians that can hop. Some have webbed feet and are good swimmers. They live mainly on land and catch worms and insects with their long, sticky tongues.

Skin colour camouflages the frog so that it is hard to see

◄ FROGSPAWN
Most frogs lay their eggs in water, because the eggs have no shell to stop them from drying out. Tadpoles hatch from the eggs and take about 12 weeks to grow and change into frogs. At first the tadpoles breathe like fish, using gills, but as they change, lungs replace the gills. The tadpoles grow back legs, front legs, a big head, and finally their tail disappears.

Large eardrum, evidence of the frog's acute hearing

Bowed front leg absorbs the shock on landing

Strong back leg enables the frog to leap long distances on land

CAECILIAN ▼
A caecilian has a long, thin body, and no legs. It cannot see well and uses its good sense of smell to find food. A small tentacle below each eye collects chemical information, which the caecilian uses in its hunt for earthworms. It is not easy to see caecilians because they live mainly in the soil.

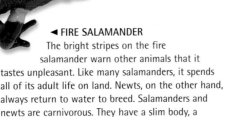

amphibians

◄ FIRE SALAMANDER
The bright stripes on the fire salamander warn other animals that it tastes unpleasant. Like many salamanders, it spends all of its adult life on land. Newts, on the other hand, always return to water to breed. Salamanders and newts are carnivorous. They have a slim body, a long tail, and four legs of about the same size.

FIND OUT MORE ▸▸ Behaviour Cycles **322–323** • Courtship **306–307** • Defence **320–321**

REPTILES

Reptiles are ectothermic (cold-blooded) animals. They cannot generate their own body heat and many bask in the sun to get warm. Tough, dry scales cover their skin, providing a waterproof covering and protecting the reptile from harm. Reptiles with legs have a sprawling gait because their limbs are jointed to the side, not below as in birds or mammals.

▼ GREEN IGUANA

This large lizard lives in trees. Its green, scale-covered skin provides camouflage among the leaves and its long, clawed fingers and toes help the iguana climb well. If attacked, it uses its long tail as a whip. Young iguanas can shed their tails and regrow the missing part.

Head with well developed eyes and nostrils

Crest of modified scales runs along the middle of the back

Green colour helps the lizard blend in with its surroundings

reptiles

Long, whip-like tail, used in defence

Dewlap of male iguana is inflated for display

Long toes give a good grip when climbing

Scales are small, a characteristic of lizard skin

Back leg, at right-angles to the body, gives a splayed gait

CLASS: REPTILIA

Reptiles have a tough skin, with scales made of keratin. They lay soft-shelled eggs, usually on land. There are nearly 8,000 species.

	Order: Squamata (lizards, snakes) **Features:** snakes do not have eyelids, most lizards do
	Order: Crocodilia (alligators, caiman, crocodiles, gharials) **Features:** semi-aquatic, sharp teeth
	Order: Testudines (tortoises, turtles, terrapins) **Features:** hard outer shell, cut food with sharp jaws

POISONOUS FANGS ▶

A rattlesnake bites its prey and injects poison through long fangs. The fangs fold back against the roof of the mouth when they are not in use, and drop down automatically as the mouth opens. Two heat-sensitive pits between the eyes and nose help the rattlesnake to locate its prey.

▼ HATCHING OUT

The leopard tortoise, like all reptiles, emerges fully formed from its egg and has to fend for itself immediately. For some reptiles, the sex of the young depends on the temperature of the eggs as the embryos develop.

Komodo dragon grows up to 3 m (10 ft) long

KOMODO DRAGON ▶

The largest of the lizards, the Komodo dragon is found on only a few islands in Indonesia. It uses its excellent sense of smell to find prey. Unlike mammals, reptiles cannot chew food – the Komodo dragon tears off chunks of flesh with its jagged teeth.

FIND OUT MORE ▶▶ Behaviour Cycles **322–323** • Courtship **306–307** • Feeding **312–313** • Growing Up **310–311**

BIRDS

Birds are endothermic (warm-blooded) animals that have feathers, beaks, and scales on their legs. They lays eggs, which they usually keep warm in nests until the young hatch. Most birds are good at flying. They have powerful wings and light, strong bones. Flight has enabled birds to colonize every habitat in the world, including remote islands and polar regions.

CLASS: AVES (BIRDS)

There are 9,000 species in a total of 29 orders. They all lay eggs that are protected by a light, strong, calcium-carbonate shell. Birds work hard to try to ensure that their chicks survive.

	Order: Passeriformes (perching birds) **Features:** grasping foot, complex songs
	Order: Falconiformes (birds of prey) **Features:** hooked bill, acute eyesight, curved talons
	Order: Piciformes (woodpeckers, toucans) **Features:** two backward, two forward toes, long, pointed bill
	Order: Anseriformes (waterfowl) **Features:** broad beaks, strongly webbed front toes
	Order: Apodiformes (hummingbirds, swifts) **Features:** nectar feeders, acrobatic fliers, rapid wing beat, can hover
	Order: Columbiformes (pigeons, doves) **Features:** plump, small bill, head bobs as bird walks
	Order: Charadriiformes (waders, gulls, auks) **Features:** mostly strong fliers, feed in or near water
	Order: Galliformes (game birds) **Features:** mainly ground-dwelling, short, broad wings

Long feathers increase the wing's surface area

Contour feathers streamline the bird's shape and keep it warm

Wing feathers spread out to slow down for landing

Flight feathers are long and stiff

Talons (long, sharp claws) seize prey firmly

Tail feathers are used for steering and braking

◄ **HAWK LANDING**
A bird of prey, like this red-tailed hawk, hunts small animals. Excellent eyesight enables it to spot animals on the ground while it is flying. Some birds of prey hover in one spot before diving to the kill. This hawk is about to land. Its wing and tail feathers fan out to slow it down.

▲ **HOLLOW BONES**
A bird's bones are mostly hollow, with no marrow. Struts, called trabeculae, strengthen the bones so that they do not break in flight. In some bones, the hollow cavities contain extensions of the air sacs from the lungs. Extensive air sacs enable the bird to get the oxygen it needs to fly quickly and easily.

◄ **FLIGHTLESS BIRDS**
A number of birds can no longer fly. The ostrich could outrun an attacking animal, so did not need flight to escape. The kiwi, in New Zealand, had no natural predators, so could adapt to life on the ground. The penguin lives at sea and swims rather than flies.

CATCHING FISH ▲
The Eurasian kingfisher is skilled at diving for fish. It folds its wings back to enter the water, catches a fish in its pointed beak, and pushes its wings down to resurface. The kingfisher can see underwater better than other birds because a clear membrane covers its eyes and protects them. The Eurasian kingfisher eats fish, but most of its relatives catch insects.

FIND OUT MORE ▶▶ Courtship **306–307** • Defence **320–321** • Growing Up **310–311** • Reproduction **308–309**

MAMMALS

All mammals are endothermic (warm-blooded), have some fur or hair on their body, and feed their young milk. They have a bony skeleton with a backbone, and their lower jaw, made of one bone, hinges directly onto the skull. Mammals breathe using lungs. A few mammals lay eggs, and some carry their young in pouches, but most have a placenta and give birth to live young. Mammals are found all over the world, on land, in the air, and in water.

Brown fur is thick and dense

Powerful shoulder muscles used when digging

Hind legs on which the bear stands upright if threatened

Strong claws, used for digging, tearing food, and climbing

CLASS: MAMMALIA

There are about 5,000 species of mammal in a total of 27 orders, of which the following are a selection.

	Class: Monotremata (duck-billed platypus, echidna) **Features:** lay eggs, short legs, small head, tiny eyes
	Class: Diprotodonta (pouched mammals) **Features:** young born at early stage and develop in pouch
	Class: Perissodactyla (odd-toed, hoofed mammals) **Features:** leg's weight on central toe
	Class: Carnassials (sharp cheek teeth) **Features:** carnassial teeth for cutting flesh
	Class: Cetartiodactyla (whales, dolphins, deer, sheep) **Features:** new grouping based on genetic and fossil evidence
	Class: Primates (lemurs, apes, monkeys, humans) **Features:** large brain, forward-facing eyes
	Class: Rodentia (rodents) **Features:** incisor teeth grow continuously, most have good sense of smell and hearing

DIVING PLATYPUS ►
The duck-billed platypus closes its eyes, ears, and nose when diving and finds its way using sense receptors around its bill. The platypus lays eggs. It does not have nipples, so when the young hatch, they suck milk from the fur around the openings of the milk glands. It lives by rivers in Australia and Tasmania.

▲ BROWN BEAR
The brown bear is an omnivore, eating plants and animals. It walks on all fours, with its heel on the ground. It is a placental mammal, which means that the young are able to develop and grow inside the female's body. The cubs look like tiny adults when born, but are helpless and stay with their mother for at least two years.

Clawed thumb

mammals

▲ BAT
Bats are the only mammals that fly. A bird's wing is made up of the whole of the forelimb, but in bats the flight membrane stretches between its very long fingers. Most bats feed at night and rest, often in large groups, during the day.

Large wing membrane for agile flight

◄ LEAPING DOLPHIN
Dolphins, like whales, spend their entire life in the water, but must still surface to breathe air through their lungs. Their fat reserves, called blubber, keep them warm in cold seas.

FIND OUT MORE ►► Growing Up **310–311** • Reproduction **308–309**

LIFE CYCLES

The life cycle of an animal consists of all the stages from the start of one generation to the beginning of the next. For many insects, it takes only a few weeks for the young to become adults and reproduce themselves, but for larger animals it can take years. Some animals reproduce once and die, but many reproduce repeatedly during their adult life. A number of animals undergo a transformation, known as **METAMORPHOSIS**, as the young animal changes, gradually or directly, into the adult form.

PARASITIC FLUKE ▶
The schistosome fluke is a parasite – it lives in another animal, known as the host. This fluke uses suckers to anchor itself in human veins and feeds on blood cells. Flukes have a complicated life cycle, with a number of larval stages that live in different hosts. The larvae often live in molluscs, but the adult stage usually lives in a vertebrate animal, often causing serious diseases.

A LONG LIFE ▲
An African elephant's gestation – the time it takes for the baby to grow in the womb before it is born – is 22 months, the longest of any mammal. When the baby elephant is born, all the herd take good care of it. Adult elephants have no natural enemies and can live to be 60 years old.

METAMORPHOSIS

Metamorphosis involves a radical change from the young animal to its adult form. The young, known as larvae, live in a different way to the adults. Incomplete metamorphosis, seen in the transformation of a tadpole to a frog, involves a number of gradual changes. Complete metamorphosis, seen in the change from a caterpillar to a butterfly, takes place inside a pupa and totally rearranges the body parts.

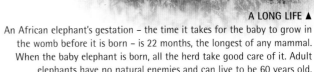
life cycles

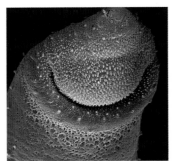

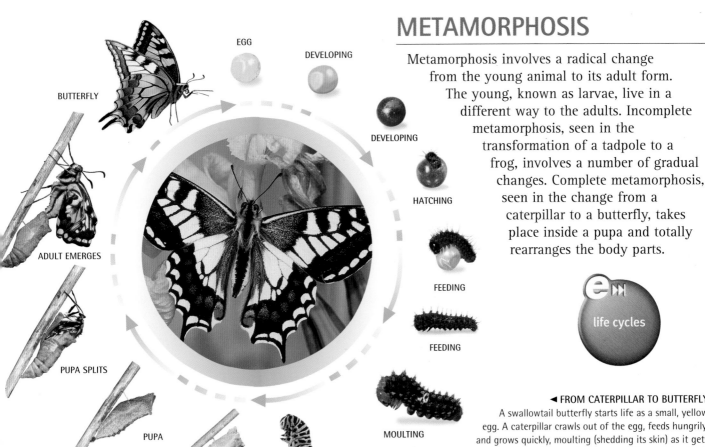

BUTTERFLY

EGG

DEVELOPING

DEVELOPING

HATCHING

FEEDING

FEEDING

MOULTING

CATERPILLAR

PUPA FORMS

PUPA

PUPA SPLITS

ADULT EMERGES

◀ FROM CATERPILLAR TO BUTTERFLY
A swallowtail butterfly starts life as a small, yellow egg. A caterpillar crawls out of the egg, feeds hungrily, and grows quickly, moulting (shedding its skin) as it gets bigger. After about four weeks, the caterpillar attaches itself with a silk thread to a plant stem. Its wriggles out of its skin, revealing a soft pupa, which hardens. A few weeks later, the pupa case splits and the adult butterfly climbs out.

FIND OUT MORE ▶▶ Courtship 306–307 • Growing Up 310–311 • Reproduction 308–309

COURTSHIP

Some animals perform complex rituals to attract a mate. These displays, performed during the breeding season, are known as courtship. Usually it is the males that perform. They may court one female or several in turn. Sometimes groups of males perform at a particular spot, called a lek, with females visiting to select a mate. Some animals have only one partner throughout their life. They do not need to perform a display, but they do need to keep a strong bond with their partners.

COLOURFUL PLUMAGE ▶
In some species, males and females look very different. Sometimes the difference is only a matter of size, but during the breeding season other differences may appear. In birds such as peacocks, the males develop elaborate tail feathers, which they fan out and quiver to attract females.

Male's tail feathers are displayed in a spectacular fan

MALE PEACOCK

FEMALE PEAHEN

▲ MOCK DUEL
In some birds, the males and females both perform a series of courtship rituals. Great blue herons raise their necks and feathers and duel with each other, shaking twigs and calling out to one another. The feathers of both sexes change to a similar colour during the breeding season, though the males' are usually more brightly coloured.

BONDING THROUGH GROOMING ▲
Golden lion tamarins mate for life so they do not need to waste energy on courtship displays. They do, however, spend time bonding with their partners by grooming (cleaning) one another. These tamarins live in family groups of about four to eight members. The males help bring up the young, and older siblings also help out so they can learn about parenting.

▲ MATING CALLS
Male frogs and toads call out to attract females to their breeding pond or stream. Each species has its own call, which helps a female to find a mate of the same species in the breeding pool if it is used by several species at once. Many species, such as the Brazilian torrent frog, have expanding vocal pouches which make their calls loud and clear. These frogs also kick their legs out during courtship displays.

courtship

MIRROR DANCING ▲
In some courtship displays, males and females copy each other's movements as if dancing. Here two butterfly fish swim alongside each other through coral, showing off their colourful bodies to one another. As well as being part of a bonding ritual, this dance allows the fish to confirm each other's identity so that they do not try to mate with the wrong species.

ANOLE LIZARD ►
Like birds, many male lizards become more brightly coloured during the breeding season, despite the fact they may be more easily seen by predators. However, the male anole lizards are different. They have permanent colourful dewlaps under their throats that remain hidden unless they are being used to attract females.

▲ RUTTING RED DEER
In autumn, stags (male deer) start to gather harems (groups) of females to mate with. They fiercely defend these harems from rival males. Usually the larger males with the bigger antlers have their pick of the females. Males of the same size battle to decide which of them will remain with the females and which males must retreat.

▲ LEAVING A SCENT
Male orchid bees attract females by marking a spot, or lek, with their particular scent. The females that are attracted by the smell fly to the lek, and mate with a male. Birds that attract females by singing or displays may also use leks. Some hoofed mammals use leks when they mark their territory with urine and faeces.

FIND OUT MORE ►► Communication 318–319 • Reproduction 308–309 • Senses 316–317

REPRODUCTION

Animals reproduce in one of two ways. In asexual reproduction, animals produce young, which are identical to themselves, without mating with another animal. Most creatures that reproduce in this way do not live very long but can reproduce in large numbers rapidly. In sexual reproduction, a female animal's egg unites with a male's sperm cell after mating, in a process known as **FERTILIZATION**. The offspring inherit features, called traits, from both parents. These animals tend to develop more slowly and many have parental care after birth.

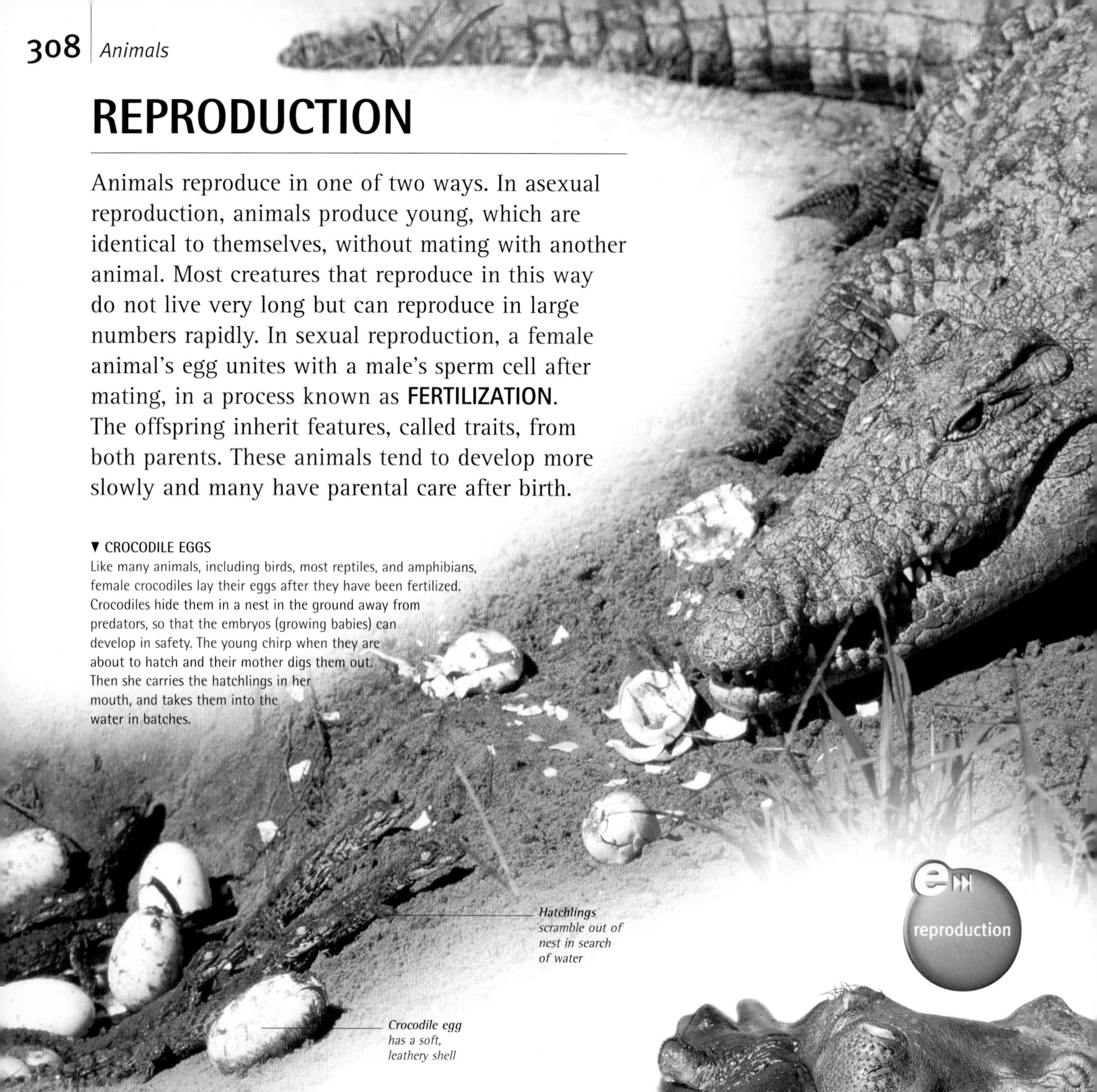

▼ CROCODILE EGGS

Like many animals, including birds, most reptiles, and amphibians, female crocodiles lay their eggs after they have been fertilized. Crocodiles hide them in a nest in the ground away from predators, so that the embryos (growing babies) can develop in safety. The young chirp when they are about to hatch and their mother digs them out. Then she carries the hatchlings in her mouth, and takes them into the water in batches.

Hatchlings scramble out of nest in search of water

Crocodile egg has a soft, leathery shell

reproduction

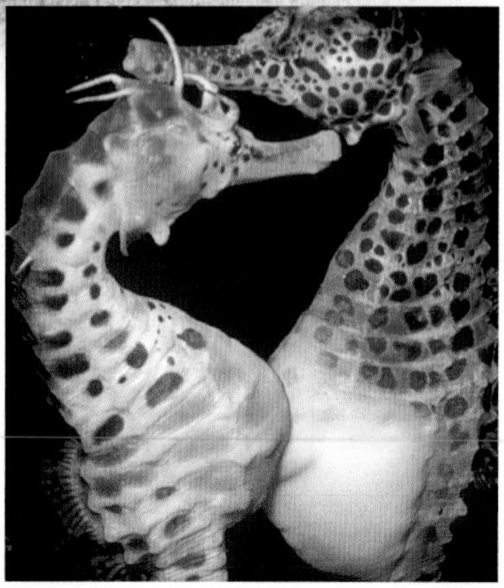

◄ ROLE REVERSAL

In many animals, the male contribution to reproduction ends straight after he has fertilized the female's eggs. However, with sea horses it is the male who looks after the eggs. The female deposits them in a special pouch on the male's abdomen, where they are fertilized. He then carries the eggs in his pouch until they hatch 2–6 weeks later.

FERTILIZATION

During fertilization, a male sex cell (sperm) and a female sex cell (ovum) unite to produce a cell that will grow into a new animal. Sex cells have half the number of chromosomes (chemicals which tell the cell how to grow into another individual) than other body cells. When the sex cells unite, the full amount of chromosomes is restored.

Male's sperm cells travel through a tube into the female's body

Fertilization occurs in the female's oviducts (tubes) where her eggs are

Small insect wrapped in strands of silk by the male

INTERNAL FERTILIZATION ▲
Many animals reproduce through internal fertilization. A male and female pair up and mate, and the female's eggs are fertilized inside her body. Empid flies mate in this way. The male empid fly is smaller than the female and risks being eaten during mating. To protect himself, he presents the female with a small insect to distract her.

Female hippopotamus suckles her calf for about 8 months

Single bud is several days old

New bud is forming

Adult hydra from which new buds grow

ASEXUAL BUDDING ▶
Simple animals, such as hydra, reproduce asexually by forming new growths from their bodies. This is called budding. Each bud eventually breaks off and forms another animal. Other animals, such as sponges, can reproduce by breaking off parts of their bodies. This is called fragmentation.

◀ LIVE YOUNG
Most mammals, some reptiles and amphibians, and several invertebrates give birth to live young. Once the egg is fertilized inside the female it stays there while it develops. This is known as the gestation period. In hippopotamuses, gestation lasts for about 240 days. During this time, the developing calf is protected in a constant environment and obtains nourishment directly from its mother's body.

▲ EGG RELEASE IN EXTERNAL FERTILIZATION
Fertilization outside a female's body is a random process. Some eggs are not fertilized, and sex cells and eggs can be easily eaten by predators. Corals simply release eggs and sperm into the water. To increase the chances of fertilization occurring, corals of the same species all spawn at the same time. That way a predator could not possibly eat all the sex cells and eggs.

FIND OUT MORE ▸▸ Growing Up 310–311 • Life Cycles 305 • Reproductive Systems 362–363

◄ POUCH BABY
The babies of pouched mammals are born very small and undeveloped and then grow in their mother's pouch. They gain all the nourishment they need from their mother's milk. A baby red kangaroo does not leave the pouch for 190 days and will stay with its mother for at least a year afterwards.

◄ CICHLID FISH
In many animals, parental care simply means protection. The young are safe as long as they remain close to their guardian. Female cichlid fish have a pouch in their throats to carry their eggs. When the eggs hatch, the young fish remain in the mother's mouth until they can fend for themselves.

◄ WOODLICE EXOSKELETON
Animals with exoskeletons can only grow by moulting (shedding) their outer skin. Woodlice are unusual because, unlike other crustaceans, they shed half their shell at a time. The exoskeleton breaks in the middle and the back half is shed first. A few days later the front half breaks off. During moulting, the lice are at risk and often hide away from predators.

GROWING UP

As animals grow they can change in both their form and their behaviour. Some animals change suddenly and drastically – for instance, caterpillars metamorphose into butterflies in a matter of weeks. Most animals develop gradually. Some young animals are cared for by parents. This allows them to learn about life from an experienced adult. Others are left to fend for themselves and have to rely on their instinct.

Calves are able to stand up a few minutes after their birth

WILDEBEEST ▲
All female wildebeest have calves over the same two weeks to reduce the number of opportunities that predators have to attack their young. The calves can stand and run 20 minutes after birth – an essential adaptation for their survival. They also follow the first moving thing that they see – usually their mother – in a process called imprinting. It is a form of learning that ensures the calves stick close by their mothers as they move around and graze.

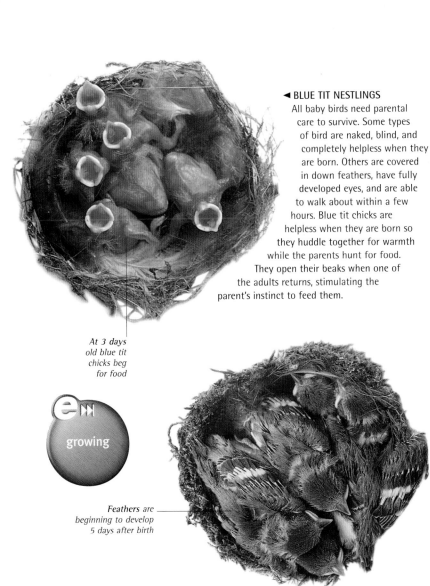

◄ BLUE TIT NESTLINGS
All baby birds need parental care to survive. Some types of bird are naked, blind, and completely helpless when they are born. Others are covered in down feathers, have fully developed eyes, and are able to walk about within a few hours. Blue tit chicks are helpless when they are born so they huddle together for warmth while the parents hunt for food. They open their beaks when one of the adults returns, stimulating the parent's instinct to feed them.

At 3 days old blue tit chicks beg for food

e ▶▶
growing

Feathers are beginning to develop 5 days after birth

LOGGERHEAD TURTLE HATCHLINGS ▲
Some babies are left to fend for themselves from the start. The female loggerhead turtle buries her eggs in the sand and then abandons them. When the young turtles hatch, they dig their way to the surface and instinctively head for the sea. Many of the hatchlings are eaten by birds and other predators as they make their way over the sand.

▼ ELEPHANTS
Elephants live in close-knit family groups. From the moment a calf is born, it has the benefit of its mother's guidance and is protected by all the females in the group. Elephants live up to 60 years and mature slowly. Calves stay with their mothers for several years to learn all they need to know to survive, such as where all the water holes are.

Female relatives, such as aunts, help raise a calf

FIND OUT MORE ▶▶ Defence 320–321 • Growth 366–367 • Life Cycles 305

FEEDING

Unlike plants, animals cannot make food from sunlight. Animals have to feed to produce the energy they need to grow, move, and reproduce. Some animals eat only plants, others eat meat, and some eat both. Most animals have a gut with a mouth at one end and an anus at the other. Food is broken down as it passes from one end of the gut to the other. The nutrients are absorbed into the body in a process called **DIGESTION**.

PREDATOR AND PREY ▶
Ospreys are birds of prey that feed entirely on fish. Like all predators, they have specially adapted features to help them catch prey. Ospreys have powerful wings and feet that enable them to swoop down and grab fish. Birds do not have teeth but a muscular organ in their digestive system called a gizzard for grinding up food. This organ also prevents harmful bones from passing into the birds' intestines.

Tentacles are fanned to capture food

Tube is made by the worm. The worm lives inside

Talons (claws) on powerful feet for grabbing prey

◀ FAN WORMS FILTER FEEDING
Many aquatic animals, including whales and basking sharks, take in food by filter feeding. Fan worms that live on seabeds use the tentacles around their mouths to filter food. As they draw water towards them with their tentacles, food particles are trapped in their tentacles with mucus. The food is then passed into their mouths.

Lure contains luminous bacteria to attract prey

◀ ANGLER FISH
Some deep-sea predators, such as angler fish, use lures to catch their prey. In deep water, food is often scarce, so attracting prey rather than chasing it saves energy. The angler fish's lure glows in the dark. Once the prey is lured near enough, the enormous jaws of the fish snap shut.

DIGESTION

During digestion, animals break down food into pieces small enough to be absorbed. The process is made quicker by chemicals in the gut called enzymes. In some invertebrate animals, digestion begins outside the body, in others food is taken in through the mouth and then digested. In mammals, food is sometimes chewed before swallowing.

feeding

OMNIVORES ▼
Animals that eat both plants and animals are called omnivores. They are highly adaptable and use whatever food they can get hold of. Racoons, with their skilful hands, are particularly successful at finding food. Their diet ranges from fish and young birds to shoots and berries. Some omnivores are at home in urban areas, surviving on garden produce and waste.

SPIDER CATCHING FLY ►
Many spiders trap their prey in webs, paralysing any animal that becomes stuck with a venomous bite. Spiders pour digestive enzymes into the prey and then suck in the resulting liquid. They can live for weeks without eating because they are able to store nutrients.

◄ TIGER'S TEETH
The teeth of a tiger are highly specialized for gripping and eating meat. The long, pointed canine teeth are used for biting into an animal's neck. The carnassial (cheek) teeth pull flesh off bones or slice up meat with a scissor-like action.

Carnassial tooth for slicing

Large canines for biting into flesh

HERBIVORE RHINO ►
Plant-eating animals are known as herbivores. Plants may be easy to find but are sometimes low in nutrients. To survive, herbivores, like this rhinoceros, have to spend much of the day eating. They also have special digestive systems that get the maximum nutrition from food.

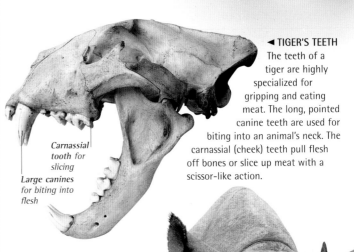

Abdomen is swollen, filled with honey and nectar

▼ HONEYPOT ANT
Animals have found remarkable ways to hoard food. Honeypot ants create larders by selecting newly hatched worker ants to act as storage vessels. They are fed a mixture of honey and nectar, causing their abdomens to swell. In the dry season, when there is less food about, the other ants entice them to regurgitate droplets of food so they can all drink.

FIND OUT MORE ▸▸ Behaviour Cycles **322–323** • Digestive System **358–359** • Ecology **326–327**

MOVEMENT

All animals are mobile for at least some part of their lives because they need to find food. Most movement is controlled by a nervous system that causes **MUSCLES** to contract and relax in a co-ordinated way. The **SKELETON** provides anchorage for these muscles. To move efficiently through water, land, and air, animals have special adaptations, such as fins, legs, and wings.

▲ RUNNING GAZELLE

Ungulates (animals with hooves) are hunted by many predators. Gazelles use their speed and endurance to escape capture. Their lower legs are very long, which lengthens their stride. They also have two toes instead of five, which needs less muscle and so saves energy.

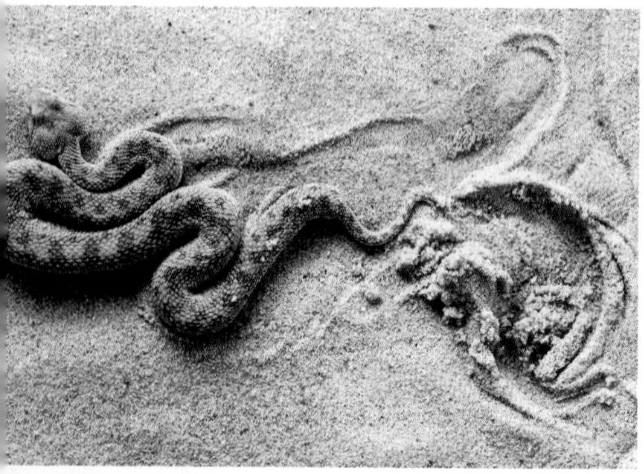

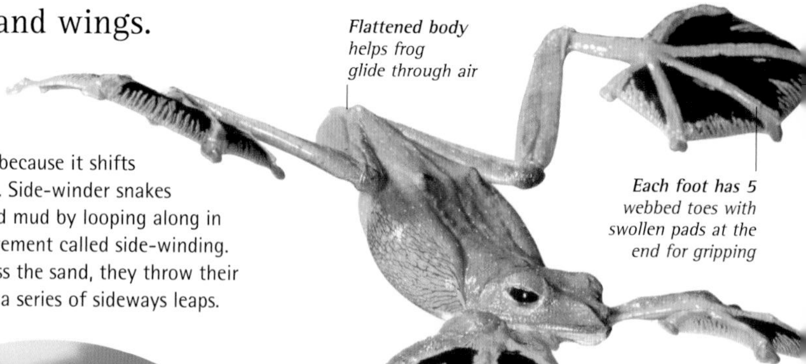

Flattened body helps frog glide through air

◄ SIDE-WINDER SNAKE

Sand is not easy to cross because it shifts under an animal's weight. Side-winder snakes move across soft sand and mud by looping along in S-shaped curves in a movement called side-winding. Instead of slithering across the sand, they throw their bodies through the air in a series of sideways leaps.

Each foot has 5 webbed toes with swollen pads at the end for gripping

▲ GLIDER

Several tree-living animals glide from tree to tree using flaps of skin like parachutes. Flying frogs have large, webbed feet that they hold out when they leap, so they can glide farther between the trees.

Wings beat 100–400 times per second

◄ JET PROPULSION

Although fish are strong swimmers, many other marine animals drift along at the mercy of the ocean currents. Jellyfish are able to control their movement to a limited extent. They have a ring of muscle around the edge of their bell-shaped body, which can be contracted and relaxed, like an umbrella opening and closing. This pushes the water backwards, making the jellyfish move in the opposite direction.

◄ FLYING

Insects are the smallest animals capable of powered flight. Four-winged insects, such as butterflies, use muscles directly attached to the base of their wings to move the wings up and down. Bees fly by using muscles attached to the top and bottom of their body. When the muscles contract, the wings move upward; when they relax, the wings drop down.

movement

MUSCLES

Muscles are bundles of fibres that provide the power for animals to move. When a nerve stimulates a muscle into action, the muscles contract (pull back), causing movement. In simple animals, such as snails, muscles contract in waves from one end of the body to the other, pushing the animal along. In vertebrates, such as the horse, muscles work in pairs and pull against bones. The area where different bones meet is called a joint.

Resilin pad in the hind leg expands just before the flea jumps

◄ HIGH JUMP
Fleas need to jump around in order to find an animal from which to suck blood. They can leap an amazing 33 cm (13 in), using muscle energy that is stored in a pad of springy material, called resilin, in their legs. When the leg muscles are triggered to jump, the flea is catapulted into the air.

SKELETON

Many animals have a rigid skeleton to support their bodies and some have jointed legs, which allow them to move rapidly. Mammals have the most complex skeletons of all animals. They have backbones made up of many small bones called vertebrae and limbs with several joint types. This complicated skeleton enables them to make lots of different movements.

Muscles lie inside the hollow leg segments

EXOSKELETON MOVEMENT ▲
Animals with exoskeletons, such as crabs, have several pairs of jointed legs. Each pair is made up of a series of hollow sections joined together at joints. Pairs of muscles joined to the inner surface of the joint allow the crab to scuttle sideways quickly.

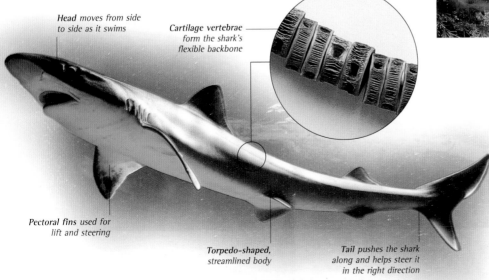

Head moves from side to side as it swims

Cartilage vertebrae form the shark's flexible backbone

Pectoral fins used for lift and steering

Torpedo-shaped, streamlined body

Tail pushes the shark along and helps steer it in the right direction

◄ STREAMLINED SWIMMER
Sharks bodies are specialized for moving fast through water. They have skeletons made from a firm elastic substance called cartilage. Cartilage is lighter than bone, enabling sharks to swim efficiently. Using rhythmic contractions of their body muscle, and with additional push from their tail, they reach speeds of 30–50 kph (19–31 mph).

SENSES

To find food, mate, avoid danger, and communicate, animals rely on information gathered by their senses. Information is processed by their nervous system, which tells their bodies how to respond to stimulation from their surroundings. Some animals have groups of sensory cells that do little more than register light. More advanced animals use a combination of **VISION, HEARING, SMELL,** and touch.

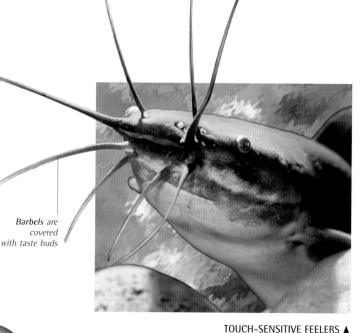

Barbels are covered with taste buds

▼ ON THE ALERT

Tigers have remarkable senses. Their acute sense of hearing is used to locate prey in dense undergrowth. Binocular vision (vision with two forward-facing eyes) gives them the ability to judge distance accurately, and a reflecting membrane at the back of their eyes helps them to see well in the dark. Their whiskers act as touch detectors, which help them find their way when hunting at night. Tigers also have an excellent sense of smell.

senses

TOUCH-SENSITIVE FEELERS ▲

Catfish are so called because of their whisker-like, fleshy barbels. They use them to feel their way around in murky river water to find food. Most other animals have touch-sensitive receptors all over their bodies. Antelopes, for example, twitch when even the tiniest of insects land on them.

External ear part is funnel-shaped to gather sound

Whiskers are sensitive to air movement

Smell is used to sense territorial boundaries and mates

VISION

Animal eyes may look very different from one another but they all respond to light. Earthworms have light sensitive receptors, which help them move away from light and stay underground. Other animals have eyes with complex structures, which allow the animals to focus and see distinct images. The position of the eyes is also important. Being able to see all around is vital if you are a prey animal such as a rabbit, whereas being able to judge distances is important for hunters and tree dwellers.

Eyes have all-round vision

COMPOUND EYES ►
Insects have large compound eyes made up of many lenses. Each lens sees an individual image. The brain puts together this information to make a complete mosaic image. These eyes are very sensitive to movement but have poor focus.

Light-sensitive cells are arranged in a circular pattern

HEARING

Hearing is the ability to detect sound waves. It is important for communication, finding a mate, and hunting for prey. The main organ concerned with hearing is the ear. An important part of the ear is a tightly stretched membrane called an eardrum, which vibrates when it picks up sound. Some animals interpret vibrations through other parts of their bodies. Snakes are able to sense sound vibrations.

GROUND VIBRATIONS ►
Many lizards, such as this green iguana have an eardrum just behind the eye that picks up sounds in the air. They also have a special bone in the jaw that picks up sound vibrations from the ground. This is also how snakes and other reptiles that lack an eardrum are able to "hear".

Eardrum used for detecting airborne vibrations

EXTERNAL EARS ►
Mammals have external ear parts called pinnae. They draw in sound waves, focusing them on the eardrum inside. The fennec fox is the smallest of the dog family, but has the largest ears because it needs to hear insects, which it feeds on at night.

SMELL

Smell is one of the two chemical senses, the other being taste. Humans have a relatively poor sense of smell, but it is a vital means of communication to many creatures. Scent can be used to mark territory, and to attract a mate. It also allows animals to track and find food.

Antennae of beetle are long for detecting pheromones

◄ A NOSE FOR FOOD
A good sense of smell is necessary for many animals to find food. Fruit bats navigate by sight but locate ripening fruit, such as mangoes, by their scent. Many carnivores track their prey by smell. Wolves can tell their prey's sex, age, and state of health from its smell.

◄ SMELL DETECTORS
Some insects, including ants, beetles, and moths, smell using antennae. Females release chemicals called pheromones to attract a mate. Males can home in on the female from distances of 8 km (5 miles) or more. In an emergency, some ants can produce an alarm pheromone to get other ants to assist them.

FIND OUT MORE ►► Communication 318–319 • Defence 320–321 • Feeding 312–313

Howling sound can travel approximately 104 sq km (40 sq m)

◄ HOWLING WOLVES
Wolves live in social groups called packs, consisting of a dominant male and female and their offspring. They communicate using body language, sounds, and scent. They use their ears, tails, and facial expressions to convey dominance and submission depending on their position in the pack. Wolves whine to greet one another, and howl to let others know they are there.

COMMUNICATION

Sounds, signals, and gestures make up an animal's **LANGUAGE**, and are essential to survival. The method of communication often depends on how close together the animals are. Sound is effective over long distances and in the dark, whereas body language and light are visual signals that are generally used at close quarters. Smell is used to communicate breeding times and to indicate territorial boundaries. Animals usually communicate with members of their own species using a code that only they can understand.

JANE GOODALL
British, 1934–

Jane Goodall has been studying the behaviour of chimpanzees for over 40 years. She was the first person to record that chimps make and use tools, a skill previously attributed only to humans. Her study methods – monitoring a family of chimpanzees in their natural environment – revolutionized research on ape behaviour.

communication

▲ FEAR
When they are afraid, chimps open their lips but keep their teeth together, rather like a forced smile. Chimps use this expression when they are approached by a chimp of higher rank.

▲ SUBMISSION
When chimps pout with their mouths slightly open, they are indicating submission to a higher-ranking chimp, possibly after a dispute of some kind. They may whimper at the same time.

▲ EXCITEMENT
An open mouth suggests excitement. Young chimps use this face when they are playing. It is accompanied by grunts and screams. The more excited the chimps, the louder the grunts are.

BIG FIN REEF SQUID ▲
Many animals in the deep ocean, ranging from squid to plankton, produce shimmering light to communicate. This light is referred to as bioluminescence. Some animals produce it themselves in organs called photophores. Others have sacs of bacteria in their skin that produce light. Animals use light to find mates and food, for defence, and camouflage.

BIOLUMINESCENCE ON PLANKTON

LANGUAGE

Unlike human language, animal language is not dominated by vocal signals. They use combinations of behaviour to talk to one another. Some animals communicate using high- and low- frequency sounds, which humans cannot hear, while others communicate using light that is invisible to people. Some animals use smell to communicate with one another.

Tiny pegs on hind legs rub against wings to produce chirping sound

▲ CHIRPING GRASSHOPPER
Many male insects produce sound by rubbing together certain hard parts of their bodies. Grasshoppers and crickets produce chirping sounds called stridulation to attract females. Some grasshoppers rub their hind legs across their forewings. Crickets rub the top part of their hind legs against their abdomen.

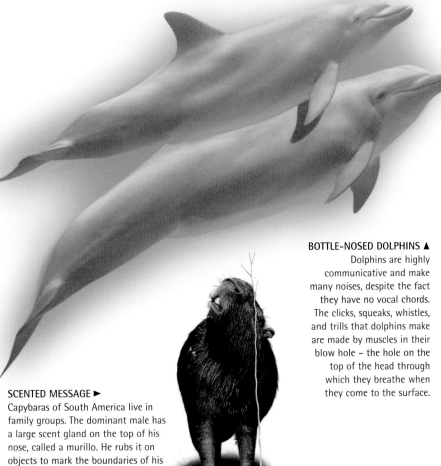

BOTTLE-NOSED DOLPHINS ▲
Dolphins are highly communicative and make many noises, despite the fact they have no vocal chords. The clicks, squeaks, whistles, and trills that dolphins make are made by muscles in their blow hole – the hole on the top of the head through which they breathe when they come to the surface.

SCENTED MESSAGE ▶
Capybaras of South America live in family groups. The dominant male has a large scent gland on the top of his nose, called a murillo. He rubs it on objects to mark the boundaries of his territory and to warn off intruders. The scent message remains until he returns to remark the spot. Many animals communicate through scent.

LOW-FREQUENCY SOUNDS ▲
Elephants produce many sounds, including some that humans cannot hear. These low-frequency rumbles travel over long distances through the air and under the ground. Elephants detect the vibrations with their feet and the tip of their trunks. These sounds may explain how lone male elephants find females and how family members communicate when they are a long way from each other.

FIND OUT MORE ▶▶ Senses **316–317** • Populations **324**

DEFENCE

Animals have evolved many ways of protecting themselves from **PREDATORS**. Most use their keen senses to detect an attack and make a dash for safety. Some animals are able to camouflage themselves. Others have sharp spines, produce poisons, or make themselves look bigger than they actually are. A few animals lose limbs or tails deliberately as a means of escape, only to regenerate them later. Nevertheless, predators have also evolved – and are better at catching prey.

▲ MIMIC BUTTERFLY
Many butterflies have markings known as false eyes on their wings. Those of the owl butterfly mimic the eyes of an owl and so frighten off small birds that might normally be eaten by owls.

POISON-DART FROG ►
Some amphibians defend themselves by producing poisons. Predators of poison-dart frogs, such as snakes and spiders, are not harmed by mild poisons. So these frogs have evolved highly poisonous skin secretions that are unpleasant or even lethal to their predators. These frogs have also developed colourful markings to warn predators that they are dangerous.

Skin is covered in a shiny, poisonous secretion

OUTWITTING THE ENEMY ►
Several animals, including porcupines and sea urchins, have spines for defence. The spiny puffer fish has spines but also another form of defence. When attacked, it inflates its body by swallowing large amounts of water. As its body swells, the spines stand up. The swollen fish sticks in the throat of any attacker and cannot be swallowed.

Fish gulps water and becomes too large to swallow

◄ CAMOUFLAGE
Many flatfish that rest on the sea floor, such as plaice, are able to change the colour of their skin to match the ground they are lying on. This makes them almost invisible to predators. Just under the surface of their skin, there are special cells that can become lighter or darker. Other animals that can change colour to match their surroundings include chameleons and squid.

TREE SKINK TRICK

Some lizards, such as the skink, can shed part of their tail if they are attacked from behind. The tail continues to move in the attacker's mouth, drawing attention away from the escaping skink. The skink is then able to grow another tail.

Tail has broken off at a special breaking point

New tail is beginning to grow from the stump

New tail is a different colour

PREDATORS

Animals that kill other animals for food are called predators. Predation is the main reason why a creature of one species attacks a member of another species. Animals being attacked may defend their young or family group, but in most cases they try and escape capture. Some creatures eat members of their own species – usually the eggs or younger members. Predators may actively hunt for food or lie in wait to catch it.

Venon-injecting fangs are at the front of the mouth

LETHAL WEAPONS ▶

Snakes have two methods for killing large prey. Venomous snakes bite their quarry, injecting venom that affects the quarry's nervous or circulatory systems. Constricting snakes grab their prey and suffocate it by throwing coils of their body around it. They strike with great speed but may take several hours to eat a large animal.

Gap in jaw expands to accommodate large prey

▼ PACK ATTACK

African hunting dogs take it in turns to head the chase so they can wear out the prey before they wear out themselves. When the prey eventually tires, they all move in together for the kill. Several types of mammal predators hunt in groups. This allows them to tackle prey that would be too big, dangerous, or tiring for them to tackle alone.

protection

FIND OUT MORE ▶▶ Evolution **328–329** • Feeding **312–313** • Plant Survival **274–275**

BEHAVIOUR CYCLES

Animals have many instinctive behaviours that are linked to climate. Seasonal changes trigger journeys known as migrations. Populations of animals move from one area to another and back again. They may cover thousands of miles to find food or to breed. Other animals stay put during harsh weather, entering a phase of **DORMANCY**. Some creatures are active at night and others during the day.

▼ MIGRATING REINDEER
In summer, reindeer eat and put on as much weight as possible to see them through the harsh winter. They move about continually, grazing in one area before moving on to a fresh patch. With the first heavy fall of snow in autumn, they migrate southward to their winter feeding grounds, where the weather is milder and food more easily accessible. In spring, the females lead the way back to northern lands.

Dense woolly undercoat for protection from cold

Bodies are strong and built for endurance

Large hooved feet prevent reindeer from sinking into the snow

▲ LONG JOURNEY HOME
Salmon live most of their lives at sea, but return to the streams where they were born to spawn (produce eggs). How they find these streams is uncertain, but it is thought they use their sense of smell to navigate. The effort of swimming upstream and spawning exhausts the fish and few survive to make the journey back.

Well-developed ears for excellent hearing

Large eyes to see insects in the dark

Woolly coat protects against cold nights

NOCTURNAL GALAGO ▶
Nocturnal creatures, such as this galago, which lives in the forests of Senegal, Africa, are adapted to living in the dark. During the day this small animal would be vulnerable to attack from predators, and would have to compete with many other insect-eaters. By sleeping away the daylight hours and hunting at night, it avoids these problems.

LONG-DISTANCE NAVIGATION ▶
As summer moves into autumn, the days shorten, and birds such as snow geese gather in large numbers before migrating to warmer climates. Some birds are able to fly vast distances non-stop to their destinations. Birds navigate using the Sun and stars, and familiar landmarks, such as rivers and coastlines.

◀ **MONARCH MARATHON**
Each year, up to 100 million monarch butterflies migrate south from all over the USA to the relative warmth of California or Mexico. Falling temperatures in late summer trigger the migration as they cannot fly at temperatures below 12.8°C (55°F). At their winter roosts, they cluster together for warmth, each butterfly hanging its wings over the one below.

body clocks

Snow geese migrate a distance of 4,800 km (3,000 miles)

DORMANCY

Hibernation is a form of dormancy (inactivity) in which animals build up body reserves to sustain them through winter. They enter a deep sleep and their body systems slow right down. In torpor, another form of dormancy, animals become sluggish, but are easily aroused. In aestivation, animals retreat into sand or mud burrows during dry seasons, until rain returns.

HIBERNATION ▲
Dormice lay down fat so they can survive their long period of hibernation – up to 7 months for European species. Before winter they eat huge quantities of seeds and berries. At other times of the year they eat insects as well as berries. The change in their diet makes them sleepy and ready to hibernate. During the end of their hibernation, they rouse occasionally and begin to come out of their deep sleep. As soon as dormice wake up they mate.

▲ **SPADE-FOOT TOAD**
Spade-foot toads survive periods of drought by digging a burrow deep in the ground. They can bury themselves over a metre down before shedding several layers of skin to make a covering that stops them from drying out. They remain safely cocooned until it rains again. The increase in moisture causes the protective covering to break open and the toads dig their way back to the surface.

FIND OUT MORE ▶▶ Climate **236–237** • Feeding **312–313** • Populations **324**

POPULATIONS

Animals of the same animal species, living in the same area, and interbreeding with one another, are referred to as a population. The size of a population and the area it occupies may vary over time due to disease and competition from other animals. The term population includes animals that live on their own, sometimes roaming over a wide area, animals that form a family group, or a larger group, such as a colony. Some groups are temporary and only form during the breeding season.

populations

▲ UNITED PROTECTION
Meerkats are mongooses that live in groups called troops. Each troop contains several family units and occupies a territory, which they defend together from neighbouring meerkats and predators. One or two individuals act as sentinels. They climb onto the nearest high place and look out for danger.

◄ COLONIAL ORGANIZATION
Termites live in colonies of over a million members, in large mounds. The members are all descended from a single queen and her mate and they only survive if they work as a team. A queen termite lays some 30,000 eggs each day. She is dependent on the worker termites to look after her and her eggs. Other termites, the soldiers, defend the mound from attack, using their powerful jaws to bite any intruders.

Chambers in the middle of the nest house the termites

▲ FAMILY MEMBERS
Cheetahs have complex societies. Males may roam from place to place, or claim their own territory. They live on their own or in pairs. These pairs, which are often brothers, live and hunt together throughout their lifetime. They sometimes claim territory that overlaps that of females, marking the boundaries with their urine on trees. Male and female cheetahs mix only when they mate.

Mound-building is carried out by worker ants

FIND OUT MORE ▸▸ Defence 320–321 • Reproduction 308–309

COMMUNITIES

A community consists of a number of different populations that interact with one another. Because environments vary considerably in size and complexity, so too do the communities that occupy them. Within any community, different animals interact with one another. Some of these interactions benefit both partners, others benefit only one. However, intense competition between species has no advantage to either animal.

▼ LIVING OFF LEFTOVERS

This humphead wrasse fish is swimming with a remora fish just below it. The remora has a disk on the top of its head so it can attach itself to large fish. By hitching a ride, the remora saves energy, is protected from attack, and can pick up any food scraps left by its carrier. The wrasse fish neither gains nor loses from the relationship.

TWO-WAY BENEFIT ▲

A relationship between two species from which both benefit is known as mutualism. This oxpecker bird is feeding on the lice and ticks that infest an African gazelle. The bird has the benefit of a constant food supply. The gazelle gains from having insects removed that would otherwise suck its blood.

communities

ONE-WAY BENEFIT ►

Some relationships only benefit one of the animals, and are harmful to the other. Deer ticks are parasites that live by sucking the blood of deer. They also suck the blood of other creatures, including humans, and can pass on diseases to the animals they live off.

Lionfish and many other life forms inhabit a coral reef

MARINE COMMUNITY ▼

Coral reefs are formed by many corals living together. They provide a very rich environment for a multitude of animals. However, the corals, which form the basis of the reef, need more food than is available around them to survive. They obtain the additional food they need from microscopic algae, which live inside them. The algae need sunlight to manufacture their food so coral reefs are only found in clear, shallow water.

Nooks and crannies in the reef provide shelter for animals

FIND OUT MORE ▸▸ Ecology 326–327 • Habitats 246–247 • Parasitic Plants 270

ECOLOGY

Ecology examines the relationship between living things, and between living things and their environment. Animals adapt to an environment and take on a specific role, such as predator or prey. This role is known as their ecological niche. There may be herbivores that eat plants, carnivores that eat herbivores, and omnivores that eat both. This progression from plants to carnivores is called a **FOOD CHAIN**.

ECOSYSTEM ▶
Communities of animals and the environments with which they interact are called ecosystems. They include entire food chains. The grassland of the African savanna is a large ecosystem. There are plants, grazers such as zebra and wildebeest that eat the plants, and carnivores such as lion and leopard that prey on the grazers.

ECOLOGICAL NICHE ▲
The harpy eagle's niche is that of a predator in the forests of South America. It has special adaptations, such as short, broad wings so it can fly between trees. There are many niches in a particular environment, but all animals have to compete with other members of the community for resources, such as food. An animal may not be able to dominate a niche forever.

Seaweed is anchored to rocks to prevent it from being washed away

Limpet clings tightly to rocks, without moving, to conserve water when the tide is out

▲ HARSH HABITAT
A habitat is an area, such as a seashore or a woodland, that is home to certain types of animals. Some habitats support a wide variety of living things. Others have fewer niches and therefore support fewer species, though they may gather in large numbers. King penguins are one of the few animals that can survive in the harsh, cold conditions of the Antarctic region.

Common starfish rests in safe crevice

Velvet crabs move to the bottom of the pool when the tide goes out

Beadlet anemones pull their tentacles in if uncovered when the tide goes out

Snakelock anemone cannot pull in its tentacles so must stay submerged

▲ ROCK POOL
A rock pool has many niches for animals to colonize. Some parts of the rock pool are exposed at low tide, some face the waves, while others are more sheltered. Some also have drastic changes in temperature. The animals and seaweeds in the rock pool have to adapt to these conditions.

FOOD CHAIN

Animals obtain energy and nutrients by eating other living things. The flow of energy from one living thing to another is called a food chain.

Owl *may eat the weasel or the rabbit*

TERTIARY CONSUMER

SECONDARY CONSUMER

PRIMARY CONSUMER

PRODUCER

FOOD PYRAMID ▶

Plants get their energy directly from the sun and so are at the bottom of almost all food chains. Called producers, they provide energy to the herbivores that eat them. Herbivores are called primary consumers. They are eaten by carnivores (secondary consumers). The animals that eat the secondary consumers are called tertiary consumers. Many carnivores eat herbivores and smaller carnivores, so can be secondary consumers and tertiary consumers. There are fewer animals at the top of the food chain than at the bottom.

FOOD WEB ▶

Within animal communities, there are many food chains. Many animals, such as foxes, eat a variety of foods, so chains can be interconnected, creating a food web. Even when animals die, they become part of a food chain. They decay, releasing nutrients which become food for a living thing.

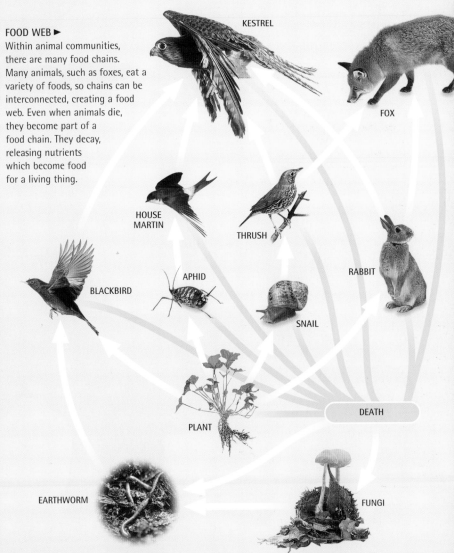

KESTREL

FOX

HOUSE MARTIN

THRUSH

RABBIT

BLACKBIRD

APHID

SNAIL

PLANT

DEATH

EARTHWORM

FUNGI

FIND OUT MORE ▶▶ Communities 325 • Feeding 312–313 • Habitats 246–247

SPECIALIZED FOR SURVIVAL

Although the forelimbs of mammals, birds, and reptiles are modified in different ways, the basic design is the same, suggesting they all descended from a common ancestor. The basic design includes one upper arm bone, two lower arm bones, and five fingers.

Elbow joint lets wing fold

First finger

Second finger elongated and fused

Bird A bird's wing has similar bones to a mammal forelimb but the numbers of fingers is reduced from five to two. The same basic pattern of bones is seen in all land vertebrates, including amphibians.

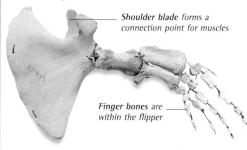

Shoulder blade forms a connection point for muscles

Finger bones are within the flipper

Dolphin This mammal's limb has evolved into a flipper, an adaptation to life in water. In life the individual fingers are hidden beneath flesh but in the skeleton they can be clearly seen.

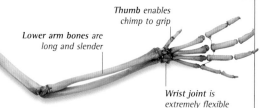

Thumb enables chimp to grip

Lower arm bones are long and slender

Wrist joint is extremely flexible

Chimpanzee This arm is modified for climbing and grasping. The thumb opposes the four fingers, allowing the animal to grip, and the elbow and wrist rotate providing additional dexterity.

EVOLUTION

The process by which changes occur in living things over time is known as evolution. The changes are passed from one generation to the next in genes. **NATURAL SELECTION** is one process by which evolution may occur. In nature, individuals with an **ADAPTATION** that helps them survive are more likely to reproduce. More of these individuals pass on their genes than their rivals, so the adaptation is more common in the next generation and builds up in the species.

e⏭ evolution

Bony plates provide armour for the head

◀ **DUNKLEOSTEUS**
Some animals become extinct as a result of evolution. They are replaced by other animals that are better able to survive. *Dunkleosteus* was an armoured fish with powerful jaws that lived about 350 million years ago. It may have become extinct as larger, faster sharks evolved, out-competing it for the fish they both hunted.

▼ **EVOLUTION OF THE ELEPHANT**
Today's elephants are the result of a long process of evolution. Over millions of years, small changes were passed from one generation to the next. The first fossil elephant species were small, but over time they increased both in size and weight. The three species alive today are the sole survivors of a once much more widespread group.

Asian elephant is one of the world's largest land animals

Phiomia was not much larger than a cow

Gomphotherium probably used its long lower teeth to dig up aquatic vegetation

Moeritherium lived in rivers and fed on water plants

Deinotherium existed until about 2 million years ago

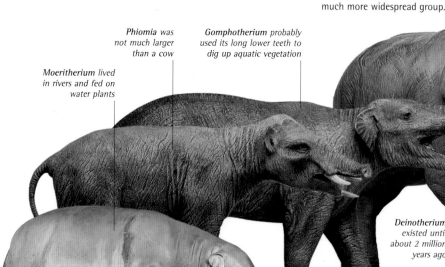

MOERITHERIUM   PHIOMIA   GOMPHOTHERIUM   DEINOTHERIUM   ASIAN ELEPHANT

NATURAL SELECTION

Not all offspring survive to become adults. Those with favourable variations, such as long, thick fur in a cold environment, are more likely to survive than those without. This effect of different characteristics on survival is what the scientist Charles Darwin called natural selection. Natural selection is a cause of evolution but it is not the only cause.

PEPPERED MOTHS ►
During the Industrial Revolution in the 1880s, pollution blackened trees in parts of England. Previously rare black peppered moths began to increase, as they were harder for birds to spot than their speckled counterparts. By 1900, most moths in industrial areas were black. Now, with pollution controlled, the black population has fallen again.

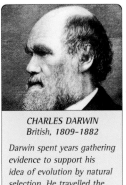

CHARLES DARWIN
British, 1809-1882

Darwin spent years gathering evidence to support his idea of evolution by natural selection. He travelled the world on expeditions aboard the ship, HMS Beagle. When he reached the Galapagos Islands in the Pacific, he was inspired by the number of unique species he found there.

The akiapola'au forages for insects, often under bark

The iiwi feeds on nectar from ohia flowers

The 'Apapane feeds on insects and ohia nectar

The Maui parrotbill tears back bark in search of beetles

The Nihoa finch uses its heavy bill to crush seeds

The original species, now extinct, probably ate insects and nectar

The Amakihi is a nectar-feeder, like the iiwi

◄ HONEY CREEPERS
Natural selection can create new species. These Hawaiian honeycreepers all evolved from a single ancestor, which arrived on the islands long ago. With no other birds for competition, the honeycreepers began to feed on different foods. Over many generations, their bills changed to cope with their new diets.

ADAPTATION

Adaptation is an outcome of natural selection. It is the gradual matching of an animal to its environment over time. It applies to everything about an animal from its anatomy and behaviour to its life cycle. It is important in evolutionary terms because the better adapted an animal is, the more likely it is to survive and produce offspring.

ECHIDNA ▲
The short-nosed echidna of Australia and Tasmania is well adapted to its diet of ants and termites. It has powerful claws to break into ant nests and termite mounds, and a long, sticky tongue to collect its prey. The short-nosed echidna also has spines to protect itself. It cannot roll up like a hedgehog – instead, when threatened, it digs quickly downwards to protect its soft underbelly.

◄ MARINE IGUANAS
The marine iguana lives on the Galapagos Islands and feeds exclusively on seaweed. It shows a number of adaptations to this lifestyle. Because it feeds underwater, the marine iguana is a good swimmer and has a long tail, flattened from side to side, to help propel it through the water. It also has special glandular structures in its nose to help it get rid of excess salt.

FIND OUT MORE ►► Extinction 334 • Genetics 364-365

PREHISTORIC LIFE

Since life began, more than 3.5 billion years ago, evolution has produced an enormous variety of living things. The earliest living things were simple, microscopic forms, such as bacteria. They evolved into increasingly complex creatures, eventually developing into the animals we know today. However, some animals, such as **DINOSAURS**, have become extinct.

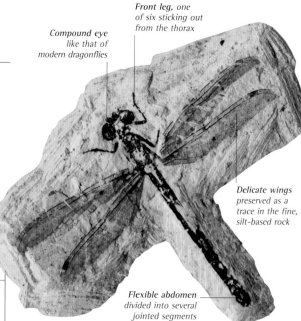

Compound eye like that of modern dragonflies

Front leg, one of six sticking out from the thorax

Delicate wings preserved as a trace in the fine, silt-based rock

Flexible abdomen divided into several jointed segments

FOSSIL DRAGONFLY ▲
This prehistoric dragonfly has been preserved as a fossil. Like many insects, dragonflies have barely changed since they first evolved. Insects are a remarkably successful group of animals that first appeared around 400 million years ago. About 320 million years ago, some insects evolved wings, making them the earliest animals to fly.

ERAS OF LIFE ON EARTH	
Proterozoic	More than 540 mya, dating back to when life began.
Palaeozoic	Beginning 540 mya, this era lasted for 290 million years. Fish appeared in the seas and rivers around 500 mya. Over time, some of these developed legs and lungs, giving rise to amphibians. Later, reptiles, such as *Dimetrodon*, evolved from these air-breathing amphibians. **DRAGONFLY** **CHEIROLEPIDID FISH** **DIMETRODON**
Mesozoic	This era lasted from 250 to 65 mya. It was dominated by dinosaurs, pterosaurs, and giant sea reptiles. Birds evolved about 150 mya. Placental mammals evolved from more primitive mammals at the end of the era. **ARCHELON** **STEGOSAURUS** **ELEPHANT SHREW**
Cenozoic	This era began 65 mya and continues today. The ancestors of most modern mammal groups appeared. Our own ancestors, the first upright hominids, evolved about 5 mya. **WOOLLY MAMMOTH** **AUTRALOPITHECUS HABILIS**

▼ ELASMOSAURUS
Between about 206 and 65 million years ago, long necked reptiles called plesiosaurs lived in the oceans. *Elasmosaurus* was one of the largest. As plesiosaurs have no living relatives, scientists can only guess how they lived and moved. The huge flippers may have worked like the oars of a boat, pulling the animal along, or they might have provided propulsion by moving up and down.

DINOSAURS

An extinct group of land-living reptiles, dinosaurs had erect legs set under their bodies rather than out to the side. All dinosaurs fall into one of two types, depending on the shape of their hip bones. Their closest living relatives are birds, which evolved from saurischian (lizard-hipped) dinosaurs.

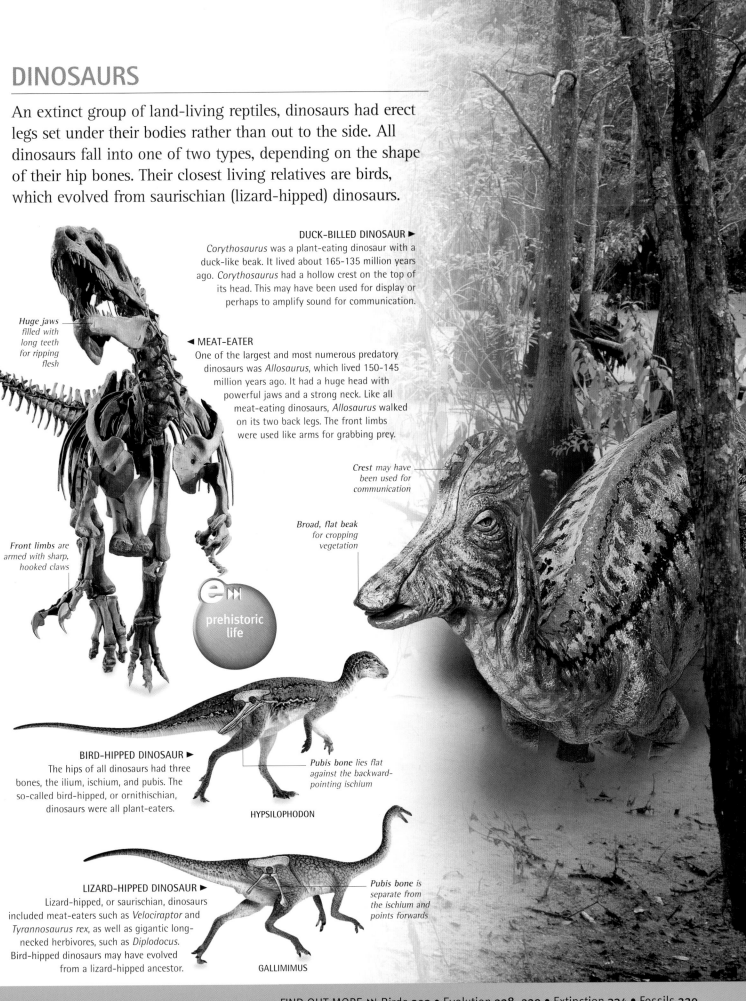

DUCK-BILLED DINOSAUR ▶
Corythosaurus was a plant-eating dinosaur with a duck-like beak. It lived about 165-135 million years ago. *Corythosaurus* had a hollow crest on the top of its head. This may have been used for display or perhaps to amplify sound for communication.

Huge jaws filled with long teeth for ripping flesh

◀ MEAT-EATER
One of the largest and most numerous predatory dinosaurs was *Allosaurus*, which lived 150-145 million years ago. It had a huge head with powerful jaws and a strong neck. Like all meat-eating dinosaurs, *Allosaurus* walked on its two back legs. The front limbs were used like arms for grabbing prey.

Crest may have been used for communication

Front limbs are armed with sharp, hooked claws

Broad, flat beak for cropping vegetation

prehistoric life

BIRD-HIPPED DINOSAUR ▶
The hips of all dinosaurs had three bones, the ilium, ischium, and pubis. The so-called bird-hipped, or ornithischian, dinosaurs were all plant-eaters.

Pubis bone lies flat against the backward-pointing ischium

HYPSILOPHODON

LIZARD-HIPPED DINOSAUR ▶
Lizard-hipped, or saurischian, dinosaurs included meat-eaters such as *Velociraptor* and *Tyrannosaurus rex*, as well as gigantic long-necked herbivores, such as *Diplodocus*. Bird-hipped dinosaurs may have evolved from a lizard-hipped ancestor.

Pubis bone is separate from the ischium and points forwards

GALLIMIMUS

FIND OUT MORE ▶▶ Birds **303** • Evolution **328–329** • Extinction **334** • Fossils **220**

PALAEONTOLOGY

The study of ancient life, known from **FOSSILS**, is called palaeontology and scientists who work in this field are known as palaeontologists. By looking closely at fossil skeletons and comparing them with those of living animals, palaeontologists try to work out what extinct creatures looked like and how they might have lived. They also use other clues from so-called "trace fossils" – preserved footprints and other remains of prehistoric animal activity.

Skull has large holes to make it light in weight, and a horny beak

Long fingers were used to grasp foliage and pull it towards the mouth

HETERODONTOSAURUS SKELETON ▶
Complete dinosaur skeletons are rare. It is more usual to find just a few teeth or bones. Palaeontologists identify isolated fossils by comparing and matching them with bones from better-preserved specimens. To flesh out dinosaur skeletons, scientists calculate the size and position of muscles from the marks left where they attached to the bones.

Toes had small claws to provide grip, like the spikes on a running shoe

Jaws were relatively small and contained simple teeth for feeding on leaves

HETERODONTOSAURUS

Colour can only be guessed at – skin pigments do not fossilize

▲ DINOSAUR FOOTPRINTS
Palaeontologists study more than just bones. Prehistoric creatures such as dinosaurs left other evidence of their lives behind. By looking at fossil footprints, palaeontologists can work out how dinosaurs moved and even how quickly they ran. As well as footprints, fossil dinosaur droppings have been found. These tell palaeontologists what these creatures ate.

Legs were well muscled, enabling Heterodontosaurus to run quite fast

Tail was held out as the dinosaur moved, to counterbalance its body

◀ SABRE-TOOTHED CAT
Some fossils can tell us about a prehistoric animal's diet. The fossil of a sabre-toothed cat, for instance, shows the sharp, cutting teeth of a meat-eater. It also shows a powerful jaw needed to hold prey. Some palaeontologists think the large canines were used for killing – others suggest they may have been for display.

Skull has a huge zygomatic arch to accomodate the massive jaw muscles

Canine teeth are massive and extend well beyond the lower jaw

Lower jaw moved up and down but not from side to side

DINOSAUR EGGS ▶
In recent decades, large numbers of dinosaur eggs have been discovered. Some are preserved in clutches of a dozen or more, suggesting that dinosaurs made nests. There is even one example of a dinosaur preserved with a clutch of eggs – evidence that there may have been some kind of parental care.

FOSSILS

Fossils are the only evidence of prehistoric life. They occur in sedimentary rocks, which form from compacted sediments, such as silt and sand. Rapid burial in sediment prevents the animal from breaking up or being eaten by scavengers. Over time what remains may be bone, or replaced by minerals, or dissolved out, leaving a mould of the animal's shape. Fossils are exposed by water or weather wearing away the rock they are in.

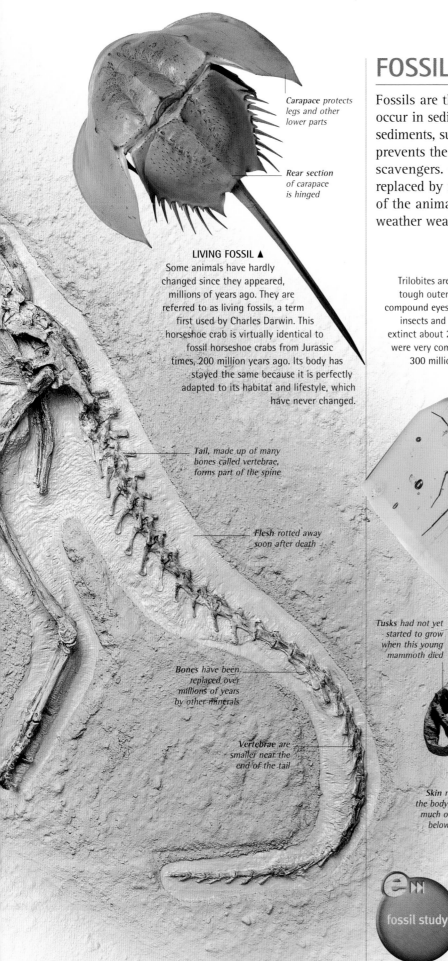

Carapace protects legs and other lower parts

Rear section of carapace is hinged

LIVING FOSSIL ▲
Some animals have hardly changed since they appeared, millions of years ago. They are referred to as living fossils, a term first used by Charles Darwin. This horseshoe crab is virtually identical to fossil horseshoe crabs from Jurassic times, 200 million years ago. Its body has stayed the same because it is perfectly adapted to its habitat and lifestyle, which have never changed.

Tail, made up of many bones called vertebrae, forms part of the spine

Flesh rotted away soon after death

Bones have been replaced over millions of years by other minerals

Vertebrae are smaller near the end of the tail

SET IN STONE ▶
Trilobites are extinct animals that had a tough outer skeleton, jointed legs, and compound eyes – features seen in today's insects and crustaceans. They became extinct about 248 million years ago, but were very common sea creatures in the 300 million years before that time.

◀ TRAPPED IN AMBER
Amber is fossilized tree resin. Sometimes it contains the remains of insects and other small animals that became trapped in it as it oozed from the tree. Evidence from these almost perfect fossils provides rare information about soft tissues and delicate structures. Attempts have been made to extract DNA from these fossils, but so far none has been successful.

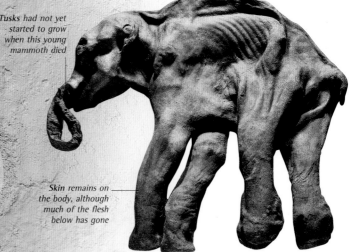

Tusks had not yet started to grow when this young mammoth died

Skin remains on the body, although much of the flesh below has gone

fossil study

PRESERVED IN ICE ▲
Occasionally, prehistoric animals are preserved intact. This baby woolly mammoth died thousands of years ago but quickly froze solid and was covered by snow and ice, which protected it from scavengers. Although most of its hair has gone, other features remain, including the tiny ears that set it apart from modern elephants.

FIND OUT MORE ▶▶ Evolution **328–329** • Extinction **334–335** • Fossils **220–221** • Prehistoric Life **330–331**

EXTINCTION

Since life on Earth began, a huge number of animals have appeared, flourished and then disappeared again. These disappearances are called extinctions. Individual species become extinct for a variety of reasons, including competition and habitat changes. At least five times in the past there have been mass extinctions, where large numbers of animal species have died out in a short period.

CHICXILUB
CRATER

TYRANNOSAURUS
REX SKULL

extinction

METEOR STRIKE ▲
Dinosaurs became extinct 65 million years ago, at the end of the Mesozoic era. Their disappearance has been linked to a massive meteor strike, which left a vast crater beneath the Gulf of Mexico. The volcanic activity in the Western Ghats of India is also thought to have played a part. Scientists think the gas and dust generated filled the atmosphere, blotting out the sun for centuries. It killed 70 per cent of all animal life, including the dinosaurs.

HUNTED TO DEATH ▶
Today, over-hunting is a major threat to many animals. In the past, it has contributed to many extinctions, including that of the dodo, a flightless pigeon from the island of Mauritius. Discovered in 1600, the dodo was easy to catch because it could not fly and was not afraid of humans. Sailors killed large numbers for food and lamp oil, and introduced animals such as rats, which destroyed their nests. By 1680, the dodo was extinct.

◀ ISLAND ISOLATION
Many islands have unique species, found nowhere else. If new predators are introduced, they have no way of escaping. This makes them especially vulnerable to extinction. One way of protecting island animals is to make their homes nature reserves. This giant tortoise is from the Galapagos Islands, which are protected by the government of Ecuador.

FIND OUT MORE ▶▶ Asteroids **184** • Prehistoric Life **330–331**

CONSERVATION

Wildlife conservation is becoming increasingly important. Not many animals can evolve quickly enough to survive human-induced change, and few can adapt to live close to people. The best way to conserve wild animals is to protect their habitats. Some habitats support more species than others. The total number of species is a measure of their **BIODIVERSITY**. Areas with very high numbers are called hot spots.

RECREATING THE QUAGGA ▶
The quagga was hunted to extinction in the late 19th century. Although once considered a separate species, it is now known to have been a subspecies of the plains zebra. This discovery led scientists to try to recreate the quagga by selectively breeding from plains zebras with reduced striping and a browner coats. The resulting animals look remarkably like the quaggas seen in museums.

conservation

BIODIVERSITY

The variety of life within habitats is known as biodiversity. Biodiversity is measured in terms of species numbers, which depends, over time, on the rate at which species evolve compared with the rate at which they become extinct. Biodiversity varies naturally between different habitats. For example, habitats near the poles, such as tundra, have much lower biodiversity than those near the equator, such as tropical rainforest.

▲ CAPTIVE BREEDING
Animals facing extinction in the wild can be saved by increasing their numbers in zoos. Pandas do not usually breed well in captivity, but in recent years the number of captive births has risen considerably. This is partly due to increased cooperation between zoos, with more loaning out their male or female pandas to form new pairs around the world.

▲ COCK-OF-THE-ROCK
Many wild creatures are closely linked to particular habitats. The Andean cock-of-the-rock is found only in mountain forests in the north of South America's Andes range. If those forests were to be cut down, this bird would become extinct in the wild.

▲ CAPUCHIN MONKEY
The Amazon rain forest has the highest biodiversity of any habitat. This capuchin represents just one of countless species that live there. A single tree may host over 1,000 insect species and there may be 300 tree species in a single hectare (2½ acres).

FIND OUT MORE ▶▶ Evolution 328–329 • Habitats 246–247

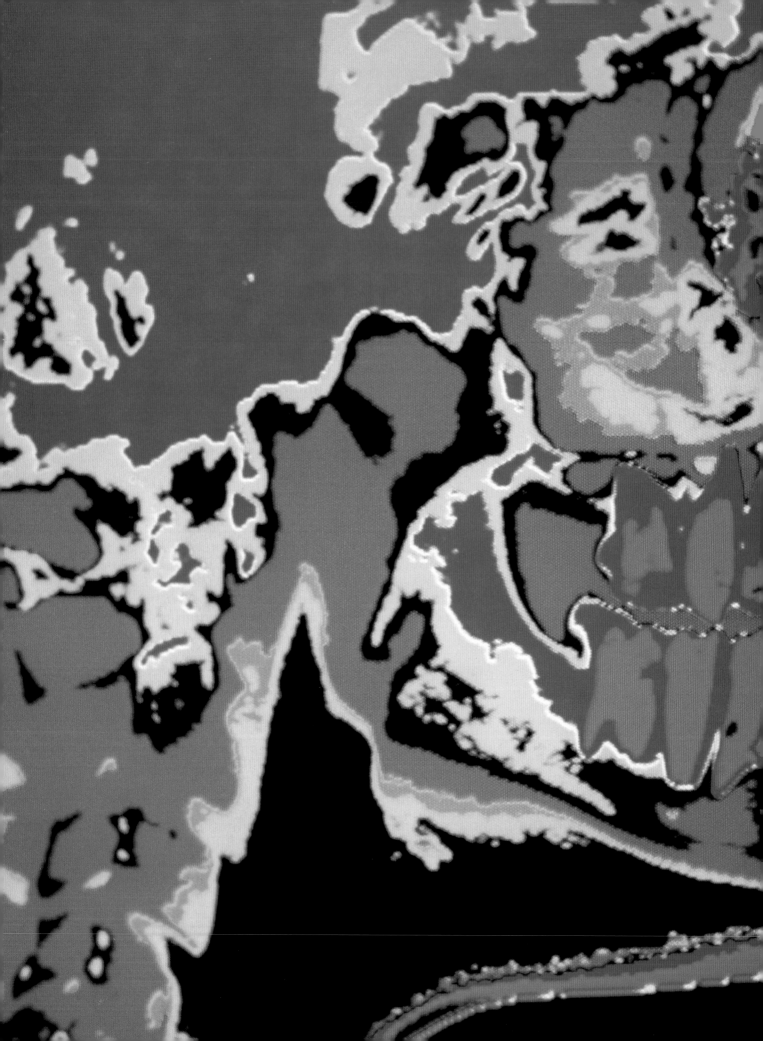

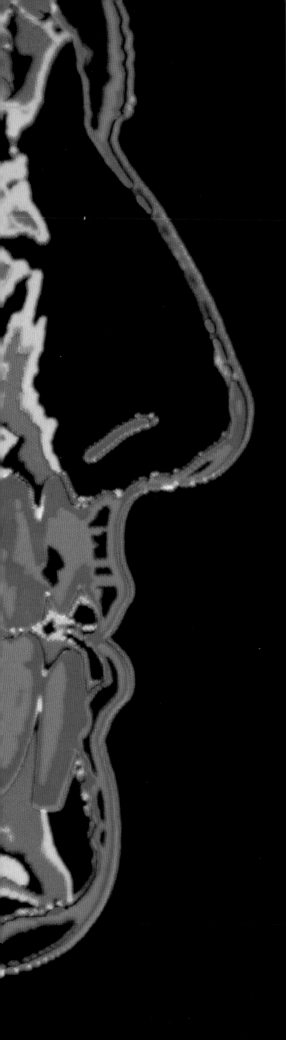

HUMAN BODY

BODY SYSTEMS

Our body structures are arranged into several different systems, each with its own specific function. The smallest units in the body are **CELLS**, which share certain characteristics. These tiny structures are collected into **TISSUES**, which are themselves arranged into **ORGANS**. Different body systems consist of collections of cells, tissues, and organs with a common purpose.

◄ **INTEGUMENTARY SYSTEM**
The skin, hair, and nails form the body's outer covering, or integument. They help to protect the body's internal parts from damage and provide a barrier to invasion by infectious organisms. An adult's skin covers an area of about 2 m² or 22 sq ft.

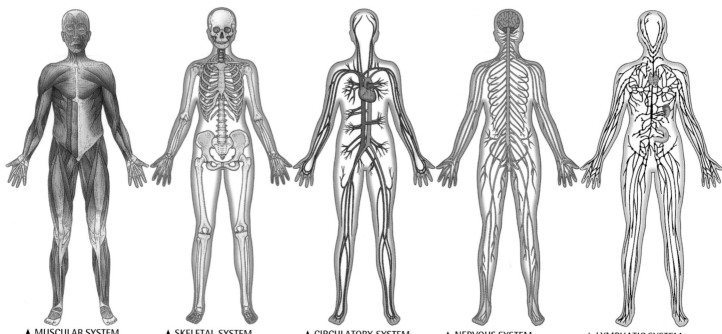

▲ **MUSCULAR SYSTEM**
The muscular system consists of layers of muscles that cover the bones of the skeleton, extend across joints, and can contract and relax to produce movement.

▲ **SKELETAL SYSTEM**
The skeleton is a strong yet flexible framework of bones and connective tissue. It provides support for the body and protection for many of its internal parts.

▲ **CIRCULATORY SYSTEM**
This system consists of the heart and a network of vessels that carry blood. It supplies oxygen and nutrients to the body's cells and removes waste products.

▲ **NERVOUS SYSTEM**
The nervous system is the body's main control system. It consists of the brain, the spinal cord, and a network of nerves that extend out to the rest of the body.

▲ **LYMPHATIC SYSTEM**
This system is a network of vessels that collects fluid from tissues and returns it to the blood. It also contains groups of cells that protect the body against infection.

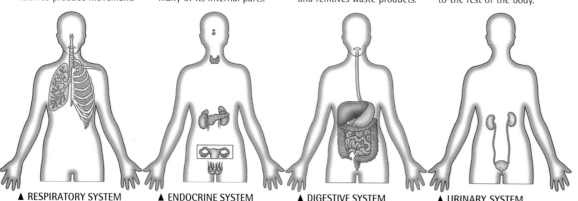

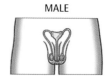

FEMALE

MALE

▲ **RESPIRATORY SYSTEM**
The respiratory system is centred on the lungs, which work to get life-giving oxygen into the blood. They also rid the body of a waste product, carbon dioxide.

▲ **ENDOCRINE SYSTEM**
Many body processes, such as growth and energy production, are directed by hormones. These chemicals are released by the glands of the endocrine system.

▲ **DIGESTIVE SYSTEM**
The digestive system takes in the food the body needs to fuel its activities. It breaks the food down into units called nutrients and absorbs the nutrients into the blood.

▲ **URINARY SYSTEM**
The body's cells produce waste products, many of which are eliminated in urine. The job of the urinary system is to make urine and expel it from the body.

▲ **REPRODUCTIVE SYSTEM**
The male and female parts of the reproductive system produce the sperm and eggs needed to create a new person. They also bring these tiny cells together.

CELLS

The basic building blocks of the body are tiny structures called cells. The human body contains trillions of cells, which fall into several types – nerve cells, muscle cells, fat cells, liver cells, and so on – each with a different function. A typical cell has a central nucleus surrounded by some jellylike material called cytoplasm. Covering the cytoplasm is the plasma membrane. This controls the movement of substances into and out of the cell.

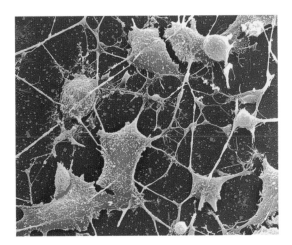

▲ NERVE CELLS
Nerve cells, or neurons, are one of the most numerous types of body cell. Each nerve cell has a central body, containing the cell nucleus, and fibrelike projections, which can be up to 1 m (3⅓ ft) long. The nervous system contains billions of neurons, which collect and transmit information around the body. The adult brain alone may contain as many as 25 billion neurons.

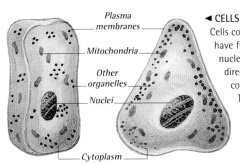

Plasma membranes — Mitochondria — Other organelles — Nuclei — Cytoplasm

◄ CELLS
Cells come in different shapes and sizes, but all have features in common. Most cells have a nucleus. This contains genetic material, which directs the cell's activities. The cytoplasm contains small structures called organelles. There are several types of organelle, each with a specific job. Mitochondria, for example, produce energy for the cell.

TISSUES

Cells group together to form tissues, each with specific functions. Connective tissue is the most widespread; it separates and supports other tissues and organs, and includes cartilage and bone. Adipose tissue is packed with fat cells, which provide energy storage and insulation. Epithelial tissue protects and lines the surfaces of many body organs. Other types include muscle and nervous tissue.

body systems

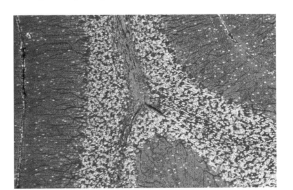

◄ NERVOUS TISSUE
Nervous tissue contains neurons and supporting cells called glial cells. This micrograph shows some tissue in the cerebellum, a part of the brain that helps smooth out and coordinate your body movements. This tissue contains layers, visible as variations in its appearance when viewed through a microscope. The lighter, speckled areas contain nerve cell bodies. The smoother blue areas are richer in nerve cell fibres.

ORGANS

Tissues are grouped together in the body to form organs. These include the brain, heart, lungs, kidneys, and liver. Each body organ has a specific shape and is made up of different types of tissue that work together. For example, the heart consists mainly of a specialized type of muscle tissue, which contracts rhythmically to provide the heart's pumping action. But it also contains nervous tissue, which carries the electrical signals that bring about the contractions, and is lined with epithelial tissue.

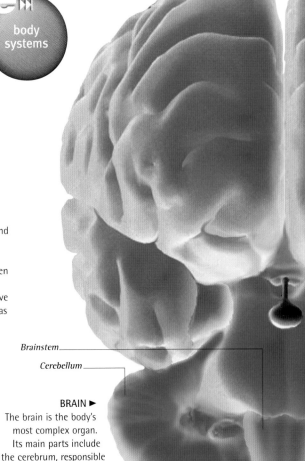

Cerebrum — Brainstem — Cerebellum —

BRAIN ►
The brain is the body's most complex organ. Its main parts include the cerebrum, responsible for thought and reasoning, and the brainstem, which controls vital processes such as breathing. The brain consists mainly of nervous tissue.

FIND OUT MORE ▸▸ Circulatory System **352–353** • Muscular System **342–343** • Nervous System **344–345**

SKELETAL SYSTEM

The body is supported and its internal parts protected by a strong yet flexible framework of **BONES** called the skeleton. These bones meet at **JOINTS**, most of which allow movement between the bones they connect. As well as protection and movement, bones provide a store for the mineral calcium, which is vital to the working of nerves and muscles. They also contain bone marrow, which makes blood cells and stores fat.

THE HUMAN SKELETON ▶

The skeleton contains 206 bones. Babies have over 270, but by adulthood many of these have fused together. Some of the main individual bones, and groups of bones, are labelled here. They fall into two groups: the axial skeleton, made up of the bones of the head, spine, ribs, and breastbone; and the appendicular skeleton, containing the bones of the limbs, the pelvis, the shoulder blades, and the collarbones.

Cranium

Cervical vertebra

Scapula (shoulder blade)

Mandible (lower jaw)

Thoracic vertebra

Phalanx (finger bone), one of 14 in each hand

Shoulder joint

Clavicle (collarbone)

Sternum (breastbone)

Humerus

Metacarpal, one of five in each hand

Radius

Carpals (wrist bones)

Lumbar vertebra

Rib, one of 24

Ulna

Sacrum

Elbow joint

Ilium (hip bone), part of pelvis

Coccyx

Hip joint

Ischium, part of pelvis

Pubis (pubic bone), part of pelvis

Femur (thigh bone)

◀ THE SPINE

This highly flexible structure, also called the vertebral column, supports the head and body. It also protects the delicate tissues of the spinal cord. It is made up of 33 bones called vertebrae, separated by intervertebral discs, which act as shock absorbers. The bones of the spine are kept in place and supported by attached ligaments and muscles.

Cervical spine (7 bones)

Knee joint

Thoracic spine (12 bones)

Tarsals (ankle bones)

Tibia (shin bone)

Lumbar spine (5 bones)

Intervertebral disc

Phalanx (toe bone), one of 14 in each foot

Patella (knee cap)

Fibula

skeleton

Sacrum (5 fused bones)

Coccyx (4 fused bones)

Calcaneus (heel bone)

Metatarsal, one of five in each foot

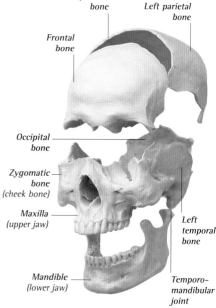

Right parietal bone

Left parietal bone

Frontal bone

Occipital bone

Zygomatic bone (cheek bone)

Maxilla (upper jaw)

Left temporal bone

Mandible (lower jaw)

Temporo-mandibular joint

THE SKULL ▲

The skull consists of 22 bones (excluding the three bones in each middle ear). All the larger skull bones are shown in this exploded view. They fall into two main groups. One group (including the frontal, parietal, and temporal bones) surrounds the brain and is fused together to form the cranium. The remainder of the bones form the face.

BONES

Bones are relatively light, yet five times stronger than steel. They contain cells, minerals, protein, and water. Bones are composed of two types of tissue: cancellous (spongy) and compact bone. These are living tissues that are constantly broken down and rebuilt by the cells they contain.

Shaft of a long bone consists mainly of hard, dense, compact bone

Compact bone

Cancellous (spongy) bone

BONE STRUCTURE ▶
Bones have an outer layer of compact bone, one of the body's hardest materials. On the inside is an area of cancellous bone, which may contain red bone marrow. In adult long bones, like this femur, the shaft is compact bone overlaying an area that may contain yellow bone marrow (a fatty tissue).

Bone end has a thin layer of compact bone overlaying cancellous bone

◀ COMPACT BONE
Compact bone consists of units called osteons, each about 1 mm (¹⁄₂₅ in) across. One osteon is shown here. It is made up of numerous tiny rings of a hard tissue arranged around a central canal, through which blood vessels and nerves pass.

CANCELLOUS BONE ▶
In cancellous (spongy) bone, struts of rigid bone tissue called trabeculae connect up to form a honeycomblike structure. Cancellous bone is less dense than compact bone but is still very strong.

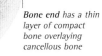

Spaces may contain red bone marrow

Trabeculae

◀ RED BONE MARROW
Red bone marrow is the site where the body's red blood cells and some white blood cells are made. This microscopic view shows one red cell surrounded by white cells. With age, the red marrow in the long bones is gradually replaced by fat cells.

JOINTS

Joints are the parts of the body where bones meet. Some, such as the joints in the cranium, allow no movement between the bones. Others, such as the joints in the spine, allow limited movement. A few, such as the hip joints, permit a wide range of movement. The bones of many joints are held in place by muscles and bands of tissue called ligaments.

Scapula

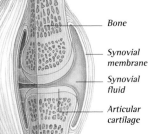

BALL-AND-SOCKET JOINT ▶
This colour-enhanced X-ray shows the shoulder joint, which, like the hip, is a ball-and-socket joint. The rounded upper end of the humerus fits into the cup-shaped socket of the scapula. This allows the humerus to rotate freely. The joint is kept in place by surrounding muscles and ligaments.

Humerus

Humerus

BALL-AND-SOCKET JOINT

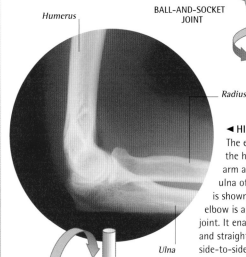

Radius

◀ HINGE JOINT
The elbow joint, where the humerus of the upper arm and the radius and ulna of the lower arm meet, is shown on this X-ray. The elbow is an example of a hinge joint. It enables the arm to bend and straighten, but it allows little side-to-side movement.

Ulna

HINGE JOINT

PARTS OF A SYNOVIAL JOINT ▶
All free-moving joints, such as the finger, hip, knee, and elbow joints, are called synovial joints and have a similar structure. The synovial membrane that lines the joint produces a fluid that lubricates movement. The bone ends are covered by a layer of articular cartilage, which is smooth and so minimizes friction. The joint is kept in place by a fibrous capsule, which encases the joint completely.

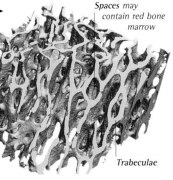

Bone

Synovial membrane

Synovial fluid

Articular cartilage

Ligament forming fibrous capsule

FIND OUT MORE ▶▶ Growth **366–367** • Movement **314–315** • Muscular System **342–343**

MUSCULAR SYSTEM

The skeleton is covered by layers of skeletal muscle. Each muscle is attached to two or more bones so that when the muscle contracts (shortens) it produces **MOVEMENT**. Skeletal muscle makes up about 40 per cent of body weight. As well as producing movement, some muscles remain partially contracted for long periods to maintain the body's posture.

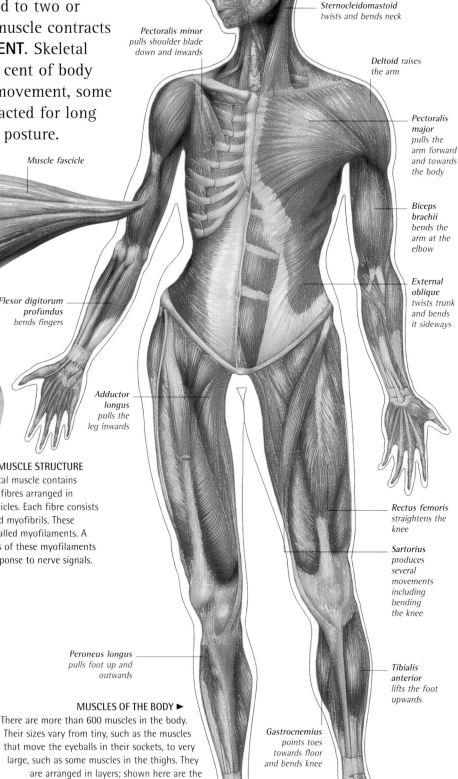

Frontalis wrinkles forehead and lifts eyebrows

Orbicularis oculi closes eyelid

Sternocleidomastoid twists and bends neck

Pectoralis minor pulls shoulder blade down and inwards

Deltoid raises the arm

Pectoralis major pulls the arm forward and towards the body

Biceps brachii bends the arm at the elbow

External oblique twists trunk and bends it sideways

Muscle fascicle

Sheath

Blood vessel

Single muscle fibre

Single myofibril

Myofilament

Flexor digitorum profundus bends fingers

Adductor longus pulls the leg inwards

Rectus femoris straightens the knee

Sartorius produces several movements including bending the knee

◀ **MUSCLE STRUCTURE**
A skeletal muscle contains many long fibres arranged in bundles called fascicles. Each fibre consists of smaller strands, called myofibrils. These contain yet smaller parts called myofilaments. A muscle contracts when sets of these myofilaments slide past each other in response to nerve signals.

Peroneus longus pulls foot up and outwards

Tibialis anterior lifts the foot upwards

Gastrocnemius points toes towards floor and bends knee

MUSCLES OF THE BODY ▶
There are more than 600 muscles in the body. Their sizes vary from tiny, such as the muscles that move the eyeballs in their sockets, to very large, such as some muscles in the thighs. They are arranged in layers; shown here are the superficial (outer) muscles at the front of the body and, on this side, some of the deeper muscles.

DEEP MUSCLES

SUPERFICIAL MUSCLES

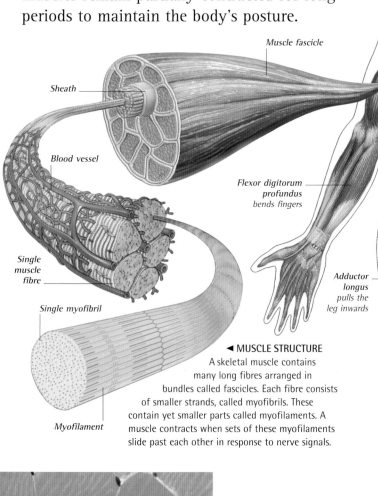

▲ **SKELETAL MUSCLE**
Skeletal muscle is also called striated or striped muscle. The stripes, which can be seen clearly when a piece of muscle is viewed under a microscope, are caused by the arrangement of myofilaments in individual muscle fibres. These lead to the appearance of alternating light and dark bands.

MOVEMENT

Skeletal muscles cross joints and are attached to the bones on either side by tough cords called tendons. They contract, to produce movement, as a result of nerve signals sent from the brain and spinal cord. Although our movements are under our conscious control, the brain can learn patterns of movements so that we can perform certain tasks, such as walking, without thinking.

Rectus femoris _and other muscles at the front of the thigh contract to straighten the knee_

Biceps femoris _and other muscles at the back of the thigh relax to allow the knee to straighten_

Gastrocnemius _relaxes to allow the knee to straighten_

MUSCLE ACTION IN MOVEMENT ▶
To straighten the knee, one group of muscles at the front of the thigh contracts, while other muscles at the back of the leg relax. Two groups of muscles such as this are called opposing groups. Contractions of opposing groups have opposite effects, such as knee straightening and bending.

OTHER MUSCLE TYPES

Skeletal muscle is not the only type of muscle in the body. There are two other types: smooth muscle and cardiac (heart) muscle. Unlike skeletal muscle, these muscles are not under our conscious control.

SMOOTH MUSCLE
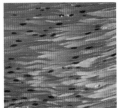

CARDIAC MUSCLE

Smooth muscle is found in the walls of many organs, such as the bladder, the womb, and the intestines, where it contracts to propel food along. It has short, spindle-shaped fibres.

Cardiac muscle contracts tirelessly throughout life to pump blood from the heart to the lungs and around the body. It is made up of a network of branching muscle fibres.

Lower leg _moves forward when knee straightens_

Nerve cell fibre

Terminal branch _of nerve cell fibre_

Muscle fibre _contracts when signal reaches it from nerve cell fibre_

End plate _is a pad at the end of a branch of a nerve cell fibre_

muscles

◀ NEUROMUSCULAR JUNCTION
To bring about a movement, the brain sends a series of signals instructing specific muscles to contract, via a network of nerve cell fibres. Each individual fibre divides into several branches before it reaches the muscle, and each branch connects to a single muscle fibre. The region where the nerve and muscle fibres meet is called a neuromuscular junction.

FIND OUT MORE ▶▶ Movement 314–315 • Nervous System 344–345 • Skeletal System 340–341

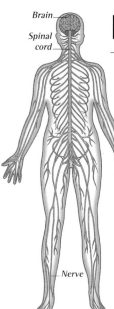

Brain
Spinal cord
Nerve

NERVOUS SYSTEM

The nervous system is the body's main control system. It is made up of the **CENTRAL NERVOUS SYSTEM** (or CNS) and a network of **NERVES** that extend from the CNS to all parts of the body. The nervous system regulates both voluntary activities, such as walking and talking, and involuntary activities, such as breathing, which you make no conscious decisions about.

▲ PARTS OF THE SYSTEM
The CNS consists of the brain and spinal cord. The rest of the nervous system, called the peripheral nervous system, consists of nerves. These include 12 pairs of nerves that branch from the brain (cranial nerves) and 31 pairs that branch from the spinal cord (spinal nerves).

Axon carries electrical signals

Synapse is a junction between two neurons and contains a gap across which electrical signals can pass

Axon's protective sheath

Cell nucleus

Dendrite is a branching projection from the cell body

Cell body

▲ NEURONS
The nervous system contains billions of neurons (nerve cells). A neuron has a cell body, arms called dendrites, and a long projecting fibre, the axon. Electrical signals – up to 2,500 per second – can pass along axons. They can also jump between neurons by means of chemicals that pass across the gaps in synapses (neuron junctions).

Cerebrum has two hemispheres (halves) and is responsible for the brain's most complex functions

Hypothalamus helps to control the body's endocrine (hormone) system

Pituitary gland produces hormones and controls the release of other hormones around the body

NERVES

Nerves are made up of bundles of the axons of nerve cells. Some of these carry information picked up by sensory receptors around the body to the CNS for processing. Other axons carry messages from the CNS to muscles, causing movement, or to the body's glands, causing the release of hormones. Many axons are surrounded by a protective sheath containing a fatty substance called myelin. This acts to insulate the axons electrically.

Myelin sheath

Fat cells

Axon

Blood vessels

Fascicle

Epineurium surrounds the entire nerve

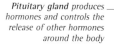

nerves

◄ NERVE STRUCTURE
Most nerves consist of several axon bundles, called fascicles. The speed at which individual nerves transmit signals varies depending on their thickness and whether or not their axons have myelin sheaths; fatter, myelinated axons transmit signals faster, at up to 350 kph (218 mph).

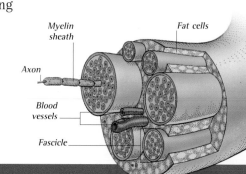

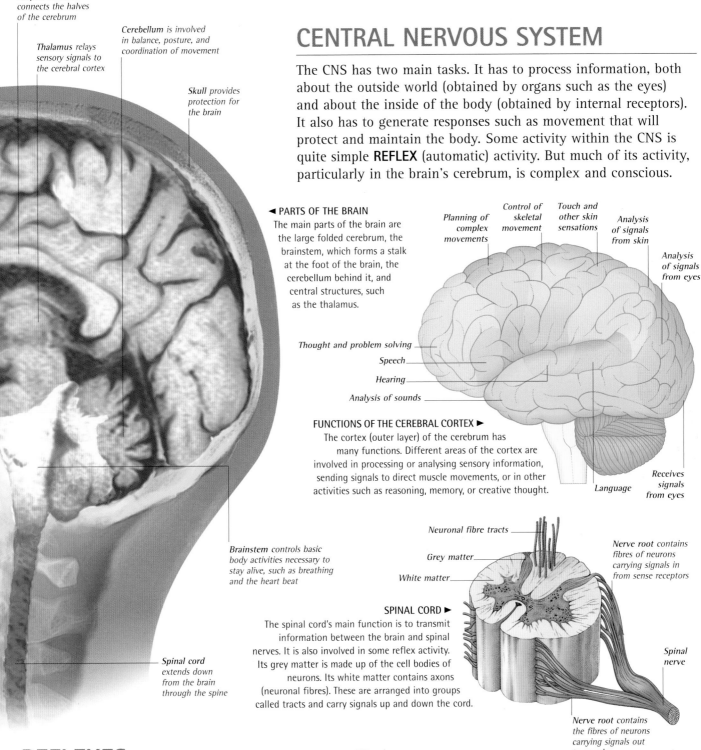

Corpus callosum connects the halves of the cerebrum

Thalamus relays sensory signals to the cerebral cortex

Cerebellum is involved in balance, posture, and coordination of movement

Skull provides protection for the brain

Brainstem controls basic body activities necessary to stay alive, such as breathing and the heart beat

Spinal cord extends down from the brain through the spine

CENTRAL NERVOUS SYSTEM

The CNS has two main tasks. It has to process information, both about the outside world (obtained by organs such as the eyes) and about the inside of the body (obtained by internal receptors). It also has to generate responses such as movement that will protect and maintain the body. Some activity within the CNS is quite simple **REFLEX** (automatic) activity. But much of its activity, particularly in the brain's cerebrum, is complex and conscious.

◀ PARTS OF THE BRAIN
The main parts of the brain are the large folded cerebrum, the brainstem, which forms a stalk at the foot of the brain, the cerebellum behind it, and central structures, such as the thalamus.

Planning of complex movements

Control of skeletal movement

Touch and other skin sensations

Analysis of signals from skin

Analysis of signals from eyes

Thought and problem solving

Speech

Hearing

Analysis of sounds

Language

Receives signals from eyes

FUNCTIONS OF THE CEREBRAL CORTEX ▶
The cortex (outer layer) of the cerebrum has many functions. Different areas of the cortex are involved in processing or analysing sensory information, sending signals to direct muscle movements, or in other activities such as reasoning, memory, or creative thought.

Neuronal fibre tracts

Grey matter

White matter

Nerve root contains fibres of neurons carrying signals in from sense receptors

Spinal nerve

SPINAL CORD ▶
The spinal cord's main function is to transmit information between the brain and spinal nerves. It is also involved in some reflex activity. Its grey matter is made up of the cell bodies of neurons. Its white matter contains axons (neuronal fibres). These are arranged into groups called tracts and carry signals up and down the cord.

Nerve root contains the fibres of neurons carrying signals out to muscles

REFLEXES

In its simplest sense, a reflex is an emergency reaction of the nervous system to a threat such as a hot object touching the skin. In a wider sense, reflexes are automatic responses to a wide range of situations in the body and are key to many internal activities, such as the heart beat. A division of the nervous system called the autonomic nervous system is in overall control of these internal activities.

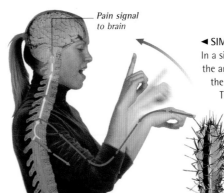

Pain signal to brain

◀ SIMPLE REFLEX ACTION
In a simple reflex, information passes from the area affected, in this case the finger, to the central nervous system (red pathway). This triggers an immediate response, in this case the contraction of a muscle (blue pathway) to withdraw the finger. Here, the reflex action involves only two nerves and the spinal cord. However, a signal also passes to the brain, which registers the pain.

FIND OUT MORE ▶▶ Balance 347 • Endocrine System 356 • Hearing 347 • Sight 348–349 • Smell 346 • Taste 346 • Touch 350

SMELL

Humans have a very keen sense of smell: we can detect thousands of different smells. This ability relies on the presence of special sensory receptors in the upper part of the nose. When stimulated by odour molecules, these receptors send signals along nerves to the brain for processing. Sometimes odour molecules do not reach the sensory area, but sniffing will help get them there.

TASTE

We can taste substances in food and drink thanks to the 10,000 or so taste buds located on structures, called papillae, on the surface of our tongues. These receptors send signals along nerves to the brain for interpretation. Four main tastes – sweet, salty, sour, and bitter – are detected by the taste buds in four areas of the tongue. The senses of taste and smell combine to analyse flavours.

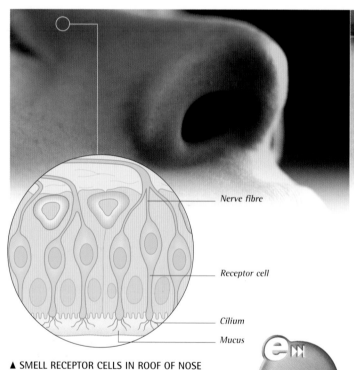

Nerve fibre

Receptor cell

Cilium

Mucus

▲ SMELL RECEPTOR CELLS IN ROOF OF NOSE
Smell receptors are specialized nerve cells. Each bears many tiny cilia (hairs), which project into the space in the upper part of the nose. A nerve fibre extends from the other end of each cell. This joins other fibres to form the olfactory nerves, which carry signals to the brain.

smell

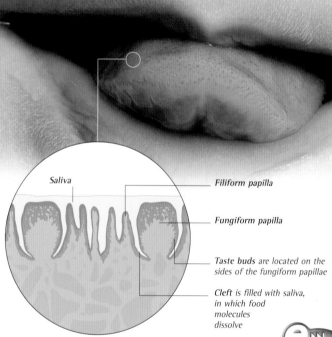

Saliva

Filiform papilla

Fungiform papilla

Taste buds are located on the sides of the fungiform papillae

Cleft is filled with saliva, in which food molecules dissolve

▲ PAPILLAE ON SURFACE OF TONGUE
Papillae are tiny protrusions on the surface of the tongue. The fungiform papillae and some other types of papillae contain taste buds. The smaller, more numerous, filiform papillae do not contain taste buds but give the tongue a rough surface, which helps it move food around the mouth.

taste

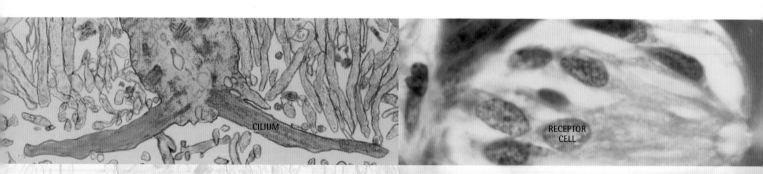

CILIUM

RECEPTOR CELL

▲ CILIA OF SMELL RECEPTOR CELL
The cilia can detect tiny amounts of substances in the air, though molecules of those substances must first be absorbed by the mucus layer. There they interact with the cilia to trigger nerve impulses.

▲ TASTE RECEPTOR CELLS IN A TASTE BUD
This taste bud contains many receptor cells. Hairs emerge from each cell. Food and drink molecules must dissolve in saliva before they can interact with these hairs and trigger signals to the brain.

FIND OUT MORE ▸▸ Nervous System 344–345

FIND OUT MORE ▸▸ Digestive System 358–359

HEARING

Our ears allow us to detect sounds, which pass through the air as waves of varying pressure. On reaching the ear, the waves travel through several structures to the cochlea in the inner ear. There, receptor cells produce signals that go to the brain. The human ear can detect sounds over a very wide range of pitch and loudness, from the high-pitched squeaks of a mouse to the roar of a passenger jet.

BALANCE

Balance is an internal sense and relies on sensory receptors that monitor the position of the head and body. Whether we are still or moving, balance is essential for maintaining our posture and stopping us falling over. The vestibule and semicircular canals of the inner ear provide information on the position and movements of the head. Combined with signals from the eyes, this helps us balance.

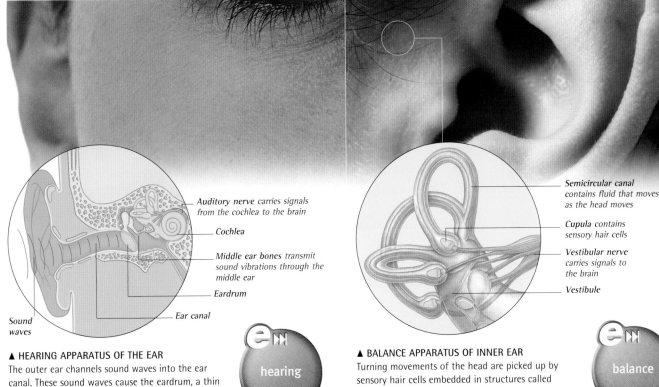

Auditory nerve carries signals from the cochlea to the brain

Cochlea

Middle ear bones transmit sound vibrations through the middle ear

Eardrum

Ear canal

Sound waves

Semicircular canal contains fluid that moves as the head moves

Cupula contains sensory hair cells

Vestibular nerve carries signals to the brain

Vestibule

▲ HEARING APPARATUS OF THE EAR
The outer ear channels sound waves into the ear canal. These sound waves cause the eardrum, a thin membrane at the end of the ear canal, to vibrate. The vibrations are transmitted via three tiny bones in the middle ear to the cochlea in the inner ear.

hearing

▲ BALANCE APPARATUS OF INNER EAR
Turning movements of the head are picked up by sensory hair cells embedded in structures called cupulae in the semicircular canals. Tilting movements of the head, and its position, are monitored by hair cells within structures in the vestibule.

balance

SENSORY HAIRS

OTOLITH

▲ SENSORY HAIRS IN COCHLEA
Inside the cochlea, sound vibrations make these sensory hairs move, which triggers signals in attached receptor cells. The signals pass to the brain, which works out the pitch and loudness of the sound.

▲ OTOLITHS IN VESTIBULE
These tiny crystals, called otoliths, are attached to sensory hair cells in the vestibule. When the head tilts, the otoliths move, causing the hair cells to bend and nerve signals to be sent to the brain.

FIND OUT MORE ▶▶ Pitch **103** • Sound **100–101**

FIND OUT MORE ▶▶ Sight **348–349**

SIGHT

Whenever we are awake, our eyes work constantly to collect information about the world. As this data is analysed by the brain, we are supplied with a detailed picture of our surroundings. We can judge distance, see in dim and bright light, and experience **COLOUR VISION**. For us to see, light rays reflected by objects around us must meet at the back of the eye. There they trigger electrical signals that are sent to the brain for interpretation.

CHANGING PUPIL SIZE

CONSTRICTED	DILATED

In bright light or when viewing close objects, the pupils of our eyes constrict (narrow). This is caused by tightening of circular muscles within the iris, the coloured region of tissue that surrounds the pupil. The constriction of the pupil reduces the number of light rays entering the eye.

In dim light or when we are viewing distant objects, our pupils dilate (widen). This is due to tightening of a different set of iris muscles that are arranged like spokes in a wheel around the pupil. Full widening of the pupil allows the maximum number of light rays to enter the eye.

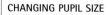

sight

Vitreous humour is the clear, jellylike fluid in the back part of the eye

Ciliary muscle can contract or relax to alter the lens's thickness

Cornea is a clear structure that partly focuses light rays

Conjunctiva is a thin, clear covering over the cornea

Lens can change shape to focus light rays on the retina

PARTS OF THE EYE ▶
The eyes sit in two bony cavities in the skull. Light rays entering the eye pass through the cornea, lens, and vitreous humour, before reaching the retina, the light-sensitive area at the back of the eye. Signals generated in the retina leave the eye along the optic nerve and go to the brain. Around each eye lie six tiny muscles, which enable the eye to turn and swivel in its socket.

Pupil is the black hole in the centre of the iris

Iris controls the amount of light entering the eye and also gives the eye its colour

Inferior oblique muscle pulls front of eyeball upwards and outwards

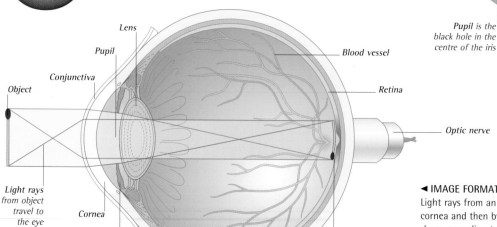

Lens

Pupil

Conjunctiva

Object

Blood vessel

Retina

Optic nerve

Light rays from object travel to the eye

Cornea

Iris

Choroid

Image of object is upside down

◀ IMAGE FORMATION
Light rays from an object are refracted (bent) first by the cornea and then by the lens, which can be made to change shape according to the distance of the object from the eye. The refraction of the rays ensure that they meet on the retina. There, images are formed upside down, but the brain makes sense of this information, so we see objects the right way up.

Superior rectus muscle pulls front of eyeball upwards

Macula is the most sensitive part of the retina and contains the highest concentration of cone cells

Optic disc (blind spot) has no light-sensitive cells

Optic nerve leaves the eye at the optic disk

Sclera is the tough, white, outer layer of the eyeball

Choroid is a dark membrane containing blood vessels

Retina is the inner layer of the eye and contains millions of light-sensitive cells

Medial rectus muscle pulls front of eyeball inwards, towards the nose

FROM EYES TO BRAIN ▶

Nerve signals leave the eyes in the optic nerves, which meet at the optic chiasma. There, fibres from the inner side of each retina cross so that each side of the brain receives information from each eye. The signals pass along the optic tracts to linked areas at the back of the brain. This part of the brain, called the visual cortex, forms a three-dimensional image of the object being viewed.

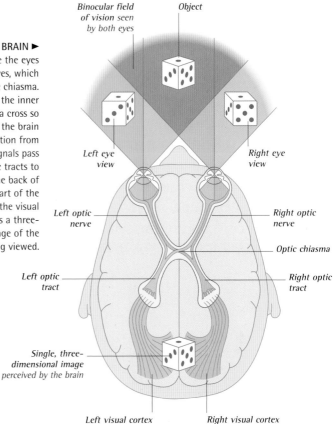

Binocular field of vision seen by both eyes

Object

Left eye view

Right eye view

Left optic nerve

Right optic nerve

Optic chiasma

Left optic tract

Right optic tract

Single, three-dimensional image perceived by the brain

Left visual cortex

Right visual cortex

COLOUR VISION

The retina houses two types of light-sensitive cells: rods and cones. The cones give us colour vision. There are three different types of cone, each sensitive to light within a different range of light wavelengths (colours). Signals are sent from the cones to the brain. From the overall pattern of signals, the brain can work out the colour of every tiny point in the scene being viewed.

Optic nerve fibres

Cone

Signals to brain

Light

Connecting nerve cells

Rod

◀ HOW THE RETINA RESPONDS TO LIGHT

When light rays reach the retina, they trigger chemical changes in different light-absorbing substances in the rod and cone cells. These changes trigger electrical signals in the cells. The rods and cones link to a system of connecting nerve cells. These perform some initial processing of the signals and then transmit them along optic nerve fibres to the brain.

▲ RODS AND CONES

In each retina, the rods (seen here coloured grey) outnumber the cones (coloured orange) by about 17 to 1. The cones only work in bright light, whereas rods respond to dim light. Unlike cones, rods are all of the same type. They are responsible for the black-and-white vision we experience in semi-darkness.

FIND OUT MORE ➤➤ Colour **122–123** • Lenses **115** • Light **110–111** • Nervous System **344–345** • Refraction **114**

TOUCH

Your sense of touch works by means of special sensory receptors scattered all over your body's surface. These receptors allow you to feel an amazing range of sensations, from the pain of touching a searing hot iron to the tickling of a feather as it brushes against your skin. The receptors send messages along nerves to the spinal cord and brain, where the information is processed.

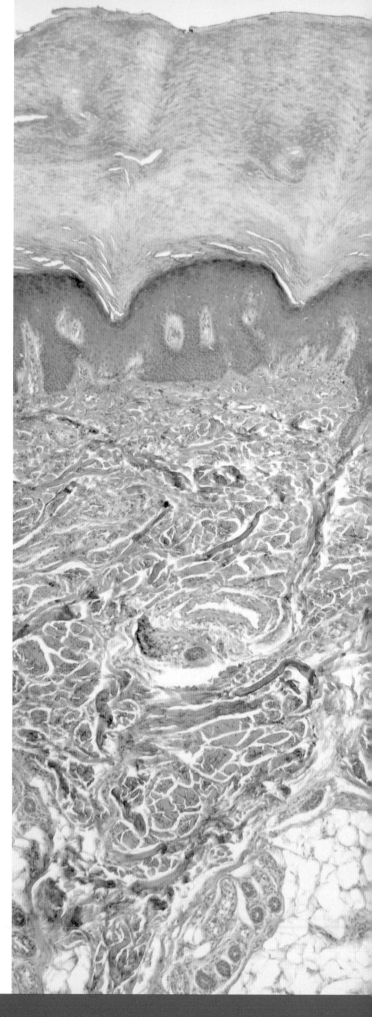

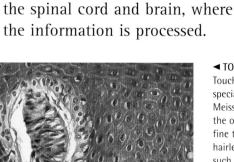

◄ TOUCH RECEPTOR
Touch receptors are types of specialized nerve ending. Meissner's corpuscles, like the one shown here, detect fine touch and are found in hairless parts of the body, such as the lips, palms, and fingertips. Other types of receptor are sensitive to pressure, stretching of the skin, vibration, or hair movements.

FINGER RIDGES ►
Some areas of the skin, such as the fingertips and palms, are folded into ridges. These help improve both touch sensitivity (as they hold more receptors) and grip. The pattern of ridges and grooves provides a means of identification, because everyone has their own unique ridge pattern. In this close-up of a finger, a basic type of ridge pattern called a loop can be seen.

▼ BRAILLE
Developed in the 19th century by a Frenchman, Louis Braille, the Braille system allows blind people to read. Words are represented by a series of raised dots, which the reader recognizes by running his or her fingers over the page. The ability to read Braille relies on the extreme sensitivity of the fingertips to touch.

SKIN

The skin, along with hair and nails, provides the body with a protective outer covering that shields it, for example, from harmful solar rays. It also provides our first line of defence against infection, helps control water loss from the body, plays an important role in **TEMPERATURE CONTROL**, and contains the receptors that provide the sense of touch.

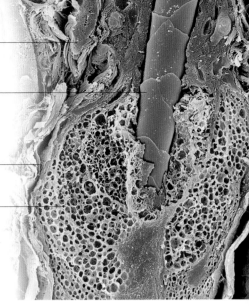

Arrector pili muscle contracts to pull the hair upright

Root of hair shaft is made of the tough fibrous protein, keratin

Lining of hair follicle

Sebaceous gland produces the oily substance sebum, which lubricates the hair and skin

◄ SKIN LAYERS
The skin has two main layers, called the epidermis and dermis. The epidermis consists of an upper layer of dead cells and a lower living layer, which replaces cells as they are lost from the upper layer. Beneath the epidermis is the thicker dermis, which overlies an insulating layer of fatty tissue.

HAIR FOLLICLE ▲
Hair grows from follicles, pockets of epidermal tissue that extend down into the dermis. This false-coloured micrograph shows a single hair follicle, magnified 200 times. Hair has a cycle of growth, rest, and then loss, when the new hair pushes the old hair out of the follicle. About 100 hairs are lost and replaced in a person's scalp every day.

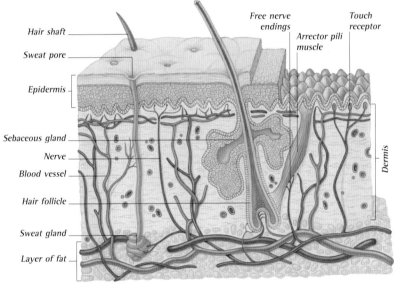

Hair shaft

Sweat pore

Epidermis

Sebaceous gland

Nerve

Blood vessel

Hair follicle

Sweat gland

Layer of fat

Free nerve endings

Touch receptor

Arrector pili muscle

Dermis

▲ STRUCTURES IN THE SKIN
Many different structures exist in the dermis, including blood vessels, nerves, hair follicles, and sweat glands. The nerve fibres terminate in free (uncovered) endings, in touch receptors, or in other types of receptors sensitive to pressure, vibration, or temperature change.

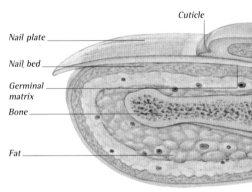

Cuticle

Nail plate

Nail bed

Germinal matrix

Bone

Fat

FINGERNAIL ▲
The ends of the fingers and toes are covered by nails. These plates of tough protective tissue are made mainly of keratin, a protein also found in hair and skin. Nails grow from a region of living cells called the germinal matrix, which lies underneath a fold of skin called the cuticle.

TEMPERATURE CONTROL

The blood vessels, hairs, and sweat glands of the skin work together to help control body temperature. If we get too hot, our sweat production increases and blood vessels widen to allow more blood to reach the skin's surface, where it cools. If we get too cold, these processes go into reverse. In addition, tiny muscles attached to the hair follicles pull the hairs erect, trapping an insulating layer of air next to the skin.

SWEAT PORE ►
Sweat, a salty liquid, reaches the surface of the skin through pores like this. The pore is surrounded by dead epidermal cells. Sweat evaporates from the surface of the skin and so helps to lower the body temperature. Sweating also rids the body of excess water and some waste products.

FIND OUT MORE ▶▶ Heat Transfer **82–83** • Muscular System **342–343** • Nervous System **344–345**

CIRCULATORY SYSTEM

The circulatory system is centred on the **HEART**, a muscular organ that rhythmically pumps **BLOOD** around a complex network of **BLOOD VESSELS** extending to every part of the body. Blood carries the oxygen and nutrients needed to fuel the activities of the body's tissues and organs, and it plays a vital role in removing the body's waste products. An average-sized adult carries about 5 litres (9 pints) of blood.

HEART

The heart contracts tirelessly – more than 2.5 billion times over an average lifetime – to pump blood around the body. These contractions are triggered by electrical impulses that originate in a specialized area of heart tissue. The signals spread through the muscle in the wall of the heart via a network of conducting fibres.

INSIDE THE HEART ▶
The heart has two upper chambers, called atria, and two lower chambers, called ventricles. Blood from the body arrives in the right atrium. This blood is low in oxygen, and is shown here in blue. The blood passes to the right ventricle, which pumps it to the lungs to pick up more oxygen. The left atrium receives oxygen-rich blood (red) back from the lungs. This passes to the left ventricle, which pumps it by way of the aorta to the body.

FROM HEAD AND ARMS

TO HEAD AND ARMS

Aorta

Pulmonary artery

TO LUNG

FROM LUNG

Pulmonary vein

Left atrium

Right atrium

Left ventricle

Right ventricle

Inferior vena cava

Heart muscle

FROM LOWER BODY

TO LOWER BODY

Blood passes through

Backflow prevented

VALVE OPEN

VALVE CLOSED

◀ HEART VALVES
At the exit of each heart chamber lies a valve, which ensures the one-way flow of blood through the heart and into the circulation. These valves are made of flaps that open to allow blood to pass through but snap tightly shut to prevent backflow. The valves have three flaps, except for the valve between the left atrium and left ventricle, which has two.

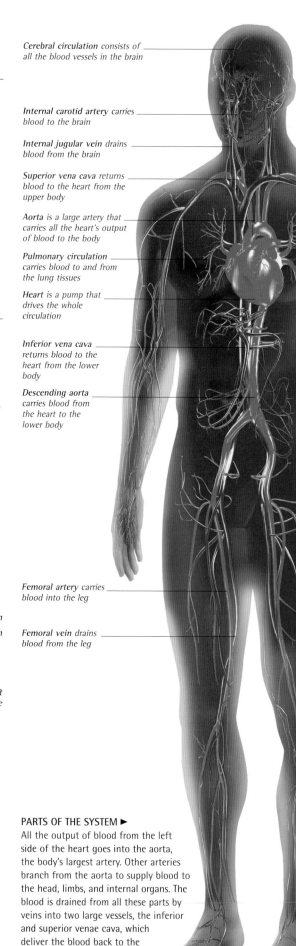

Cerebral circulation consists of all the blood vessels in the brain

Internal carotid artery carries blood to the brain

Internal jugular vein drains blood from the brain

Superior vena cava returns blood to the heart from the upper body

Aorta is a large artery that carries all the heart's output of blood to the body

Pulmonary circulation carries blood to and from the lung tissues

Heart is a pump that drives the whole circulation

Inferior vena cava returns blood to the heart from the lower body

Descending aorta carries blood from the heart to the lower body

Femoral artery carries blood into the leg

Femoral vein drains blood from the leg

PARTS OF THE SYSTEM ▶
All the output of blood from the left side of the heart goes into the aorta, the body's largest artery. Other arteries branch from the aorta to supply blood to the head, limbs, and internal organs. The blood is drained from all these parts by veins into two large vessels, the inferior and superior venae cava, which deliver the blood back to the right side of the heart.

BLOOD VESSELS

If an adult's blood vessels were laid end to end, they would stretch out over 100,000 km (62,500 miles). There are three main types of vessel. Arteries carry blood from the heart to the body's tissues, while veins carry blood back from the tissues to the heart. Small arteries are called arterioles and small veins are referred to as venules. The third and smallest type of vessel, capillaries, form a network connecting the smallest arterioles with the smallest venules.

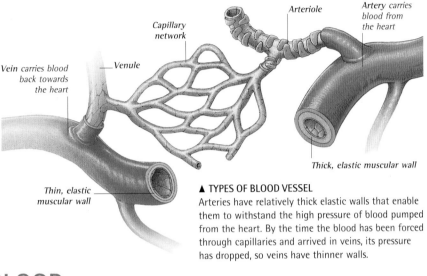

Capillary network · *Arteriole* · *Artery carries blood from the heart*

Vein carries blood back towards the heart · *Venule*

Thick, elastic muscular wall

Thin, elastic muscular wall

▲ TYPES OF BLOOD VESSEL
Arteries have relatively thick elastic walls that enable them to withstand the high pressure of blood pumped from the heart. By the time the blood has been forced through capillaries and arrived in veins, its pressure has dropped, so veins have thinner walls.

CAPILLARIES IN GALL BLADDER ▲
The largest blood vessel visible in this micrograph is a small arteriole. It is surrounded by a network of fine capillaries. Gaps in the thin walls of capillaries allow substances to pass to and from surrounding cells. They are the site where oxygen and nutrients enter the body's tissues, and waste products are absorbed from the tissues.

BLOOD

Blood is composed of a straw-coloured fluid, plasma, and huge numbers of blood cells that float in the plasma. Of the two main types of blood cell, red blood cells carry oxygen to the body's tissues, and white blood cells help defend the body against infection. Blood also transports nutrients, proteins needed for blood clotting, and waste products.

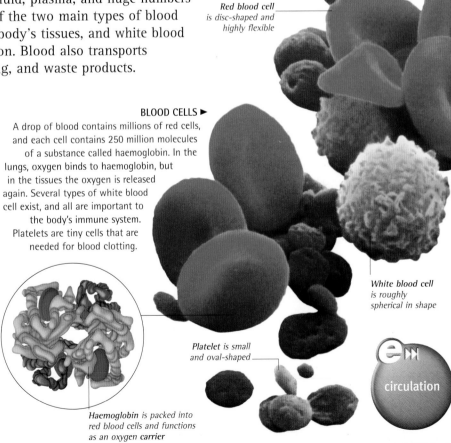

Red blood cell is disc-shaped and highly flexible

White blood cell is roughly spherical in shape

Platelet is small and oval-shaped

BLOOD CELLS ▶
A drop of blood contains millions of red cells, and each cell contains 250 million molecules of a substance called haemoglobin. In the lungs, oxygen binds to haemoglobin, but in the tissues the oxygen is released again. Several types of white blood cell exist, and all are important to the body's immune system. Platelets are tiny cells that are needed for blood clotting.

Haemoglobin is packed into red blood cells and functions as an oxygen carrier

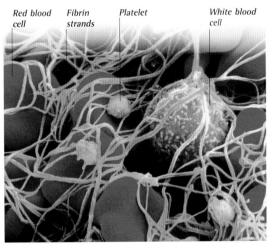

Red blood cell · *Fibrin strands* · *Platelet* · *White blood cell*

▲ BLOOD CLOTTING
If a blood vessel is damaged, a clot forms to stop blood leaking. First, platelets stick together to form a plug that stops the leak. At the same time, a complex sequence of chemical events in the blood leads to the production of long strands of a protein called fibrin. These bind the blood cells and debris together to form a gel-like clot that gradually solidifies. The solid clot remains until the blood vessel has been repaired.

circulation

FIND OUT MORE ▶▶ Immune System 357 • Muscular System 342–343 • Respiratory System 354–355

RESPIRATORY SYSTEM

The respiratory system provides your body cells with the oxygen they need and rids them of carbon dioxide, a waste product. Playing a central part in this process are the lungs, two organs in the chest that work closely with the blood circulation. **BREATHING** is the cycle of moving air into and out of the lungs. Structures in the respiratory system are also vital to **SPEECH**.

▼ RESIN CAST OF LUNG AIRWAYS
When air is breathed in, it passes down the trachea. This divides into two airways called main bronchi, which go to the two lungs. Each splits into smaller bronchi, which then split into bronchioles. These terminate in groups of tiny air sacs called alveoli.

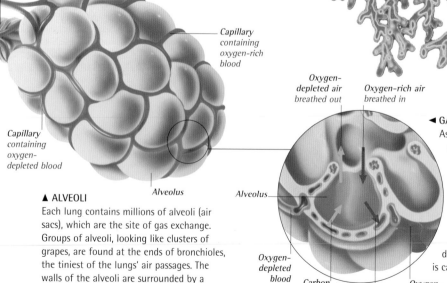

Cartilage rings keep the walls of the trachea open to allow air to pass

Trachea (windpipe) is a wide tube in the upper chest

Left main bronchus

Right main bronchus is one of two large airways that connect the trachea to the lungs

Small bronchi branch off the main bronchus

Group of alveoli

Bronchioles are the tiniest and most numerous air passages in the lungs, branching off small bronchi

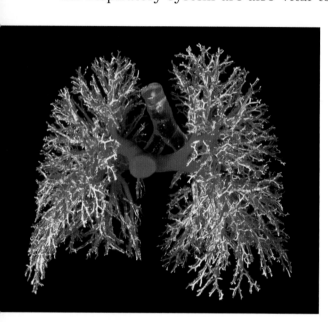

▲ BLOOD SUPPLY
Blood is carried from the heart to the lungs by the pulmonary arteries. They are the thick red vessels at the centre of this photograph of a resin cast of two lungs. The pulmonary arteries split into many branches, forming an intricate network of vessels that carry blood to the lungs' alveoli. There, oxygen enters the blood, and carbon dioxide leaves it.

End of a bronchiole

Capillary containing oxygen-rich blood

Capillary containing oxygen-depleted blood

Alveolus

▲ ALVEOLI
Each lung contains millions of alveoli (air sacs), which are the site of gas exchange. Groups of alveoli, looking like clusters of grapes, are found at the ends of bronchioles, the tiniest of the lungs' air passages. The walls of the alveoli are surrounded by a dense network of capillaries carrying blood.

Oxygen-depleted air breathed out

Oxygen-rich air breathed in

Alveolus

Oxygen-depleted blood

Carbon dioxide

Oxygen

Oxygen-rich blood

◄ GAS EXCHANGE IN ALVEOLI
As oxygen-depleted blood passes close to the wall of an alveolus, carbon dioxide passes from the blood into the alveolus. At the same time, oxygen passes from the alveolus into the blood, where it binds with haemoglobin in red blood cells. The swapping of carbon dioxide for oxygen in the lungs is called gas exchange.

JOHN SCOTT HALDANE
Scottish, 1860-1936

In 1905, the scientist J S Haldane made the important discovery that the urge to breathe is caused by a build-up of carbon dioxide in the blood. As the blood level of carbon dioxide rises, this is detected by a small region in the brain, which triggers quicker breathing.

SPEECH

Our ability to speak relies on the presence of two folds of tissue called vocal cords in the larynx (voice box) at the top of the trachea. As air passes between the cords when we breathe out, they vibrate. During speech, a centre in the brain sends signals to tiny muscles that alter the position and length of the cords, producing different sounds. These are modified into meaningful speech by movements of the lips, cheeks, and tongue.

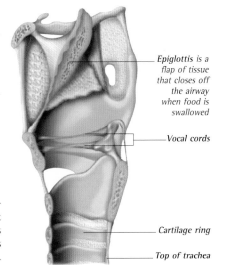

Epiglottis is a flap of tissue that closes off the airway when food is swallowed

Vocal cords

Cartilage ring

Top of trachea

CROSS-SECTION THROUGH THE LARYNX ▶
The larynx lies between the back of the pharynx (throat) and the top of the trachea. The vocal cords stretch across the larynx. When air from the lungs passes through them, they vibrate to produce sounds.

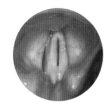

CLOSED VOCAL CORDS

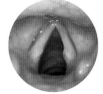

OPEN VOCAL CORDS

BREATHING

Breathing is the process of drawing air into the lungs and then expelling it again. Adults breathe at a rate of around 12-15 times per minute at rest but at a faster rate during exercise. With each breath, the lungs take in around 0.5 litres (1 pint) of air.

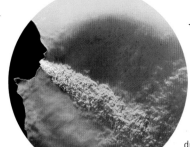

◀ COUGHING
If dust or germs enter the respiratory system, they can irritate the larynx, trachea, or bronchi. This may trigger coughing. When you cough, muscles in the chest and abdomen contract suddenly, increasing air pressure within the lungs. As a result, a spray of liquid drops containing dust and other unwanted material is forced out. Germs can pass from one person to another in this way.

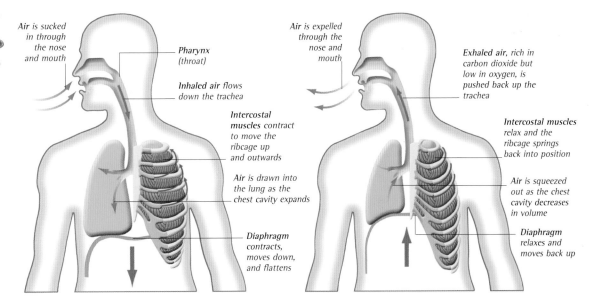

Air is sucked in through the nose and mouth

Pharynx (throat)

Inhaled air flows down the trachea

Intercostal muscles contract to move the ribcage up and outwards

Air is drawn into the lung as the chest cavity expands

Diaphragm contracts, moves down, and flattens

Air is expelled through the nose and mouth

Exhaled air, rich in carbon dioxide but low in oxygen, is pushed back up the trachea

Intercostal muscles relax and the ribcage springs back into position

Air is squeezed out as the chest cavity decreases in volume

Diaphragm relaxes and moves back up

respiration

▲ INHALATION
During inhalation, the intercostal muscles between the ribs contract. So does the diaphragm, a muscular sheet at the base of the chest cavity. The ribcage expands, and the diaphragm flattens, which increases the size of the chest cavity. This increase in chest volume causes the pressure of air in the lungs to be lower than the pressure of the air outside the body. As a result, air is drawn down the trachea into the lungs.

▲ EXHALATION
During exhalation, the intercostal muscles relax, and so does the diaphragm. The ribs move downwards and inwards, causing the ribcage to contract, and the diaphragm moves up. As the volume of the chest cavity decreases, the pressure of air within the lungs becomes higher than the pressure in the air outside the body. As a result, air moves back up the trachea and is expelled to the outside through the nose and mouth.

FIND OUT MORE ▶▶ Circulatory System 352–353 • Nervous System 344–345 • Oxygen 39 • Pressure 74–75

ENDOCRINE SYSTEM

Many body processes are influenced by hormones, chemical messengers produced by glands of the endocrine system. These glands release their hormones into the blood. The hormones are then carried to the parts of the body whose activities they influence. The endocrine system works closely with the nervous system to maintain the body in a stable state (homeostasis).

endocrine system

ENDOCRINE GLANDS ►
As well as helping to maintain homeostasis, hormones play roles in metabolism (chemical processes throughout the body), reproduction, growth, and response to stress. The production of many hormones is controlled by a feedback system; glands are kept informed of what is happening in the body and adjust the amount of hormone they produce appropriately.

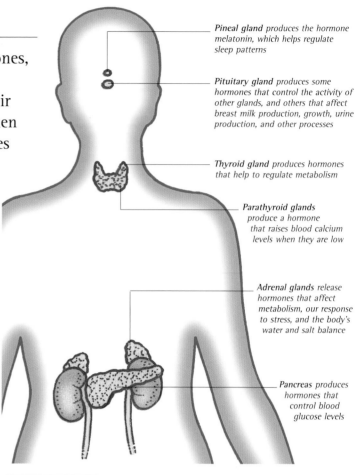

Pineal gland produces the hormone melatonin, which helps regulate sleep patterns

Pituitary gland produces some hormones that control the activity of other glands, and others that affect breast milk production, growth, urine production, and other processes

Thyroid gland produces hormones that help to regulate metabolism

Parathyroid glands produce a hormone that raises blood calcium levels when they are low

Adrenal glands release hormones that affect metabolism, our response to stress, and the body's water and salt balance

Pancreas produces hormones that control blood glucose levels

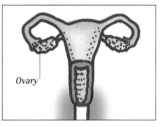

Ovary

Testis

▲ OVARIES
The ovaries produce oestrogen and progesterone. They are involved in the development of female sexual characteristics at puberty.

▲ TESTES
The testes produce testosterone, which stimulates the production of sperm and the development of male sexual characteristics at puberty.

Beta cell secretes insulin, which promotes absorption of glucose by body cells

Alpha cell secretes glucagon, which works to increase blood glucose levels

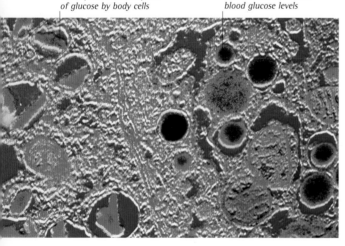

▲ HORMONE FACTORY
This false-colour micrograph shows a small region of the pancreas. Two types of cell in the pancreas release hormones that control the concentration of glucose (a simple type of sugar) in the blood. When the glucose levels are too high, the beta cells release insulin. When glucose levels are too low, the alpha cells release glucagon.

AN EXAMPLE OF HOW HORMONES WORK

The diagram below gives an example of what hormones do in the body. One hormone made by the pituitary is thyroid-stimulating hormone (TSH). TSH travels to the thyroid gland, where it stimulates release of another hormone, thyroxine. Thyroxine acts on body cells to divide and release energy, but it also controls its own level in the blood.

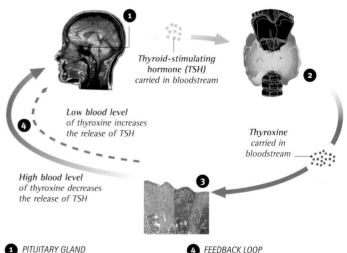

Thyroid-stimulating hormone (TSH) carried in bloodstream

Low blood level of thyroxine increases the release of TSH

High blood level of thyroxine decreases the release of TSH

Thyroxine carried in bloodstream

1 PITUITARY GLAND
secretes thyroid-stimulating hormone (TSH)

2 THYROID GLAND
is stimulated by TSH to make the hormone thyroxine

3 BODY CELLS
are stimulated by thyroxine to divide, process nutrients, and release energy

4 FEEDBACK LOOP
Thyroxine controls its own level in the blood by means of a feedback loop. If blood levels of thyroxine get too high, this slows the release of TSH by the pituitary gland. If blood levels of thyroxine drop too low, this speeds up release of TSH. In this way, the level of thyroxine in the blood is kept fairly constant all the time.

FIND OUT MORE ►► Digestive System 358–359 • Nervous System 344–345 • Reproductive System 362–363

IMMUNE SYSTEM

The role of the immune system is to protect the body against germs and cancers. When it detects substances that it recognizes as abnormal or foreign to the body, the system mounts an **IMMUNE RESPONSE**. Its chief weapons are cells called lymphocytes. Some of these are carried in lymph – excess body fluid that drains into the blood via a system of vessels and nodes.

Incoming lymph vessel brings lymph to the node

Outgoing lymph vessel drains lymph from the node

Sinus is a channel containing macrophages, which filter unwanted organisms and material out of the lymph as it passes through

Valve

immune system

Capsule of fibrous tissue surrounds the lymphatic tissue

Venule drains blood from the node

Arteriole supplies blood to the node

INSIDE A LYMPH NODE ▲
Lymph nodes are swellings of tissue found at intervals along lymph vessels. They make and store lymphocytes, which are added to lymph before it joins the blood. They also house other types of white cell called macrophages. Lymph nodes vary from 0.1 to 3 cm (¹⁄₂₅ to 1¼ in) in diameter.

Lymphocyte (pink) combats invading organisms and cancer cells

Macrophage (brown) destroys bacteria and other unwanted matter by engulfing it

THE IMMUNE RESPONSE

When the immune system detects an invading organism, it mounts a response in two main ways. Some lymphocytes attack the invader directly. Others produce substances called antibodies that promote the organism's destruction. The immune system retains a memory of different invaders and mounts a more rapid and effective response when it encounters one for the second time. This form of memorized protection is called immunity.

LYMPHOCYTE IN ACTION ►
Here, a lymphocyte (blue) is swallowing a yeast cell (yellow) that it recognizes as alien to the body. Lymphocytes, like macrophages, are types of white blood cell, though they are found in the lymphatic system as well as in blood.

Tonsils produce lymphocytes that attack infectious organisms in inhaled air and in food

Thymus, where some lymphocytes grow before going to lymph tissue around the body

Lymph nodes are collections of lymph tissue

Spleen contains lymphocytes and macrophages

Peyer's patch is a collection of lymph tissue in the lining of the small intestine

Lymph vessels transport lymph from the body's tissues to the blood

THE LYMPHATIC SYSTEM ▲
The system consists of a network of vessels and collections of tissue called lymphoid tissue. It collects excess fluid (lymph) from the body's tissues, returns this fluid to the blood, and stores and transports cells of the immune system, such as lymphocytes.

FIND OUT MORE ►► Circulatory System **352–353** • Disease **370–371**

DIGESTIVE SYSTEM

The job of the digestive system is to break down the food we eat into smaller units called nutrients. The nutrients are then absorbed into the bloodstream and fuel the body's activities. The **MOUTH** takes in food and begins the digestive process, which continues in the **STOMACH**. The food then passes to the **INTESTINES**, which complete the process. Enzymes – substances that speed up chemical reactions in the body – play a key part in breaking down food.

digestion

PARTS OF THE DIGESTIVE SYSTEM ▶
The digestive system consists mainly of a long muscular tube, the digestive tract. This starts at the mouth, continues via the oesophagus and stomach to the intestines, and ends at the anus. The system also relies on the pancreas, liver, and gall bladder to help digest food. Each day the pancreas releases about 1.5 litres (2½ pints) of enzyme-containing juice into the tract. The liver produces about 1 litre (1¾ pints) of bile, which is stored temporarily in the gall bladder.

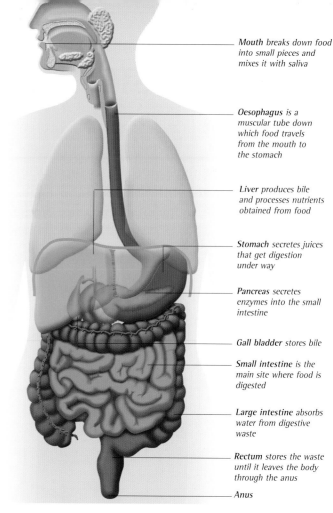

Mouth breaks down food into small pieces and mixes it with saliva

Oesophagus is a muscular tube down which food travels from the mouth to the stomach

Liver produces bile and processes nutrients obtained from food

Stomach secretes juices that get digestion under way

Pancreas secretes enzymes into the small intestine

Gall bladder stores bile

Small intestine is the main site where food is digested

Large intestine absorbs water from digestive waste

Rectum stores the waste until it leaves the body through the anus

Anus

MOUTH

The mouth is a cavity formed between the tongue at its base, cheeks at the side, hard and soft palate in its roof, and teeth at the front. The teeth tear and grind food, which is then churned through movements of the jaws and tongue. Breaking the food into smaller pieces creates a larger surface area for the action of enzymes in saliva; these begin to digest the food.

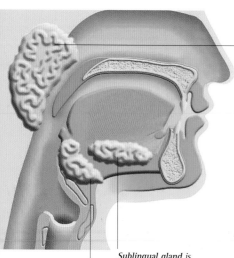

Submandibular gland is found deep in the floor of the mouth

Sublingual gland is the smallest salivary gland, situated under the tongue

Parotid gland lies in front of the ear and opens via a short tube onto the inner surface of the cheek

◄ SALIVARY GLANDS
These glands produce a mucus-rich fluid, saliva, that moistens food and so helps form it into a ball, called a bolus, that is easily swallowed. Saliva also contains enzymes that begin the digestive process. When food enters the mouth, the brain sends messages to the glands, triggering a rush of saliva. Food must mix with saliva before it can stimulate the taste buds, which check that the food is safe to eat.

Enamel

Dentine

Nerve

Pulp cavity

Root secures tooth to jawbone

◄ STRUCTURE OF A TOOTH
Each tooth has a crown, which can be seen above the gumline, and one or more roots, which are embedded in the jawbone. In the centre of the tooth lies the pulp cavity, which contains blood vessels and nerves. This is surrounded by dentine, a strong material that forms most of the tooth. The outer layer of the crown is made of enamel, the hardest substance in the body.

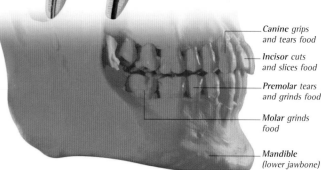

Canine grips and tears food

Incisor cuts and slices food

Premolar tears and grinds food

Molar grinds food

Mandible (lower jawbone)

▲ TYPES OF TEETH
Teeth come in four main types: incisors, canines, premolars, and molars. Each is shaped for a specific function. People have two sets of teeth during their lives: the 20 milk teeth emerge between the ages of about six months and three years. The 32 permanent teeth (like those seen here) emerge from the age of about six years into the early twenties.

STOMACH

The stomach is a muscular bag that begins the digestive process. On entering the stomach, food is mixed with gastric juices and churned by contractions of muscles in the stomach wall. Food usually stays in the stomach for about four hours. By that time, it has become a semi-liquid (called chyme) that can be released in spurts into the duodenum, the first part of the small intestine.

Oesophagus delivers food to the stomach from the mouth

INSIDE THE STOMACH ►
This computer artwork shows the stomach, part of the oesophagus, and the duodenum. When food enters the stomach, glands in its lining release an enzyme, pepsin. This begins to break down protein in food. The lining also produces hydrochloric acid, which maintains the acid environment needed for pepsin to work and kills any bacteria present.

Pyloric sphincter is an opening that can enlarge to allow chyme to enter the duodenum

STOMACH LINING ►
This micrograph shows the openings of glands in the stomach lining. These glands, which are visible as pits, release hydrochloric acid and digestive enzymes into the stomach cavity. Mucus protects the lining from the acid and enzymes. To keep it healthy, the cells of the stomach lining are replaced every few days.

Duodenum receives chyme from the stomach and digestive juices from the pancreas and gall bladder

Stomach wall has folds called rugae that allow the stomach to expand by up to 20 times to fill with food

INTESTINES

The intestines form a continuous tube that is about 8 m (26 ft) long in adults. Food is pushed along this tube by contractions of muscles in the intestinal walls. In the small intestine, muscle contractions mix chyme with enzymes produced by the pancreas and intestinal lining. It is here that nutrients are absorbed into the bloodstream. In the large intestine, water is absorbed from the digestive waste, and the remaining waste is formed into stools.

◄ VILLI OF SMALL INTESTINE
The lining of the small intestine has a huge surface area due to the folds in its lining as well as villi, tiny fingerlike structures that project from the lining. Once digested, nutrients are absorbed into tiny blood vessels within the villi. They are then carried in the blood to the liver for processing. Food can remain in the small intestine for up to five hours.

◄ INSIDE THE LARGE INTESTINE
This photograph was taken with a colonoscope, a bendy light-viewing tube inserted through the anus and up the colon, the main part of the large intestine. The triangular shape is due to the arrangement of muscles in the colon's walls. Waste may spend a day or more in the large intestine. Cells in its lining secrete a mucus that helps bind the waste to form stools.

◄ INTESTINAL BACTERIA
In this micrograph of the lining of the small intestine, some bacteria (purple) can be seen with food debris (cream-coloured). The intestines are home to more than 500 species of bacteria, which they need to function effectively. Some of these bacteria are thought to protect the intestines from disease. Certain bacteria in the large intestine make vitamin K, which the body needs for blood clotting.

FIND OUT MORE ►► Acids 32 • Bacteria 284 • Biochemistry 46–47 • Liver 360 • Taste 346

LIVER

The liver has many important jobs in the body. It produces the digestive juice bile and processes nutrients to be used by the body. It plays a role in regulating blood sugar levels and stores many important substances such as iron and some vitamins. Other functions of the liver include making proteins needed for blood clotting, breaking down old red blood cells, and removing or breaking down any toxic substances that appear in the blood.

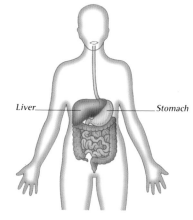

▲ LOCATION OF THE LIVER
The liver is situated mainly on the right side of your upper abdomen. One part of it partially covers the front of the stomach. It is the body's heaviest internal organ, weighing up to 1.6 kg (3½ lb) in an adult.

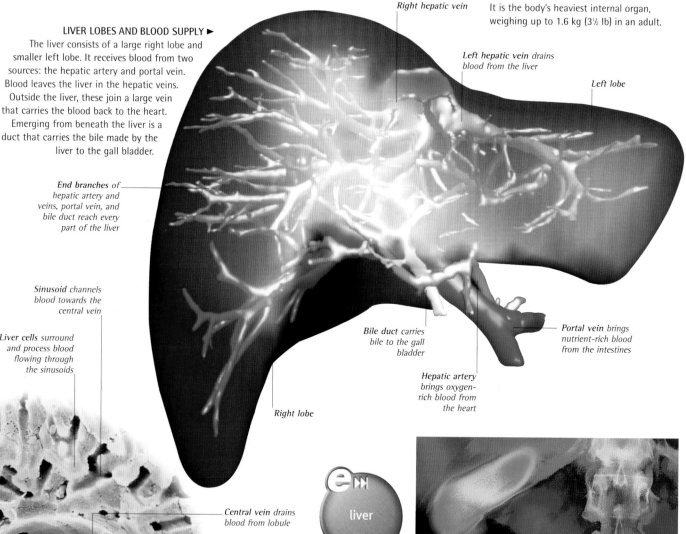

LIVER LOBES AND BLOOD SUPPLY ▶
The liver consists of a large right lobe and smaller left lobe. It receives blood from two sources: the hepatic artery and portal vein. Blood leaves the liver in the hepatic veins. Outside the liver, these join a large vein that carries the blood back to the heart. Emerging from beneath the liver is a duct that carries the bile made by the liver to the gall bladder.

End branches of hepatic artery and veins, portal vein, and bile duct reach every part of the liver

Sinusoid channels blood towards the central vein

Liver cells surround and process blood flowing through the sinusoids

Right hepatic vein

Left hepatic vein drains blood from the liver

Left lobe

Bile duct carries bile to the gall bladder

Portal vein brings nutrient-rich blood from the intestines

Hepatic artery brings oxygen-rich blood from the heart

Right lobe

Central vein drains blood from lobule

◀ LIVER LOBULE
This micrograph shows part of one of the thousands of tiny processing units, called lobules, in the liver. Blood flows through channels called sinusoids in a lobule, past groups of liver cells, and towards a central vein. As the blood flows past, the liver cells absorb some substances from it. They also release other substances into the blood.

▲ THE GALL BLADDER
This colour-enhanced X-ray shows the gall bladder (green), a small sac that lies beneath the liver. The gall bladder stores bile made by the liver and releases the bile into the small intestine when food enters the intestine from the stomach. Bile is a greenish fluid made of material produced from the breakdown of old red blood cells. It plays a key role in fat digestion.

URINARY SYSTEM

The urinary system is responsible for ridding the body of many of its waste products. It also helps to maintain a stable environment by regulating the composition of body fluids. The waste substances are removed from the blood and expelled in urine, which is produced by the kidneys. Together, these two organs receive about one-quarter of the blood pumped out by the heart, and yet they contribute less than 1 per cent to our body weight.

Renal capsule encases and protects the kidney

Blood vessels consist of an artery that brings blood to the kidney and a vein that carries blood away

Renal pelvis collects urine before it passes into the ureter

Ureter delivers urine to the bladder

Renal pyramid contains thousands of urine-collecting ducts

Renal cortex is the outer layer of the kidney, where nephrons make urine

▲ KIDNEY SCAN
This coloured CT (computed tomography) scan shows a slice through a kidney. The kidneys filter the blood they receive so that waste substances and excess water pass into the urine. Substances that the body needs stay in the blood. The kidneys each contain huge numbers of nephrons, tiny structures that make urine and fine-tune its composition to maintain a stable environment in the body.

Right kidney lies at the back of the abdomen, roughly behind the liver

Left kidney is slightly higher than the right kidney and lies behind the stomach

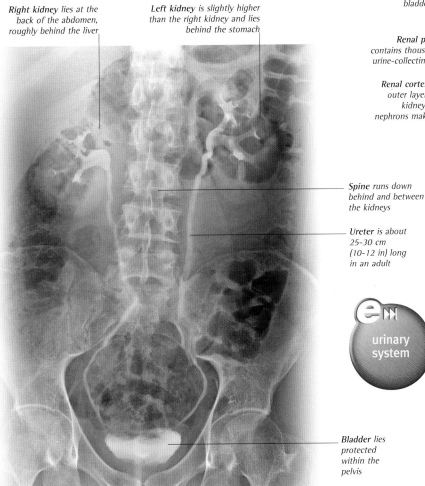

Spine runs down behind and between the kidneys

Ureter is about 25-30 cm (10-12 in) long in an adult

urinary system

Glomerulus is a tiny knot of blood vessels

Arteriole supplies blood to each glomerulus

Bladder lies protected within the pelvis

▲ PARTS OF THE URINARY SYSTEM
This colour-enhanced X-ray shows the main parts of the system. The kidneys make urine, which flows down the ureters to the bladder. The bladder stores urine and expels it from the body via the urethra, a tube that is about 20 cm (8 in) long in men and 4 cm (1½ in) long in women. Signals from the brain and spinal cord trigger emptying of the bladder.

KIDNEY GLOMERULI ▶
Each kidney contains about one million glomeruli, tiny clusters of blood capillaries. Each glomerulus forms part of a nephron, one of the kidney's filtering units. A glomerulus is the part of a nephron where water and waste products are forced out of the blood and carried away via a system of little tubes and ducts as urine.

FIND OUT MORE ▶▶ Circulatory System 352–353 • Endocrine System 356

REPRODUCTIVE SYSTEM

The male and female parts of the human reproductive system can bring together an egg and a sperm so that they join and begin the development of a new human being. The female system also protects and nourishes the developing foetus during the nine months of pregnancy. About 150,000 eggs are present in a girl's ovaries at birth; however, sperm production in boys begins only at the start of puberty.

MALE REPRODUCTIVE SYSTEM	FEMALE REPRODUCTIVE SYSTEM
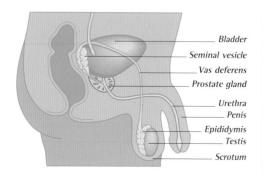 Bladder Seminal vesicle Vas deferens Prostate gland Urethra Penis Epididymis Testis Scrotum	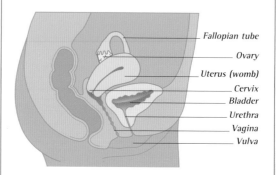 Fallopian tube Ovary Uterus (womb) Cervix Bladder Urethra Vagina Vulva
Sperm are made in the testes and mature in a coiled tube, the epididymis. At the end of sexual intercourse, they travel along the vas deferens and mix with fluids made by the prostate gland and seminal vesicle. The sperm then pass along the urethra to the tip of the penis and into the female vagina. From there they begin their journey up the female reproductive tract.	From puberty, an egg is released monthly from one of the ovaries and passes along the adjoining fallopian tube towards the uterus. The ovaries also produce female sex hormones. The uterus is the home for the foetus during pregnancy. The outlet of the uterus is called the cervix. It stretches wide open during childbirth to allow the baby to pass through.

Baby's vertebral column is curved due to its curled-up position in the uterus

Baby's neck

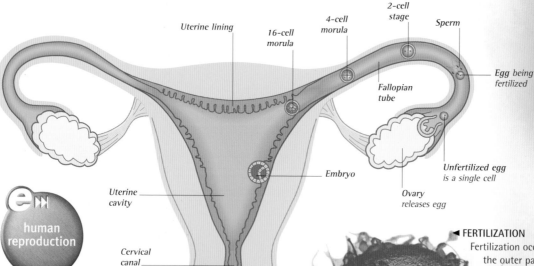

Uterine lining
16-cell morula
4-cell morula
2-cell stage
Sperm
Egg being fertilized
Fallopian tube
Embryo
Unfertilized egg is a single cell
Ovary releases egg
Uterine cavity
human reproduction
Cervical canal
Vagina

JOURNEY OF SPERM AND EGG ▲

Millions of sperm leave the penis during sexual intercourse and are deposited in the vagina. From there they travel through the cervical canal into the uterine cavity and then into the fallopian tubes. About 500 sperm reach the tubes. If one fertilizes an egg, the tiny fertilized egg begins to grow by dividing – into two cells, then four, and so on. As it divides, it travels along the fallopian tube to the uterus. There, it embeds in the uterine lining as an embryo.

◄ FERTILIZATION

Fertilization occurs in the outer part of the fallopian tube. Once they reach the egg, the sperm release substances that allow them to break through the egg's outer layers. Only one sperm penetrates the egg. The egg, and the sperm that successfully penetrates it, each provide half the genetic information needed to form a new individual.

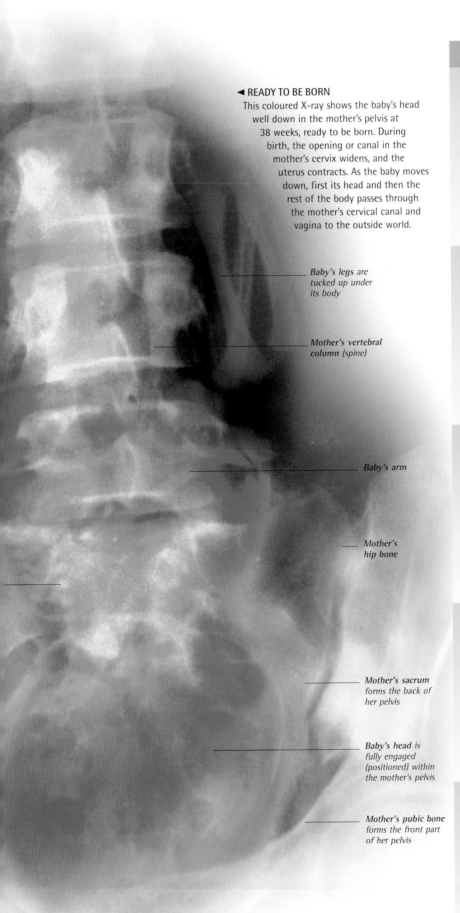

◄ READY TO BE BORN
This coloured X-ray shows the baby's head well down in the mother's pelvis at 38 weeks, ready to be born. During birth, the opening or canal in the mother's cervix widens, and the uterus contracts. As the baby moves down, first its head and then the rest of the body passes through the mother's cervical canal and vagina to the outside world.

Baby's legs are tucked up under its body

Mother's vertebral column (spine)

Baby's arm

Mother's hip bone

Mother's sacrum forms the back of her pelvis

Baby's head is fully engaged (positioned) within the mother's pelvis

Mother's pubic bone forms the front part of her pelvis

FROM FERTILIZED EGG TO NEWBORN BABY

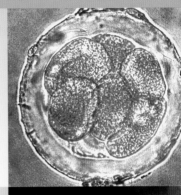

MORULA ►
A fertilized egg divides to form a ball of cells called a morula. This reaches the uterus a few days after fertilization, moved along by wafting cilia, hairlike structures in the lining of the fallopian tube. The early embryo then embeds itself in the uterine lining.

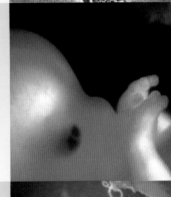

FOETUS AT 10 WEEKS ►
By this time, the embryo's cells have developed and grouped together to form tissues and organs. Most organs are formed by eight weeks, and the embryo is now called a foetus. At 10 weeks, the foetus is 5 cm (2 in) long and has facial features and limbs.

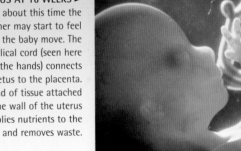

FOETUS AT 18 WEEKS ►
At about this time the mother may start to feel the baby move. The umbilical cord (seen here behind the hands) connects the foetus to the placenta. This pad of tissue attached to the wall of the uterus supplies nutrients to the foetus and removes waste.

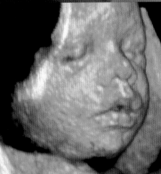

FOETUS AT 30 WEEKS ►
The baby's facial features can clearly be seen on this 3-D ultrasound image. Ultrasound scanning is often used during pregnancy to check foetal health. The baby is growing fast now and maturing in preparation for its journey into the outside world.

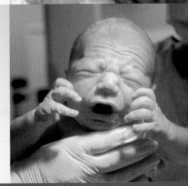

NEWBORN BABY ►
Once delivered, the baby starts breathing. The umbilical cord is clamped shut and cut, and the baby is then immediately given to the mother to suckle. The remaining stump of umbilical cord falls off a few days later, leaving the umbilicus (tummy button).

FIND OUT MORE ►► Endocrine System **356** • Genetics **364–365** • Growth **366–367** • Reproduction **308–309**

GENETICS

Genetics is the study of the instructions, contained in cells, for how our bodies develop and function. These instructions are called **GENES** and are carried by tiny objects called chromosomes in the nuclei (centres) of cells. Packed into the chromosomes is a chemical called deoxyribonucleic acid (DNA), which holds the instructions in coded form. People's genes come from their parents. The passing on of biological characteristics from parents to children via their genes is called **INHERITANCE**.

LOCATION OF CHROMOSOMES ▶
Chromosomes are tiny structures in the nucleus at the centre of each body cell. Each cell carries 46 chromosomes. Just before a cell divides, each chromosome makes a copy of itself. In this way, a complete set of chromosomes can pass into each of the two new cells that form during division.

Chromosome

Cell nucleus

DNA coil

Centromere

Chromosome arm

Gene

Gene

Gene

genetics

▲ CHROMOSOME STRUCTURE
When a chromosome copies itself just before cell division, for a short time the chromosome and its copy are joined at a point called the centromere. This gives each chromosome an X shape, with four arms extending from the centromere. Each arm consists of a long threadlike molecule of DNA, coiled up on itself. A gene is a section of the DNA molecule.

Nucleotide bases

Sugar-phosphate chain

▲ DNA DOUBLE HELIX
If the DNA molecule could be unravelled, it would reveal a structure like a twisted ladder, called the double helix. The sides of this ladder consist of a chainlike substance called sugar-phosphate. The rungs of the ladder are formed from chemicals called nucleotide bases. There are four different types of base. Their sequence within a gene (a long section of DNA) is a code that holds the instructions carried by that gene.

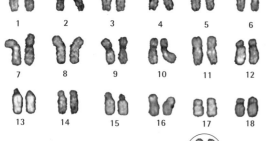

1	2	3	4	5	6
7	8	9	10	11	12
13	14	15	16	17	18
19	20	21	22		

SEX CHROMOSOMES (FEMALE)
X X

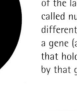

X Y

MALE VERSION

▲ CHROMOSOME SET
A full chromosome set consists of 23 pairs of chromosomes. Twenty-two of these chromosome pairs are the same in girls and boys. The last pair, called sex chromosomes, differ between the sexes. Girls have two sex chromosomes called X chromosomes. Boys have just one X chromosome and a smaller partner, called the Y chromosome. This Y chromosome contains genes that give boys their male characteristics.

GENES

A gene is a section of DNA in a chromosome. People receive half their genes from their mothers and half from their fathers, via the egg and sperm cells from which they grew. Every gene plays a part in determining how a person looks and functions. Genes influence the body through complex mechanisms in cells that translate the coded messages in DNA into the activities of each cell. The differences between people result partly from tiny variations in their genetic makeup – that is, in the sequence of bases in their DNA.

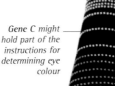

Gene A might direct the synthesis of a substance required for releasing energy from food

Gene B might be the code for a substance that protects the body against cancer

Gene C might hold part of the instructions for determining eye colour

◀ CHROMOSOME MAPPING
Different genes on a single human chromosome may hold the instructions for a wide range of body activities. Scientists are gradually building up maps that show the locations of specific genes on each chromosome. As they work out what each gene does, they add this information to the maps. Altogether, there are probably between 30,000 and 40,000 genes on the human chromosomes.

◀ DNA SEQUENCING
This photograph depicts a tiny part of the DNA sequence of one person, as viewed on a computer screen during a huge worldwide scientific study called the Human Genome Project. One aim of the project was to work out the exact sequence in which the four nucleotide bases occur in human DNA. This aim has been achieved. Scientists are now studying the variations in the sequence that make people different from each other.

GENETIC SIMILARITY

Everyone's DNA sequence is amazingly similar. The numbers below indicate in percentage terms exactly how similar. Human DNA is also very like that of chimpanzees, our closest animal relatives.

Identical twins	100%
Two brothers or sisters	99.95%
Any two unrelated humans	99.9%
Any human and a chimpanzee	99%

INHERITANCE

Family members often resemble each other in traits (characteristics) controlled by genes, such as eye colour. Some traits are determined by the combined action of just two genes, one from each parent. Each member of this gene pair exists as one of two or more forms, with differing effects. One of these forms, called the dominant form, may mask the action of others and only has to be inherited in a single dose to produce its specific effect. Other gene forms, called recessive, have no effect unless they are inherited in a double dose, one from each parent.

INHERITED TRAIT ▶
An example of a simply inherited trait is the ability to taste phenylthiocarbamide (PTC), a bitter substance found in some fruits and vegetables. Only about two-thirds of us can taste PTC. To be a PTC taster depends on inheriting at least one copy of the dominant "taster" gene for this characteristic. People are nontasters only if they inherit an alternative "nontaster" gene from both parents.

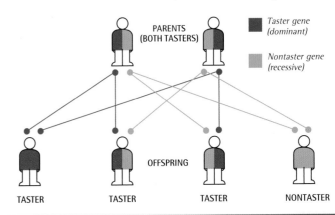

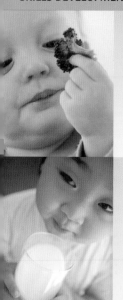

◄ 12 MONTHS
By around this age, many babies can grasp an object between forefinger and thumb, can eat with their fingers, walk holding on to furniture, and can say "dada" and "mama".

◄ 18 MONTHS
Many babies of this age have progressed to drinking from a cup. They can take off shoes and socks, turn pages, and enjoy scribbling. Some can point to their eyes, nose, and mouth.

◄ 2 YEARS
Toddlers aged around 2 years can usually build a tower of four bricks, kick a ball, point to parts of the body, eat with a fork and spoon, and undress without help. Some are able to draw a straight line.

◄ 3 YEARS
Children aged around 3 years can usually eat with a knife and fork. Most can copy a circle, talk in short sentences, pedal a tricycle, and run fast. Many know their first and last names.

◄ 4 YEARS
Many 4-year-old children are able to dress without help, and can draw a simple picture of a man. They can copy a square and a cross, can count up to 10, and can brush their own teeth.

◄ 6 YEARS
By 6 years of age, some children can tie their own shoelaces. They can bounce and catch a ball, copy a triangle, and can speak to others fluently and clearly. They can draw a detailed picture of a man.

GROWTH

The processes of growth and development continue throughout life. They include physical changes, such as the increase in height that occurs throughout childhood, and mental changes, such as the continual development of new skills from early childhood onwards. One distinct phase of growth is **PUBERTY**, the time when boys' and girls' reproductive systems mature. As life continues, the body has a constant need for new cells to repair its worn out parts. **AGEING** is a natural part of the life process that results from some slowing of this capacity for self-renewal.

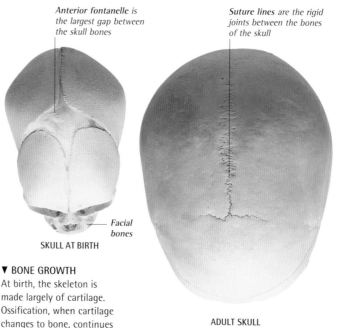

Anterior fontanelle is the largest gap between the skull bones

Suture lines are the rigid joints between the bones of the skull

Facial bones

SKULL AT BIRTH

ADULT SKULL

◄ SKULL GROWTH
The skull increases in size during childhood to accommodate the growing brain. At birth, the bones of the skull are separated by gaps, filled with fibrous tissue, called fontanelles. The anterior fontanelle closes at around 18 months. The posterior fontanelle, located farther back, closes at the age of 3 months.

▼ BONE GROWTH
At birth, the skeleton is made largely of cartilage. Ossification, when cartilage changes to bone, continues throughout childhood and adolescence. It is finished by the age of about 20 years. Cells called osteoblasts are responsible for this; they produce a substance that forms bone when calcium is added to it.

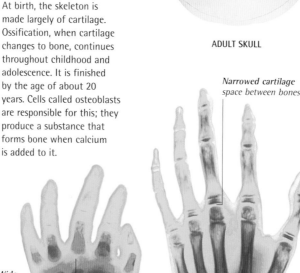

Joint has closed with bone replacing nearly all of the cartilage

Narrowed cartilage space between bones

Wide cartilage space

Wrist is made of cartilage

1-YEAR OLD

Wrist bones can now be seen

13-YEAR OLD

Ossification of wrist is complete

20-YEAR OLD

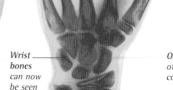

PUBERTY

Puberty occurs between the ages of about 10 and 14 in girls and between 12 and 15 in boys. Hormonal changes promote rapid growth, changes in body shape, and development of the reproductive organs. In girls, the menstrual cycle begins. In boys, the testes start to produce sperm. As childhood ends, boys and girls become more self-aware and independent.

Single acne bacterium splitting in two

◄ ACNE BACTERIA
A common condition affecting both boys and girls during puberty is a type of skin inflammation called acne. It is caused by excessive production of an oily substance, sebum, by the skin, as a result of hormonal changes. The sebum can block hair follicles, providing a site for bacteria to multiply.

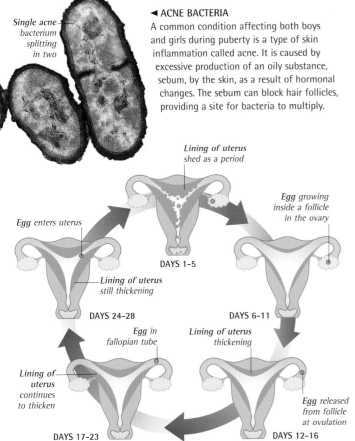

Lining of uterus shed as a period

Egg growing inside a follicle in the ovary

DAYS 1-5

Egg enters uterus

Lining of uterus still thickening

DAYS 24-28

DAYS 6-11

Egg in fallopian tube

Lining of uterus thickening

Lining of uterus continues to thicken

DAYS 17-23

Egg released from follicle at ovulation

DAYS 12-16

▲ MENSTRUAL CYCLE
During a girl's teenage years, the immature eggs present in her ovaries at birth begin to develop and are released on a monthly basis. The lining of the uterus thickens every month in preparation for a fertilized egg. However, fertilization does not usually occur and the lining is shed as a period. The menstrual cycle is controlled by hormones produced by the pituitary gland and by the ovaries.

FACIAL HAIR ►
As well as the pubic and armpit hair that appears in girls and boys at puberty, boys develop hair on the face and often the chest. This new hair growth is caused by an increase in the male hormone testosterone. In addition, a boy's voice deepens as his voice box gets bigger and the vocal cords lengthen.

AGEING

As the body ages, the turnover of its cells slows. The skin loses some of its elasticity and wrinkles develop. The bones of the skeleton slowly become less dense. The capacity of the body to repair itself is gradually reduced; wounds take longer to heal and broken bones longer to mend. The eye's capacity to focus lessens, and by the age of 50 many people need reading glasses. However, the lifelong accumulation of skills and knowledge means that many people find old age to be one of the most enjoyable parts of their lives.

▼ BONE THINNING
Loss of bone tissue is a normal part of ageing, but in some people this bone thinning is more severe and increases the risk of fractures. This X-ray shows a side view of some vertebrae (bones in the spine) affected by severe bone thinning, called osteoporosis. One vertebra has been compressed to form a wedge shape. This may result in the spine becoming curved.

growth

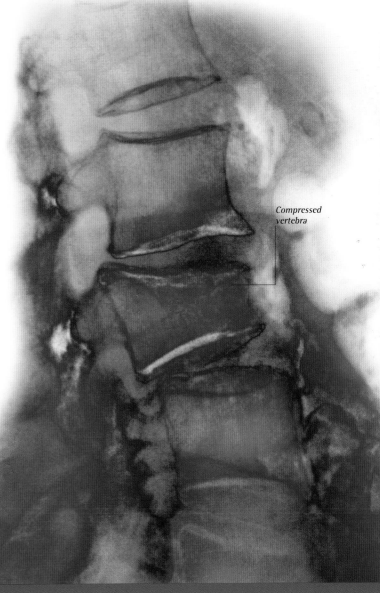

Compressed vertebra

FIND OUT MORE ►► Endocrine System 356 • Reproductive System 362–363 • Skeletal System 340–341

HEALTH

Many factors play a part in keeping your body and mind in good health. These include taking plenty of **EXERCISE**, eating a healthy **DIET**, and observing some basic rules of **HYGIENE**. Other aspects include getting enough sleep and having the health check-ups and disease prevention measures (such as vaccination) that are advised for each age group. By avoiding harmful habits such as smoking, people can also reduce their risk of many diseases.

EXERCISE

Exercising regularly to keep physically fit is a vital part of a healthy lifestyle. Even moderate exercise performed regularly helps protect against disease and prolongs life expectancy. As well as being enjoyable and improving strength and suppleness, regular exercise improves the efficiency of the muscles, the heart, and the whole circulatory system. It also helps prevent a person from becoming overweight, which has many negative effects on health.

▲ TAKING A PULSE
You can assess your fitness from some pulse measurements. Feel for your pulse as shown and count the number of beats in a minute. After five minutes exercise, you will find the pulse rate has increased. In a fit person, the rate returns to normal within a few minutes.

FITNESS BENEFITS OF EXERCISE

ACTIVITY	STAMINA	FLEXIBILITY	STRENGTH
Basketball	***	**	*
Cycling (fast)	***	*	**
Dancing	**	***	*
Hill walking	**	*	*
Jogging	***	*	*
Judo	*	***	*
Skipping (fast)	***	*	*
Soccer	***	*	*
Swimming	***	***	***
Tennis	*	**	*
Yoga	*	***	*

*KEY * GOOD EFFECT ** VERY GOOD EFFECT *** EXCELLENT EFFECT*

HYGIENE

There are two main aspects to hygiene. One is to observe some basic measures for minimizing the spread of germs. This includes washing hands before handling food and after going to the toilet or playing with a pet. The other aspect is simply a matter of keeping clean and odour-free. Taking a bath or shower regularly, and washing the body thoroughly all over using plenty of soap and water, is highly effective.

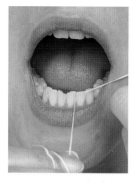

▲ FLOSSING YOUR TEETH
Maintaining your teeth and gums in good condition is important to overall health as well as for avoiding tooth decay and gum disease. Daily flossing, as well as brushing twice a day, can achieve this.

SWIMMING ▶
Swimming, cycling, jogging, and brisk walking are valuable forms of what is called aerobic exercise. This describes any activity that can be done continuously for at least 12 minutes and uses oxygen to provide energy for your muscles. Most people should aim to do some vigorous aerobic exercise at least three times a week, for at least 20 minutes each session, to stay fit and in good mental and physical shape.

DIET

A healthy diet provides all the nutrients you need to grow and function normally. These include proteins, which supply materials for growth and repair; carbohydrates and some fat, which give energy; and tiny amounts of various vitamins and minerals, which play many roles in the body. Your diet should also provide fibre (indigestible plant matter). Eating a wide mix of foods should automatically supply all the nutrients you need. For really good health, however, you should limit your intake of sugar-rich and fatty foods.

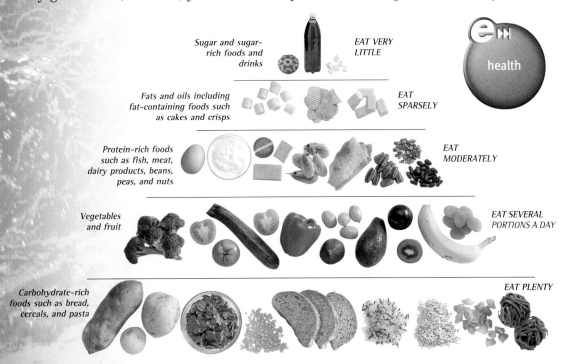

Sugar and sugar-rich foods and drinks — EAT VERY LITTLE

Fats and oils including fat-containing foods such as cakes and crisps — EAT SPARSELY

Protein-rich foods such as fish, meat, dairy products, beans, peas, and nuts — EAT MODERATELY

Vegetables and fruit — EAT SEVERAL PORTIONS A DAY

Carbohydrate-rich foods such as bread, cereals, and pasta — EAT PLENTY

▲ FOOD PYRAMID

One route to a healthy diet is to use a food pyramid. The bulk of a person's diet should consist of foods near the base of the pyramid: carbohydrate-rich foods, vegetables, and fruit. These supply energy, fibre, vitamins, and minerals. Protein-rich foods, in the middle row, should be eaten in moderate amounts. Foods at the top of the pyramid, which supply energy but are low in most nutrients, are best eaten only in small amounts.

Stir-fry ingredients include baby sweetcorn, broccoli, carrots, red chilli peppers, and asparagus

▲ FRESH FOOD

Healthy food is one that contains a mix of fresh ingredients, such as the components of this vegetable stir-fry. For a balanced meal, these might be combined with a carbohydrate-rich food, like rice, and a source of protein, such as some fried chicken. One advantage of including plenty of vegetables in the diet is that they are a good source of dietary fibre. This adds bulk to food and helps the functioning of the intestines.

SOME THREATS TO HEALTH

ACTIVITY	DANGER
Smoking	Tobacco smoking is very harmful to health. It causes many types of cancer, damages lungs, contributes to heart disease, and reduces life expectancy. Tobacco contains a substance, nicotine, that is highly addictive.
Alcohol use	Excessive use of alcohol can seriously damage some body organs and increases the risk of cancer. It can lead to addiction (difficulty abstaining) and mental problems. Limited alcohol use (up to a certain number of units per week as advised by a doctor) can, however, be good for an adult's health.
Drug use	Using drugs not prescribed or recommended by a doctor can cause some serious mental and physical health problems, depending on the drug. Addiction is common.

DISEASE

When a person has a disease, a part of that person's body fails to function properly. Diseases produce different patterns of symptoms: knowing these helps a doctor to recognize many different diseases. Three of the most common categories of disease are **HEART DISEASE, INFECTIOUS DISEASES**, and **CANCER**. Various factors can affect the risk of a disease developing; for example, smoking is known to increase the risk of many diseases, whereas regular exercise may reduce the risk.

SOME OTHER CATEGORIES OF DISEASE	
CATEGORY	*CAUSE AND EXAMPLES*
Nutritional deficiencies	Lack of an essential nutrient in the diet Example: scurvy, caused by lack of vitamin C
Genetic disorders	Inheritance of faulty genes Examples: cystic fibrosis, haemophilia
Degenerative diseases	Gradual loss of function in body parts Example: osteoarthritis
Endocrine disorders	Disturbance of the body's hormonal balance Example: diabetes mellitus
Autoimmune diseases	Immune system attacks the body's own tissues Example: rheumatoid arthritis
Allergies	Immune system's sensitivity to substances from outside the body Example: hay fever

HEART DISEASE

Heart disease is caused by narrowing of the coronary arteries, the vessels that supply blood to the heart muscle. Some factors that increase the risk of getting the disease are smoking, being overweight, a diet rich in animal fats, and a lack of exercise. Heart disease is common in rich countries. Drugs, and in some cases surgery, are used to treat it.

Atheroma, mainly consisting of cholesterol and other fatty substances

◀ NARROWED ARTERY
The narrowing of the coronary arteries in heart disease is caused by fatty deposits, called atheroma, on the walls of the arteries. This cross-section shows an artery that has been severely narrowed by atheroma. Atheroma can also develop in arteries supplying the brain. This can lead to a stroke, in which the blood supply to an area of brain is cut off.

Narrowed passage restricts blood flow

Artery wall

▲ CORONARY ARTERY DISEASE
This X-ray shows narrowing (circled) in one of the coronary arteries, which have been injected with dye. If an affected artery becomes blocked, the blood supply to an area of muscle is cut off and the affected muscle dies. This is called a heart attack and can be fatal.

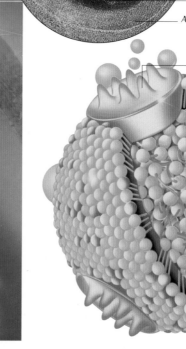

Protein molecule

Fatty acid molecule

Cholesterol molecule

▲ CHOLESTEROL CARRIER PARTICLE
A high concentration of particles like this in the blood is thought to increase the risk of atheroma. Called low-density lipoprotein particles, they are composed of fatty substances and some protein. The level of these particles in the blood tends to be higher if a person's diet is rich in fats of animal origin.

INFECTIOUS DISEASES

Infectious diseases are caused by organisms that invade body tissues or organs and affect their functioning. Infections can be passed between people and can affect any part of the body. The two main types of infectious organisms are bacteria and viruses. Other types include fungi, protozoa, and worms. Some infectious diseases can be treated with drugs.

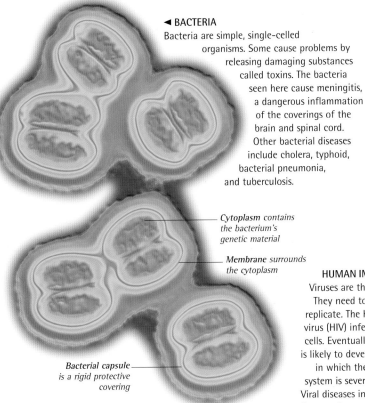

◀ BACTERIA
Bacteria are simple, single-celled organisms. Some cause problems by releasing damaging substances called toxins. The bacteria seen here cause meningitis, a dangerous inflammation of the coverings of the brain and spinal cord. Other bacterial diseases include cholera, typhoid, bacterial pneumonia, and tuberculosis.

Cytoplasm contains the bacterium's genetic material

Membrane surrounds the cytoplasm

Bacterial capsule is a rigid protective covering

Virus particle is seen here budding from the surface of an infected white blood cell

Surface of white blood cell, which is of a type known as a T-lymphocyte

HUMAN IMMUNODEFICIENCY VIRUS ▶
Viruses are the smallest infectious particles. They need to invade other living cells to replicate. The human immunodeficiency virus (HIV) infects certain white blood cells. Eventually, an infected person is likely to develop AIDS, a disease in which the body's immune system is severely weakened. Viral diseases include colds, flu, measles, and mumps.

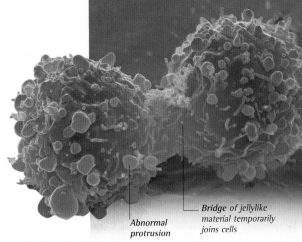

CANCER

In cancer, there is rapid and uncontrolled growth of body cells. These cells are often grouped together as a lump, called a tumour. Common sites include the lungs, large intestine, and breasts. The tumour may spread to surrounding tissues and later to other parts of the body. In some cases, cancer is fatal. However, the various treatment choices mean that it can often be cured.

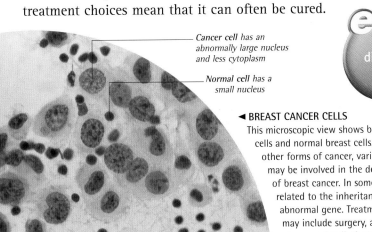

Cancer cell has an abnormally large nucleus and less cytoplasm

Normal cell has a small nucleus

disease

◀ BREAST CANCER CELLS
This microscopic view shows breast cancer cells and normal breast cells. As with other forms of cancer, various factors may be involved in the development of breast cancer. In some cases, it is related to the inheritance of an abnormal gene. Treatment options may include surgery, anticancer drugs, and radiotherapy.

Abnormal protrusion

Bridge of jellylike material temporarily joins cells

LUNG CANCER CELLS ▲
This image shows a lung cancer cell dividing to form two cells. Lung cancer is one of the commonest forms of cancer. The main cause is smoking tobacco. Exposure to high levels of air pollution may also be a factor. Treatment can include surgery, anticancer drugs, and radiotherapy, in which high-intensity radiation is used to kill cancer cells.

FIND OUT MORE ▶▶ Bacteria **284** • Circulatory System **352–353** • Fungi **282–283** • Immune System **357**

MEDICINE

Medicine is the practice of dealing with diseases and injuries to the body. One of its most important aspects is **PREVENTATIVE MEDICINE**, which recommends measures to stop diseases from developing. Two aspects of how doctors deal with existing disease are **DIAGNOSIS** – working out exactly what is wrong with an ill person – and **TREATMENT**.

PENICILLIN ▶
This micrograph shows some fungi that are the source of penicillin, an important antibiotic (natural bacteria-fighting substance). Since the discovery of penicillin in the 1920s, many antibiotics have been developed to target disease-causing bacteria. They have hugely increased life expectancy across the world.

◀ HEALTH EDUCATION
An important aspect of medicine is education about the causes and avoidance of disease. Relevant topics for education vary in different parts of the world. These people in India are learning about how to avoid insect-borne diseases. In developed countries, key areas for education include the importance of not smoking, eating a healthy diet, and taking regular exercise.

PREVENTATIVE MEDICINE

Many measures can be taken to reduce the risk of illness, from adopting a healthy lifestyle to having the appropriate vaccinations against diseases such as polio, mumps, and measles. Doctors provide advice on medical tests to have at different ages. These may include eye tests, blood pressure checks, and procedures such as mammograms (breast X-rays), which aim to detect disease at an early, treatable stage. People at special risk of a disease (one that runs in the family, for example) may be offered special tests.

VACCINATION ▶
This micrograph shows a group of polio virus particles. Polio is an infectious disease that has been eradicated from many countries through vaccination programmes. Some vaccines are offered routinely in childhood. Others may be advised when travelling to parts of the world where there is a special risk of a disease. Most are given by injection, although one form of polio vaccine is given by drops (onto the tongue), for swallowing.

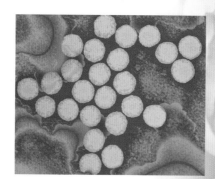

◀ CHECK-UP
Many people, particularly those at risk of having conditions such as heart disease or a stroke, see their doctors for regular check-ups. Blood pressure is often measured, as shown here. This blood pressure reading is above the normal range. If the blood pressure remains high when checked on a number of occasions, treatment is likely to be needed.

DIAGNOSIS

To make a diagnosis, the doctor first talks to the patient about his or her symptoms and other relevant matters, such as the patient's family, occupation, and lifestyle. This may be followed by a physical examination. The information gained will guide the doctor on what tests may be needed to confirm the diagnosis. These may include laboratory tests and imaging.

▲ LABORATORY TESTS
Blood tests are the most common laboratory tests and can provide much information about the functioning of the body. They can be arranged for various reasons, for example, to assess the function of the liver. Urine may also be tested, often to look for infection.

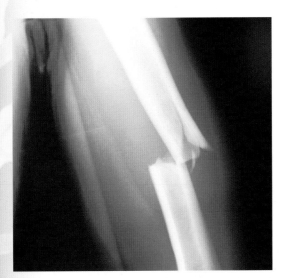

◄ X-RAY
X-rays are the most commonly used imaging method. They can be used to examine bones, often to look for and clarify the extent of a fracture, such as this one of an upper arm bone. They are also often used to examine the chest, in particular, to look for any infection in the lungs and to assess the heart.

medicine

TREATMENT

Doctors and their colleagues select from a number of different types of treatment when attempting to cure illness. These include drugs, physiotherapy, surgery, and speech therapy. Some therapies are concerned mainly with treating mental rather than physical health. These include counselling, which encourages individuals to talk through their anxieties and concerns.

DRUG TREATMENT ►
Modern drug treatment aims to relieve symptoms or to cure or control disease by correcting disturbances in body chemistry. Substances used as drugs range from relatively simple molecules, such as aspirin, to huge, complex protein molecules.

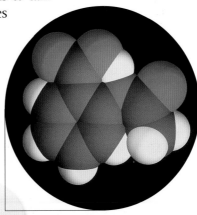

MODEL OF AN ASPIRIN MOLECULE

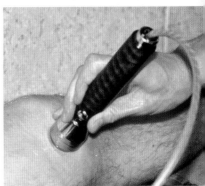

PHYSIOTHERAPY ▲
Physiotherapy uses forms of physical energy, such as heat, exercise, electricity, or sound energy, to promote healing or to improve strength and flexibility. Here, ultrasound (high frequency sound waves) is being used to speed up healing after a knee injury. Other types of physiotherapy include massage, hydrotherapy, heat treatment, and various exercise regimens.

FIND OUT MORE ►► Health **368–369** • Immune System **357** • Medical Research **376** • Medical Technology **374–375**

MEDICAL TECHNOLOGY

Technology plays an increasingly important role in modern medicine. Today, **HI-TECH IMAGING** allows doctors to view internal body structures in amazing detail, while **SURGERY** relies heavily on technologies such as lasers, robots, and computers. **ARTIFICIAL DEVICES** are commonly used either to replace diseased body parts completely, or to provide assistance to failing organs.

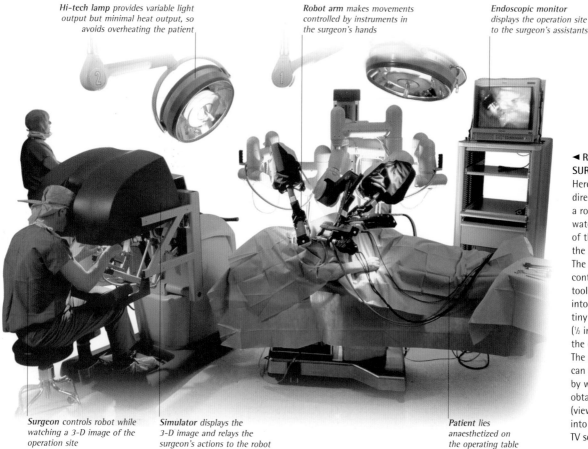

Hi-tech lamp provides variable light output but minimal heat output, so avoids overheating the patient

Robot arm makes movements controlled by instruments in the surgeon's hands

Endoscopic monitor displays the operation site to the surgeon's assistants

Surgeon controls robot while watching a 3-D image of the operation site

Simulator displays the 3-D image and relays the surgeon's actions to the robot

Patient lies anaesthetized on the operating table

◄ ROBOT-ASSISTED SURGERY
Here, a heart surgeon is directing the actions of a robot surgeon while watching a 3-D image of the operation site in the black simulator box. The system allows precise control over the surgical tools, which are inserted into the chest through a tiny incision, just 1.2 cm (½ in) wide, and held at the end of robot arms. The surgeon's assistants can view the procedure by watching images obtained by endoscopes (viewing tubes inserted into the body) on the TV screen at top right.

HI-TECH IMAGING

Modern imaging methods can provide detailed pictures of body parts, whether by injecting dyes that highlight specific structures on X-ray viewing, or by using methods that provide cross-sectional or 3-D scans. Some techniques provide information on body activity, not just structure. For example, special forms of ultrasound (high-frequency sound) can be used to monitor the flow of blood within blood vessels. The internal structures of the body can also be examined directly by means of fibreoptic endoscopes (viewing tubes).

◄ 3-D MRI BRAIN SCAN
This scan reveals structures inside the brain. MRI (magnetic resonance imaging) is an advanced imaging technique. The body part to be examined is placed inside a powerful magnet, and harmless radio waves are released towards it. A computer then builds an image by analysing the pattern of radio waves returned from the part. MRI is often used to examine the brain and other soft tissues.

Lateral ventricle (pink) is one of four fluid-filled cavities in the brain

Thalamus (orange and yellow)

SURGERY

Surgical techniques are constantly being improved and new ones developed. Today, the trend is towards using keyhole surgery, in which the cuts made into the body are kept as small as possible. Microsurgery is another important field. Here, the surgeon uses tiny instruments to repair delicate structures such as nerves, while viewing the operation site through a microscope. Transplant surgery offers hope to people with damaged internal organs – the kidney, heart, liver, small intestine, and lungs can now all be transplanted, although there is often a shortage of donor organs.

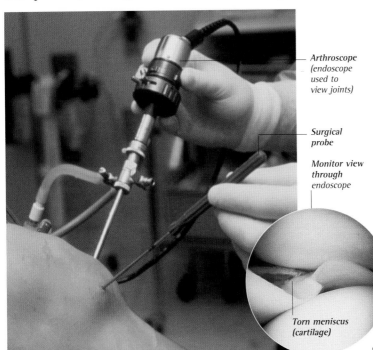

Arthroscope (endoscope used to view joints)

Surgical probe

Monitor view through endoscope

Torn meniscus (cartilage)

LASER EYE SURGERY ▶
Laser beams are used as cutting and burning tools in various types of surgery, including eye surgery. A laser can be used to treat disease of the retina at the back of the eye and to reshape the cornea for treating short-sightedness. Lasers are also used to treat some skin conditions, such as birthmarks, and to remove tattoos.

◀ KEYHOLE SURGERY
Keyhole surgery relies on using fibreoptic endoscopes, viewing instruments that may be rigid (as shown in this knee operation) or flexible. The cuts made to insert the surgical instruments and endoscope into the body are smaller than those used in other types of surgery. The main benefit is the short recovery time.

ARTIFICIAL DEVICES

The range of artificial devices used in treatment increases all the time. Devices available today include replacements for damaged hip and knee joints and an implant into the cochlea of the ear to help some types of deafness. A cataract (clouding of the lens in the eye) can now be treated by insertion of a plastic lens. It is also possible to replace defective heart valves, either with a valve constructed from human or animal tissue, or with one made out of metal and plastic.

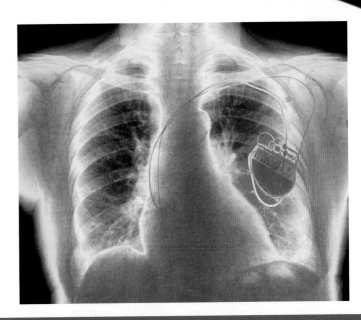

HEART PACEMAKER ▶
This chest X-ray shows an artificial pacemaker that has been inserted beneath the skin of the chest. Pacemakers are given to people whose hearts have a defective electrical system. The pacemaker delivers electrical signals to the heart via wire leads – these signals trigger contractions of the heart muscle and maintain a regular heart beat.

FIND OUT MORE ▶▶ Lasers **112** • Medical Research **376** • Robots **154–155**

MEDICAL RESEARCH

Huge resources are poured each year into studying diseases and finding new treatments. Some of the main focuses of research today are new drugs (including drugs to treat cancer and heart disease), improved artificial parts, the production of new vaccines, and the role of genes in disease. Another growing area is stem cell research. Stem cells are body cells from which all other types of cells form. Potentially, they could be laboratory-grown into a variety of human tissues, for use in body repair.

White blood cell of a different type

NANOTECHNOLOGY IN MEDICINE ▶
One day, it may be possible to introduce tiny robots, called nanorobots, into the body to carry out treatment missions. In this artist's impression, a roving nanorobot is destroying an abnormal white blood cell. Medical treatments are just one possible application of the futuristic area of research known as nanotechnology.

White blood cell being destroyed

Propeller

Red blood cell

Nanorobot introduced into bloodstream

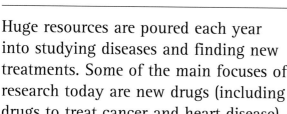

Touch-sensitive pad monitors how hard the object is being held and feeds this back to the microprocessor

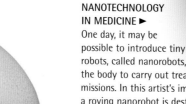

research

◀ BIONICS
This artificial hand works by means of sensors that can detect signals in the wearer's forearm muscles. A built-in microprocessor analyses the signals and orders the mechanical parts of the hand to open or close. A challenge in producing devices such as this is to get living tissue and nonliving material to work together. One focus is to find better ways of passing signals between human nerves and electronic devices.

Rotor wheel for motor that controls finger movements

Casing holds batteries and the built-in microprocessor

Swivel around which whole hand can rotate

▲ COMPUTER-AIDED DRUG DESIGN
Modern researchers often use computers to help design new medicines. For example, a scientist can use a computer model of a drug molecule and study how it interacts with a model of a target site in the body. The researcher can then make improvements to the molecule, allowing a virtual drug to be developed on computer before it is made and tested for real.

FIND OUT MORE ▶▶ Chemical Industry 50–51 • Nanotechnology 157 • Robots 154–155

INDEX

A page number in **bold** refers to the
main entry for that subject.